Ecotoxicological Characterization of Waste

Heidrun Moser • Jörg Römbke
Editors

Ecotoxicological Characterization of Waste

Results and Experiences of an International Ring Test

Springer

Editors
Heidrun Moser
Federal Environment Agency
Germany (UBA)
Dessau, Germany

Jörg Römbke
ECT Oekotoxikologie GmbH
Flörsheim, Germany

ISBN: 978-0-387-88958-0 e-ISBN: 978-0-387-88959-7
DOI: 10.1007/978-0-387-88959-7

Library of Congress Control Number: 2008937807

Printed on acid-free paper

9 8 7 6 5 4 3 2 (Corrected at 2nd printing 2010)

springer.com

Foreword

R. Wagner

The European list of wastes (LOW) (Commission Decision 2000/532/EC and updates) is the basic foundation for the implementation of waste legislation in Europe. It contains a harmonized list of different types of waste and categorizes them according to their origin or generation process. It provides for a uniform description of wastes through the application of a uniform nomenclature and the assignation of certain waste code numbers, for example, in permits or for monitoring purposes. It also labels those hazardous wastes that are subject to a number of special provisions in both European and national legislation, for example, with regard to monitoring, licenses for installations, and national obligations regarding giving notice and transfer of responsibilities. The Waste Catalogue Ordinance (AVV) transposed the European LOW into German legislation in late 2001. This ordinance encompasses 840 codes of waste in 20 main chapters, classifies 400 types of waste as hazardous, and contains mirror entries for about 200 types of waste. The latter term classifies waste either as hazardous or non-hazardous, depending on certain physical or chemical properties. The ordinance defines 14 hazard criteria (H criteria), including criterion H14 "ecotoxic," in order to classify waste with regard to properties that render it hazardous.

Unfortunately, the LOW is incomplete. Depending on the classification of hazard, specific concentrations of solids are laid down for the majority of properties relevant for health and workers' protection. There are, however, no specifications with regard to environmentally hazardous properties, in particular, criterion H14 "ecotoxic." This complicates implementation for EU Member States and thus fails to meet the target of a harmonized application of law. Concentrations of solids for ecotoxic substances contained in such wastes, which are applied to bridge this gap, do not allow reliable statements on the ecotoxicity of the respective types of waste. First, the concentration does not provide any information regarding potential release or bio-availability of the pollutants in the matrix in question, and second, the concentration of solids cannot be determined for wastes consisting of unknown or variable components.

Standardized testing procedures that are well suited for waste management practices and that will allow statements on the ecotoxic effects of wastes are a prerequisite for specifying the ecotoxicitiy of wastes in the European LOW.

Sixty-four participants from 14 European countries and the US carried out an evaluation of biological testing methods in a ring test in order to examine whether these testing procedures were methodologically sound and applicable with regard to real waste samples and eluates. The battery of tests comprised both aquatic and terrestrial testing methods. Ashes from the incineration of domestic waste, waste wood, and contaminated soil were used as test substrates. Testing thus covered a wide range of toxicity levels and different matrices of main waste streams.

The ring test to classify wastes according to their ecotoxic properties proved that with the biological testing procedures selected the hazardous impact of waste on the environment could be reliably determined. Owing to the large number of participating laboratories, the numerous results from reference testing, and the compilation of methodological parameters, comprehensive data for the battery of tests could be collected. Detailed statistical analysis of all these elements ensures that the data is sound and reliable.

Using the tested procedures allows for a deduction of concentration limit values for hazard criteria H14 "ecotoxic" of the European LOW. Methodological implementation of H14 makes it possible to assess the ecotoxicity of wastes in practice and is an important element for classifying wastes as hazardous or non-hazardous.

To expand experience and obtain a wider basis for the present results, the evaluated battery of tests will also be used for other types of waste in mirror entries.

The European collaborative study is an important contribution to establishing a European research area (6th Communication of the Commission, 2000). Through intensive scientific discussions, the organization of demonstration workshops by the Federal Environment Agency, and the exchange of documents on methodology, a high level of expertise was gained. Serving as an evaluation study of EN 14735, the results form an important element for the standardization work of CEN TC 292 "Characterization of waste."

Thus, the foundations are being laid to specify the ecotoxicity of wastes through fixed parameters, which are based on standardized biological testing methods to determine hazard classification, within the framework of the recently launched discussion on updating the European LOW.

Contents

Contributors

M.J.B. Amorim
CESAM and Department of Biology, University of Aveiro,
Aveiro, Portugal

Y. Barthel
IPL Santé Environnement durables Est, Maxéville, France

C. Bazin
INSAVALOR – POLDEN, Villeurbanne, France

R. Becker
Federal Institute for Materials Research and Testing (BAM), Berlin, Germany

A. Buchholz
Federal Institute for Materials Research and Testing (BAM), Berlin, Germany

A-M. Charissou
IPL Santé Environnement durables Est, Maxéville, France

X. Domene
CREAF, Autonomous University of Barcelona, Facultat de Ciències,
Cerdanyola del Vallès, Barcelona, Spain

G. Donnevert
University of Applied Sciences (FH) Giessen-Friedberg, Giessen, Germany

R. Eilbracht
Stuttgart, Germany

C. Firla
ECT Oekotoxikologie GmbH, Floersheim, Germany

B. Foerster
ECT Oekotoxikologie GmbH, Floersheim, Germany

C. Hafner
Hydrotox – Labor fuer Oekotoxikologie und Gewaesserschutz GmbH,
Freiburg i.Br., Germany

A. Haller
ECT Oekotoxikologie GmbH, Flörsheim, Germany

J. Hofman
1RECETOX – Research Centre for Environmental Chemistry and Ecotoxicology, Faculty of Science, Masaryk University, Brno, Czech Republic

I. Hovorková
1RECETOX – Research Centre for Environmental Chemistry and Ecotoxicology, Faculty of Science, Masaryk University, Brno, Czech Republic

T. Junker
ECT Oekotoxikologie GmbH, Floersheim, Germany

U. Kalbe
Federal Institute for Materials Research and Testing (BAM), Berlin, Germany

H. Kessler
Federal Environment Agency, Dessau; Germany

S. Kesberg
Stuttgart, Germany

R. Kuperman
Edgewood Chemical Biological Center, Aberdeen Proving Ground MD USA

J. Machát
RECETOX – Research Centre for Environmental Chemistry and Ecotoxicology, Faculty of Science, Masaryk University, Brno, Czech Republic

H. Moser
Federal Environment Agency, Dessau, Germany

T. Moser
ECT Oekotoxikologie GmbH, Floersheim, Germany

T. Natal-da-Luz
Instituto do Mar – Coimbra Interdisciplinary Centre, Department of Zoology, University of Coimbra, Coimbra, Portugal

H. Neumann-Hensel
Dr. Fintelmann und Dr. Meyer GmbH, Chemische und biologische Handelslaboratorien, Hamburg, Germany

P. Pandard
INERIS, Verneuil en Halatte, France

M. Pattard
Federal Environment Agency, Berlin, Germany

G. Persoone
MicroBioTests Inc. Mariakerke-Gent, Belgium

J.F. Postma
Grontmij|AquaSense, Amsterdam, The Netherlands
Present address: Ecofide, Lakenkopersweg 20, 1383 CS, Weesp, The Netherlands

F. Riepert
Julius Kuehn-Institut, Bundesforschungsinstitut fuer Kulturpflanzen, Berlin, Germany

C. Roeber
Dr. Fintelmann und Dr. Meyer GmbH, Chemische und biologische Handelslaboratorien, Hamburg, Germany

J. Römbke
ECT Oekotoxikologie GmbH, Floersheim, Germany

A. Scheffczyk
ECT Oekotoxikologie GmbH, Floersheim, Germany

J.P. Sousa
Instituto do Mar – Coimbra Interdisciplinary Centre, Department of Zoology, University of Coimbra, Coimbra, Portugal

S. Uhlig
quo data – Gesellschaft fuer Qualitaetsmanagement und Statistik mbH, Dresden, Germany

H.A. Van der Sloot
Energy research Centre of the Netherlands (ECN), Environmental Risk Assessment Group, Petten, The Netherlands

C. Van der Wielen
Institut Scientifique de Service Public, Liege, Belgium

A. Van Zomeren
Energy research Centre of the Netherlands (ECN), Environmental Risk Assessment Group, Petten, The Netherlands

K. Wadhia
ALcontrol Laboratories, Rotherham, South Yorkshire, United Kingdom

R. Wagner
Federal Ministry for the Environment, Nature Conservation and Nuclear Safety, Bonn, Germany

G.E. Weber
Stuttgart, Germany

R. Weltens
VITO, Environmental Toxicology, MOL Belgium

Part I
Summary and Background Information of the Ring Test

Chapter 1
Overview on the Results of the Ring Test

J. Römbke, H. Moser, and T. Moser

Abstract The assessment of waste as hazardous or non-hazardous, according to the European Waste List, includes ecotoxicological characterization. Despite being made into national law by the Waste List Ordinance 2001, no recommendations on the methodology have been provided to cover the hazard criterion (H14 "ecotoxic"), which was taken from the legislation on dangerous substances. Based on the recommendations of CEN guideline 14735 (2005), an international ring test was organized by the German Environment Agency (UBA) with the help of the Federal Institute for Materials Research and Testing (BAM), FH Giessen-Friedberg and ECT Oekotoxikologie GmbH. Sixty laboratories from 15 countries participated in the ring test. It was performed with three representative waste types: ash from an incineration plant contaminated mainly with heavy metals (INC), soil containing high concentrations of organic contaminants (PAHs) (SOI) and preserved wood waste contaminated with copper and other heavy metals (WOO). Samples were prepared (inter alia dried, sieved and homogenized) and distributed by BAM. Parallel to the biological testing, the eluates and solid samples were chemically characterized. The basic test battery used in the ring test consisted of three aquatic (Algae test, *Daphnia* acute test and Microtox test) and two terrestrial (earthworm acute and plant test with two species (oat, rape)) tests. In addition, data were submitted for ten other tests (five aquatic (including a genotoxicity test) and five terrestrial). Almost all tests were performed according to ISO guidelines, providing EC_{50} values as measurement of toxicity. Data evaluation was done following recent recommendations made by ISO (2002) and Environment Canada (2005).

In addition to data from several reference tests, 634 data sets were produced from the basic test battery, 196 data sets from additional test battery and, in parallel, 218 data sets with the respective reference substances. Only a few data sets were not acceptable (e.g. due to lack of reference data) and even fewer results were identified as statistical or biological outliers. For example, in the case of the basic test battery, the acceptance rate varied between 74.1 (Algae test) and 92.6% (*Daphnia*

J. Römbke (✉)
ECT Oekotoxikologie GmbH, Boettgerstr. 2-14, 65439 Floersheim, Germany

H. Moser and J. Römbke (eds.), *Ecotoxicological Characterization of Waste*,
DOI: 10.1007/978-0-387-88959-7_1,

test). There were no methodological problems, but further guidance on moisture determination in the terrestrial tests, as well as details concerning reference testing and data evaluation for several tests, is needed. However, changes in the procedures described by CEN 14735 (2005) for the preparation of eluates may have caused some of the variability observed in the aquatic tests.

Independently which test system is considered, SOI always had the lowest effects and WOO was the most toxic, while the EC_{50} values of INC showed an intermediate toxicity. Among the aquatic tests, daphnids and one algal species were the most sensitive ones, while plants were always more sensitive than earthworms in the terrestrial tests. A comparison of the ring test results with literature data published so far revealed good agreement, but the high sensitivity of daphnids, as well as the low sensitivity of the luminesecent bacteria, was not expected, because of the small number of data reported in the literature. Taking into account the differences between chemical/physicochemical and ecotoxicological ring tests, this variation is considered acceptable.

Based on the results from the additional tests, proposals for the modification of the existing basic test could be made. Most importantly, it was proposed that the acute earthworm test should be replaced by a more sensitive soil invertebrate test, in particular those with short durations (e.g. the Earthworm avoidance test or the Arthrobacter test, because of their combination of practicability and sensitivity). Available information indicates that besides "traditional" test versions with daphnids and algae, miniaturized microplate methods can be used. The same experience can also be used to find test methods for wastes which cannot be tested with the basic tests (e.g. the Lemna test for turbid eluates which cannot be tested in the Algae test). Finally, the ring test results also support the proposals made in the CEN guideline 14735 (2005) concerning the performance of such tests.

Further work, performed in parallel to the ring test, improves waste testing considerably. For example, special consideration should be given to the preparation of the terrestrial control and dilution medium (OECD artificial soil), as the variability of its properties may also have contributed to the overall variability of test results. Independent of the objectives of the ring test, the data collected can be used in combination with a detailed characterization of the eluates for the modeling of the bioavailable fraction of the respective contaminant, thus allowing the ecotoxicological effects observed in the tests to be understood.

In any case, more experience with different types of waste materials is necessary for all ecotoxicological test systems discussed so far, to define the range of materials which can successfully be tested. The same information is also necessary for the definition of limit values as well as for the identification of the most efficient testing strategy (e.g. the application of extended laboratory tests instead of full dose-response tests). In this context the collection of all relevant data as well as legal requirements in a data-base, together with the appropriate software like the H14-Navigator, is recommended, to improve data evaluation.

In line with recommendations made in the literature, a combination of the results of a battery of biological tests and chemical residue analysis is needed for an ecotoxicological characterization of wastes. The results and experiences from the ring test should be transferred to a harmonized approach for the implementation of the hazard criterion H14, as part of a future amendment of the European Waste List. In addition, basically the same approach - but with modifications of the test strategy and also test methods - can be used for an environmental risk assessment (ERA) of wastes. ERA is a very important prerequisite for the management of wastes, for example in re-use scenarios in which wastes come in direct contact with the environment.

Keywords Ecotoxicological laboratory tests, Solid waste, Eluate, Copper, PAH, Ring test

1.1 Introduction

1.1.1 Legal Background

The European Waste List (2001/118/EC (EC 2001)) is a harmonized list of about 850 different waste types, thus forming a consistent waste classification system across the EU. It is intended to be a catalogue of all wastes, grouped according to generic industry, process and waste type. So far, the list has been amended three times and can be revised according to the Waste Framework Directive. It includes 850 six-digit-code wastes in 20 chapters, defining 405 waste types as hazardous waste material and listing 200 waste types as so called "mirror entries". A mirror entry is defined as follows: Wastes with potential to be either hazardous or non-hazardous depending on their composition and the concentration of dangerous substances. So far, 14 hazard criteria for the characterization of hazardous waste types have been defined. The criteria H3 to H8, H10, and H11 (flammable, irritant, harmful, toxic, carcinogenic, corrosive, teratogenic, and mutagenic) are based on the concentration of dangerous substances. Criterion H14, "ecotoxic", lacks an assessment and testing strategy. In addition, no specific threshold values have been defined so far. Details of the legal background are provided in Chap. 2.

During an international workshop organized by UBA and JRC in Ispra (Italy) the participants agreed that biological test systems should be used for the ecotoxicological characterization of waste (Gawlik & Moser 2005). A distinct need for a harmonized test set was identified, to be developed and validated in the framework of CEN TC 292. In addition, a general agreement was reached that this test should address the property of ecotoxicity of waste by using test organisms as representatives for various ecosystems and trophic levels (CEN 2005). Clearly, for the validation of such a test, it is necessary to perform an international ring test. In this chapter, the procedures and results of the ring test are summarized.

1.1.2 Aims of the Ring Test

The validation of test procedures plays an essential part in the standardization process of each ecotoxicological test. The framework of European standardization (CEN) demands an internationally conducted ring test with typical test substrates, which provides valid information on the practicability and reproducibility of the results for the test procedures employed. Therefore, the main aim of the ring test was the establishment of a test battery for the ecotoxicological characterization of wastes (H14), using the CEN 14735 standard "Characterization of waste - Preparation of waste samples for ecotoxicity tests" (CEN 2005) as the basis. In this context, the following issues were addressed:

- The validation of the preparation of the test substrates according to this standard
- The assessment of the suitability of the basic tests (i.e. three aquatic and two terrestrial tests) in terms of practicability and sensitivity, including whether modifications of existing test methods are necessary
- The evaluation of the uncertainty level of the results for the various tests
- The identification of recommendations concerning tests for routine use

Finally, open questions and needs for future research had to be selected.

1.1.3 Organisation of the Ring Test

In keeping with the recommendations published by international standardisation organizations (e.g. OECD 2005) and in scientific literature (e.g. Römbke & Moser 2002), the ring test was organized in such a way that a large number of participants, who represented various countries and institutional backgrounds, and had a broad range of experience with the selected ecotoxicological tests, were selected.

1.2 Methods

1.2.1 Test Substrates

The test materials were selected after consultation with the CEN TC 292/WG7 committee. The three selected materials covered a wide range of toxicity and different matrices of the main waste flows. Their processing, characterization and distribution were conducted by BAM (Berlin). Details of their selection, process, and characterization are given in Chap. 3. The homogeneity of selected chemical parameters (heavy metals, polycyclic aromatics, Cu) was tested in accordance with the respective guidelines and was found to be satisfying with regard to the variability observed in the ring test. The whole process is shown in Fig. 1.1.

Fig. 1.1 Schematic processing of the three test substrates: from selection to dispatch

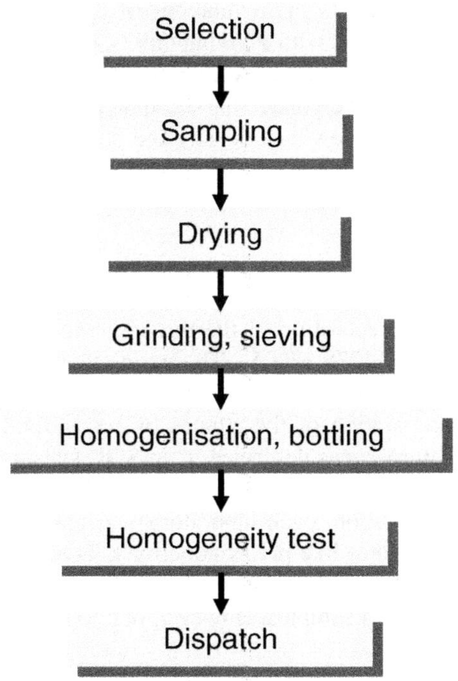

1.2.1.1 Municipal Waste Incineration Ash (INC)

The starting material (719 kg) was obtained form a Dutch incineration plant for municipal waste ("bottom ash") and was processed at BAM (drying, sieving [<4 mm], homogenization). Depending on the demand, 318 kg were bottled and distributed among the participants. Along with partly high concentrations of heavy metals (Cu 6,800 mg/kg; Zn 2,639 mg/kg; Pb 1,623 mg/kg) a high pH (about 10.5) was also observed.

1.2.1.2 PAH Contaminated Soil (SOI)

The polluted sandy soil originates from a former gasworks site in Berlin (Germany), which was dried, sieved (<4 mm) and homogenized at BAM. A total of about 680 kg of the starting material was used. Besides a high PAH content (sum of the 16 EPA-PAK: 840 mg/kg) only a minor amount of mineral oil hydrocarbons was detected (152 mg/kg) and the pH value was 8.4.

1.2.1.3 Waste Wood (WOO)

This substrate was a mixture of treated and untreated wood samples from a commercial timber processing plant, which was treated with copper-based wood preservatives according to the regulations of different European countries. The starting material was ground with a cutting mill (<4 mm). The obtained amount of 900 kg was homogenized and 617 kg were bottled in containers between 0.5 and 10 kg and dispatched to the participants. This substrate demanded a complicated homogenization procedure due to its low bulk density and poor flow ability. The copper content was high (2,110 mg/kg), while the pH was low (≈4.8).

1.2.1.4 Treatment of Samples and Preparation of the Eluates

The preparation of the three test substrates followed the provisions laid down in CEN 14735 (2005). The main steps of the procedures are shown in Fig. 1.2. As stated above, the preparation of the test substrates was performed by BAM, and the preparation of the eluates by the individual partners. The handling of the waste samples was described in an SOP, starting from the moment the samples arrived in the laboratory (including details like sample labeling, storage conditions and so on). In addition, each laboratory was asked to determine basic parameters of the test substrates like pH, conductivity, TOC and the concentration of the main pollutants for the eluates, as well as pH, organic matter content and water holding capacity for the solid samples. However, due to limited resources, this was not always possible.

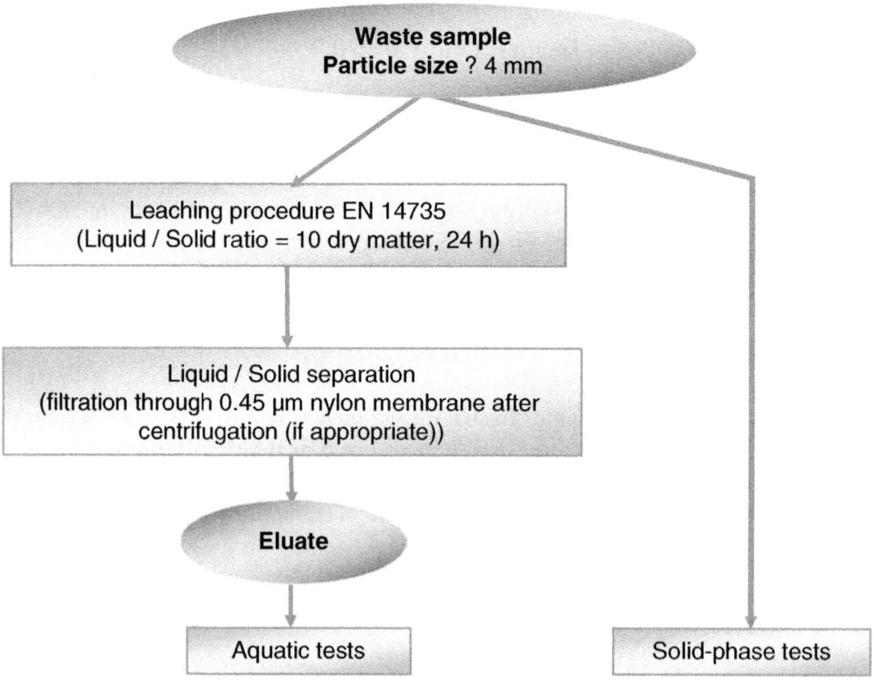

Fig. 1.2 Overview on the treatment of the three test substrates in the ring test

All the information collected was compiled in specific forms provided by the ring test organizers.

In order to perform tests following a dose-response design, the eluates as well as the waste samples had to be diluted using an appropriate dilution material (e.g. OECD reconstituted water (OECD 2004) or OECD artificial soil (OECD 1984)). Both for the control and the dilution steps, the same material had to be used. Details of the preparation of the dilutions as well as the design of the tests were laid down in the SOPs for individual tests and were part of the information given during the demonstration workshop.

1.2.2 Basic Test Methods

The ecotoxicological characterization of wastes is laid down in European standard EN 14735 (2005), which describes sample preparation and provides an informative collection of appropriate test procedures for the investigation of wastes. This collection of test procedures was condensed to a basic set of tests containing three aquatic and two terrestrial procedures, which are briefly described here (Table 1.1). Details of the performance of these five tests are given in Chaps. 7–11.

Only the tests which fulfilled the validity criteria as given in the individual guidelines (usually including a reference test) and which were performed according to the SOPs provided by the organizers, were considered acceptable for assessment. In the case of the luminescent bacteria tests, sensitivity is proved mainly by performing limit tests; therefore, the reference tests following a dose-response design were not used to check the acceptance of these tests. Since for plant tests no reference substance is formally required, the lack of such data did not cause the automatic exclusion of these data sets.

A total of 634 data sets were produced in the basic set of tests. The results of 143 Algae tests, 161 *Daphnia* tests, 154 luminescent bacteria tests, 52 earthworm tests and 124 plant tests were submitted by the participants (Table 1.2). Parallel to these tests, 172 data sets were determined in tests with the respective reference substances.

Table 1.1 Brief overview on the five tests belonging to the basic test battery

Name	Guideline	Species
Eluate (aquatic) waste tests		
Algae	ISO 8692 (2004)	*Desmodesmus subspicatus* or *Pseudo-kirchneriella subcapitata*
Daphnia	ISO 6341 (1996)	*Daphnia magna*
Luminescent bacteria	ISO 11348-1/2 (2005)	*Vibrio fischeri* (three sources, alternatively)
Solid (terrestrial) waste tests		
Earthworms (acute)	ISO 11268-1 (1997)	*Eisenia fetida* or *Eisenia andrei*
Plants	ISO 11269-2 (2004)	*Avena sativa* and *Brassica rapa*

Table 1.2 Number of accepted test data sets in percent of the total number of tests performed, separately for the five basic tests and the three test substrates

		Accepted data sets				
Test system	Total No. of tests	INC	SOI	WOO	Sum	% Total
Algae	143	35	35	36	106	74.1
Daphnia	161	47	51	51	149	92.6
Lumi. bacteria	154	45	42	53	140	90.9
Plants	124	37	35	35	107	86.3
Earthworms	52	14	15	15	44	84.6
Sum	634	178	178	190	546	85.7

1.2.3 Additional Test Methods

Ten additional test procedures (five aquatic and five terrestrial ones), regarded as being potentially appropriate for the determination of the ecotoxicity of waste, were performed in the ring test. These test methods are briefly described in Table 1.3. Details of the test performance of the additional waste eluate tests are given in Chaps. 12–15 and 21. The information provided for the additional solid waste tests is provided in Chaps. 16–20. As the experience with these additional test methods is in general lower than for the methods used in the basic set and because no SOPs were prepared for them, the acceptance of test data sets was handled on a case-by-case basis (i.e. in general no tests with a reference substance were required).

A total of 196 data sets were produced in the additional tests (Table 1.4). The results of 51 *Lemna* tests, 10 *Brachionus* tests, 12 *Pseudomonas putida* bacteria tests, 10 *Ceriodaphnia* tests and 23 umu tests were submitted by the participants for waste eluates. The respective numbers for the solid waste tests are: 21 Collembola tests, 17 earthworm reproduction tests, 12 earthworm avoidance tests, 12 enchytraeid tests and 28 *Arthrobacter* tests. In parallel to these tests, 46 data sets were determined in tests with the respective reference substances.

1.2.4 Test Data Evaluation

The statistical evaluation of the ring test data (only standard tests) was performed in a step-wise process. First, the individual test results (EC/LC50 values) were recalculated using one method (probit analysis and the ToxRat program (2006)). Secondly, they had to fulfil several acceptance criteria. Data sets which passed these acceptance criteria were evaluated in two different ways in parallel: according to the approach generally used for the validation of chemical and physicochemical methods in environmental analytics (ISO 5725-2 (2002)) and according to the warning limit approach following Environment Canada (2005).

Table 1.3 Brief overview on the ten tests belonging to the additional test battery

Name	Guideline	Species
Eluate (aquatic) waste tests		
Aquatic macrophyte	ISO 20079 (2004c)	*Lemna minor*
Rotifer	ISO/CD 20666 (2007b)	*Brachionus calyciflorus*
Sludge bacteria	ISO 10712 (1995)	*Pseudomonas putida*
Water flea	AFNOR 90-376 (2000)	*Ceriodaphnia dubia*
Umu Genotoxicity	ISO 13829 (2000)	*Salmonella choleraesius*
Solid (terrestrial) waste tests		
Earthworm reproduction	ISO 11268-2 (1998)	*Eisenia fetida* or *Eisenia andrei*
Enchytraeidae	ISO 16387 (2004d)	*Enchytraeus albidus* or *E. crypticus*
Collembola	ISO 11267 (1999)	*Folsomia candida*
Earthworm avoidance	ISO 17512-1 (2007)	*Eisenia fetida* or *Eisenia andrei*
Arthrobacter contact	DIN 38412-48 (2002)	*Arthrobacter globiformis*

Table 1.4 Number of accepted (= total number minus those without raw data) tests and EC_{50} values (% waste) of the additional test systems

	No. tests		Test substrates		
Test system	Total	Accepted	INC	SOI	WOO
Lemna	51	35	>50	>90	1.70
Rotifer	10	10	5.35	>100	0.11
Pseudomonas	12	9	11.7	>72.5	0.14
Water flea	10	10	4.78	>90	0.08
Umu-genotoxity	23	21	n.a.	n.a.	n.a.
Earthworm reproduction	17	12	16.1	>50	4.10
Enchytraeidae	12	11	31.8	>100	14.6
Collembola	21	18	26.0	47.9	5.00
Earthworm avoidance	12	12	13.7	39.8	2.90
Arthrobacter contact	28	25	22.6	12.4	0.58

However, for the evaluation of the results from ecotoxicological tests, the ISO approach had to be modified. Assuming that EC/LC_{50} values are log-normally distributed, the log-transformed EC/LC_{50} values were used, instead of the original EC/LC_{50} values, to calculate means and standard deviations. Therefore, the re-transformed total means are not arithmetic but geometric means. Those test results which were identified as statistical outliers (ISO approach) or those outside the warning limits (Environment Canada) were excluded from the calculation of the final geometrical mean EC/LC_{50}. Examples of both approaches are presented in Figs. 1.3 and 1.4. (Table 1.4). As a measure of robustness the factor between minimum and maximum EC/LC_{50} values is presented. Details of the evaluation process are provided in Chap. 4.

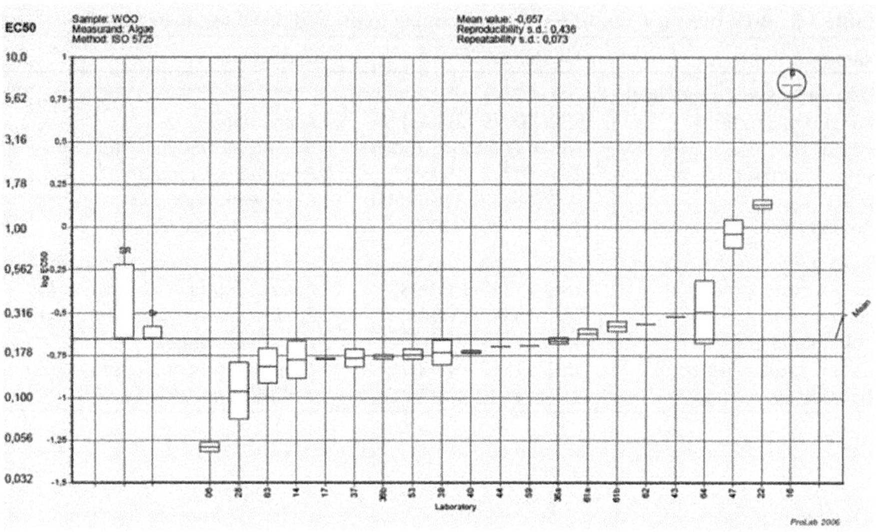

Fig. 1.3 Example for the graphical representation of the statistical evaluation according to ISO (ISO 5725-2, 2002)

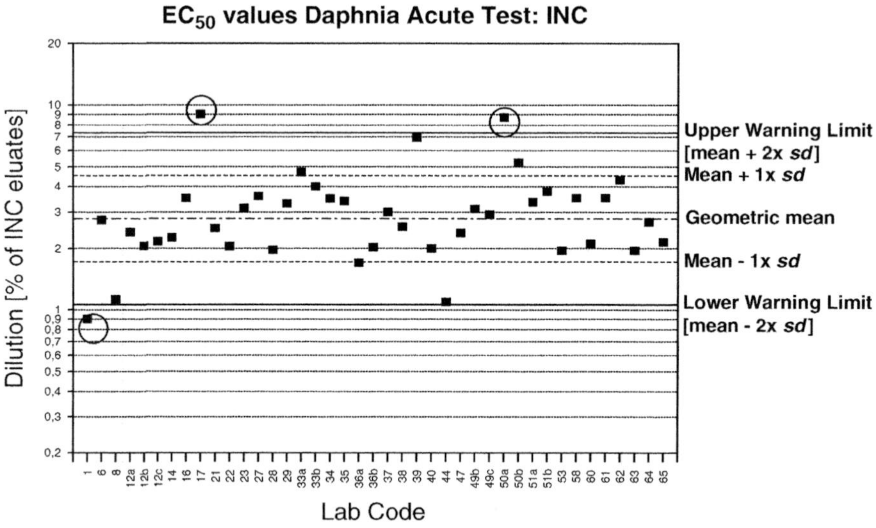

Fig. 1.4 Example for the graphical representation of the warning chart approach (Environment Canada 2005). Encircled values: statistical outliers

1.3 Results

1.3.1 Results of Reference Tests

In parallel to the tests with the three waste materials, most of the participating laboratories secured the sensitivity of the test species by performing tests with specific reference substances. These tests were evaluated separately (172 aquatic reference tests: Chap. 5 and 46 terrestrial reference tests: Chap. 6).

First of all, the data sets were large. For instance, in the case of the algae and *daphnia* tests, 45 and 76 data sets, respectively, were provided. For the earthworm acute test, the number of results with the reference substance chloroacetamide was almost as high (15) as the number of data provided in the original ring test when the range of expected results was defined (18; Edwards 1984). Secondly, only a few participants (4–13%, depending on the test system) did not perform reference tests when required. Even when such reference tests were not absolutely necessary, as in the case of the luminescent bacteria and the plant tests, a considerable number of participants (19 and 9%, respectively) provided such data. Finally, only in very few cases were the results from these tests outside the required range.

In the case of the aquatic reference tests, the number of the delivered reference tests, mainly with potassium dichromate in the standard tests, was so high that these results could be used for the re-evaluation of the data ranges listed in the respective test guidelines. Especially, the range given in the current algae test guideline, based on old ring test data, had to be changed and thus, a new range is proposed here (see Chap. 5). In the additional tests, the number of data sets provided, often with different reference substances, was much lower. Since potassium dichromate did not cause problems in the various tests, it is recommended that this substance be used as the standard reference substance in aquatic tests.

The number of the required (chloroacetamide in the earthworm test) and proposed (boric acid in the plant tests) reference tests was high enough to confirm the suitability of the substances and required/proposed ranges of LC_{50} values. Again, in the three additional tests, the number of data sets provided, often with different reference substances, was much lower. For reasons of availability (pesticides) and human toxicology (chloroacetamide) boric acid is recommended as a reference substance for all terrestrial test systems. Therefore, and despite some data being already provided (in particular for the plant growth test), more data from tests with boric acid are needed to identify the most suitable range of effects. The information gained in this ring test forms a good starting point for this purpose.

1.3.2 Basic Test Methods

As a first step, the acceptance of the data provided was determined. Using the validity criteria given in the respective guidelines, as well as the results of the reference tests as mentioned in the section 1.3.1 those data were identified as useful for further assessment. In addition, in a few cases test results were not accepted because they were considered as outliers, either due to statistical or biological reasons (EC50 < / > as the overall mean by a factor of 10). The outcome of this exercise for the five basic tests and separately for the three test substrates is given in Table 1.2. Note that the differences between the three sources of luminescent bacteria and the two plant species were too small to be considered here.

The rate of acceptance was very high and varied between 74.1 (Algae tests) and 92.6% (*Daphnia* tests). In the first attempt the acceptance rate of the algae tests was lower (slightly less than 70%), because the required range of reference results was set too low in the guidelines. Following a discussion with the ISO representatives responsible for the standardization of this test, the range was altered and the rate reached 74.1%, which is only slightly less than those rates found in the other basic tests. All further evaluation was based on the data identified as accepted here.

Before comparing the results of the individual tests, the most important methodological experiences from the five basic tests are briefly summarized. In fact, the performance of these tests did not cause problems, although most of the participating laboratories were not familiar with the biological testing of waste material or eluates. Concerning the algae tests, the most striking outcome was the clear difference in sensitivity between the two species: *Pseudokirchneriella subcapitata* is by a factor of 4 (tests with INC) to 5(tests with WOO) more sensitive than *Desmodesmus Subspicatus;* however the difference in sensitivity was not confirmed in the reference tests. No modifications were necessary in the *Daphnia* test. When performing the tests with luminescent bacteria, comparable differences were found, as in the algae tests: depending on the source (fresh/liquid vs. frozen bacteria) different sensitivities were found. In the tests with the waste substrate WOO the sensitivity differed by a factor of about 5, showing that the frozen bacteria are more sensitive. While the earthworm tests were running fine, participants asked for more guidance on how to determine the moisture of the test substrate mixtures. In addition, there was strong concern about the suitability of the currently used reference chloroacetamide, which is considered to be mutagenic. Finally, in the case of plants, the situation was similar: the test methods themselves were acceptable, but in a few cases it was difficult to find and maintain the optimal moisture level when setting up the tests. From the very beginning, it was clear that the two plant species had to be handled separately due to their different taxonomy and physiology.

Concerning the reproducibility of the accepted test results, the factor between the lowest and the highest EC50 values were compared for each test system (and species/sources) separately (Table 1.5). In addition, EC50 values outside the warning limits were not taken into consideration. However, as less than two data sets were outside the warning limits, this exclusion did not influence the evaluation considerably.

In terms of toxicity, these results can be assessed in various ways.

Table 1.5 EC_{50} values (% waste) and their minimum-maximum factor based on all accepted data minus those outside of the warning limits (see the Table A.1 in Annexure)

	LC/EC_{50} values (% waste)			EC_{50} values: Minimum-maximum factor		
Test system: Species/source	INC	SOI	WOO	INC	SOI	WOO
Algae: all	4.08	56.5	0.50	59	n.a.	138
D. subspicatus	8.80	>100	1.34	15	n.a.	2
P. subcapitata	2.42	>100	0.22	20	n.a.	138
D. magna	2.71	>100	0.38	10	n.a.	31
Lumi-Bacteria: all	35.4	65.5	2.56	8	2	122
Lumi-Bacteria: freeze	30.8	65.8	0.73	8	n.a.	20
Lumi-Bacteria: liquid/fresh	n.a.	n.a.	5.60	n.a.	n.a.	11
E. fetida/E. andrei	45.5	>100	20.1	2	n.a.	3
A. sativa	29.4	57.8[a]	10.0	17	n.a.	10
B. rapa	23.9	63.0	2.64	22	3	11

n.a. Not applicable: no EC_{50} determinable due to low toxicity
[a]Based on just six values

1.3.2.1 Toxicity of the Three Waste Substrates

Whichever exposure pathway (eluate or waste) or test system was considered, SOI always had the lowest effects and WOO was most toxic, while the EC50 values of INC were found somewhere between these extremes. A compilation of the individual results of the basic test set is given in Table A.1 in the Annexure.

1.3.2.2 Sensitivity of the Individual Test Systems in Aquatic and Terrestrial Tests

The waste substrate INC had the strongest effect on daphnids and the algae *P. subcapitata*, followed by the second algal species *D. subspicatus*, but did not have much effect on luminescent bacteria. In contrast, SOI impacted (weakly) the luminescent bacteria but showed no toxicity to algae and daphnids. The waste substrate WOO was highly toxic to all aquatic species; only the liquid/fresh luminescent bacteria reacted less strongly (in fact, freeze dried bacteria were often more sensitive than the other two sources). The terrestrial tests plants were always more impacted by the three waste substrates than earthworms, meaning that the earthworm acute test was the least sensitive. However, with the exception of WOO, which strongly affected *Brassica rapa*, the difference between the terrestrial EC_{50} values was always lower than the factor of 2. Again, with the exception of WOO, where there was a clear difference, *B. rapa* was as sensitive as *Avena sativa*.

1.3.2.3 Reproducibility of the Test Results

The difference between the lowest and highest EC_{50} values was in most cases small, which is not surprising, as the waste substrate SOI values can vary only slightly because of its low toxicity effect (actually, only by a factor of 2–3). In the waste INC tests, factors higher than ten were observed only in the algal and plant tests. The EC_{50} values for the highly toxic WOO differed in two cases by a factor less than five and in two cases by a factor clearly higher than 100. It is not known why the test results differed so much when testing the algal species *P. subcapitata* while the minimum-maximum factor for the second species *D. subspicatus* was only 2. In general, smaller factors were found in the terrestrial tests. High minimum-maximum factors were usually found when the results of two species or sources (e.g. algal tests with INC or luminescent bacteria tests with WOO) were combined, thus reflecting the different sensitivity of the species and sources, respectively.

In contrast to the terrestrial tests, the variability of the results presented in the aquatic tests integrated not only the variability of the ecotoxicity test itself, but also the variability of the entire procedure: sampling of the test portion, eluate preparation (leaching procedure and separation of the solid and liquid phases) and ecotoxicity test. A preliminary analysis of the raw data provided by the participating laboratories clearly showed that some of them did not strictly follow EN 14735 (CEN 2005) for eluate preparation (e.g., amount of waste for the leaching test or volume of eluate collected) and storage time of eluates before performing aquatic tests. This may explain partly the high variability of the results (Table 1.5), e.g. when compared with the one observed in tests with the reference chemical potassium dichromate (min-max ratio: 2.6; see Chap. 5).

The comparison of repeatability standard deviation s_r and reproducibility standard deviation s_R covers a range between 2.16 and 15.0. 42% of the biotest results are within a s_r/s_R ratio of 4 and about 85% are within a range for the s_r/s_R ratio from 2 to 8, taking into account that this range already covers influence factors like eluate preparation or the use of different test species. Considering also the differences between chemical/physicochemical and ecotoxicological ring tests, this variation is acceptable. Certainly, reproducibility and repeatability can be improved by training the test performers and a well established test quality assessment. It can be summarized that the basic test set forms a reliable methodology for waste ecotoxicity tests, which should be optimized with experiences gained in the ring test.

1.3.3 Additional Test Methods

Details of the performance of the additional waste eluate tests are given in Chaps. 12–15 and 21. The information provided by the additional solid waste tests is provided in Chaps. 16–20. As the additional test methods were not the main focus of the ring test, only the most important results will be presented here (Table 1.4). In general, the number of data sets per test system were relatively low (aquatic: 10–51; terrestrial: 10–28); thus, with the exception of the *Lemna* test (51 data sets) any discussion of the reproducibility of the test results is premature. All results were accepted except those without raw data and those violating the respective validity criteria. No methodological problems were reported by the participants.

As in the case of the basic tests, WOO had the strongest effects in all test systems and SOI the least, with INC showing intermediate results. Among the eluate tests, the tests with *Ceriodaphnia dubia* and *Brachionus calyciflorus*, followed by the test with *Pseudomonas putida*, were the most sensitive for all waste substrates. Several non-basic tests could become alternatives, considering the issue of sensitivity. While the toxicity of WOO was higher in the three non-plant additional tests, it was less compared to the *Daphnia* and one algal test with INC. The Lemna test was clearly the least sensitive method, but it could provide information about plant-related toxicity in colored or turbid waste eluates, where is not possible to apply Algae tests. Among the data provided for the Pseudomonas tests, no difference was observed between the standard and miniaturized versions. In fact, the same overall result was obtained when comparing the results of other "Toxkit microbiotests" using algae, *Daphnia* or, in much smaller numbers, rotifers (*B. calyciflorus*) (Chap. 23). However, the variation coefficients (CVs), varied substantially from one test to another and for different types of waste, ranging from 16 to as high as 92%. Considering that the CVs of tests performed with pure chemicals are usually lower, it is probable that the homogeneity of the waste samples and/or the preparation of the eluates were responsible for these differences.

Among the solid waste tests, the enchytraeid reproduction test was the least sensitive method, while the earthworm avoidance and, in particular, the Arthrobacter test were the most sensitive. This is especially interesting because these two tests are also the shortest ones (1–2 days) of all the terrestrial tests. On the other hand, the three chronic reproduction tests (28–56 days) using earthworms, enchytraeids and collembolans are more robust with minimum-maximum factors less than 10. Compared to the basic test set, almost all the zoological tests were more sensitive than the earthworm acute test. No methodological problems occurred in these tests but further information on the determination of the test substrate and the amount of moisture in the mixtures is needed. In addition, alternatives for the pesticides currently used as reference substances should be selected (e.g. boric acid).

In total, 23 umu tests were performed in the ring test. So far, the genotoxicity of waste materials has rarely been studied and it was also not part of the standard test set. No indication of water-extractable genotoxic potential could be observed in concentrations (usually) up to 66.7% eluate. However, considering the low number of participants and the restricted range of tested substrates, further testing is required. In addition, the high importance of the endpoint genotoxicity supports the inclusion of this test for waste eluates.

1.3.4 Comparison with Literature Data

So far, there have been very limited experiences with testing wastes in ecotoxicology (Kostka-Rick 2004). In addition, data are often published in "grey" reports (e.g. Deventer et al. 2004). Despite the fact that it was not the main objective to compare the results of this ring test with the few experiences published in the literature, some first examples are presented here (for details, see Chaps. 7–21).

Algal tests have already been used for the ecotoxicity evaluation of waste materials (Donat 2006) for many years now and have proved to be sensitive to different types of pollution. They have been incorporated in previous projects on the ecotoxicity of waste. While Devillers et al. (2006) could not statistically verify the need for algae within a test set for wastes, other authors included algae, mainly because they represent an important trophic level (Deventer 2006; Römbke & Moser 2007). According to these authors, algae react highly sensitively to different types of incineration ashes (in particular fresh ones), but since no EC_{50} values were determined direct comparisons are not possible.

Tests with the water flea *Daphnia magna* have been widely used for measuring the acute toxicity of waste water (e.g. Manusadzianas et al. 2003) and waste eluates (Pandard et al. 2006). Unfortunately, no effect data are given in the latter paper, but the species was not recommended for a basic test because of the higher sensitivity of *C. dubia*. This could not be confirmed in the ring test, especially because the *D. magna* tests had already been evaluated after 24 h (see Chap. 8). While no EC_{50} values are given for a study performed with twelve incineration ashes, it could be concluded from the LID values shown that this species reacts very sensitively when exposed to this type of waste (Römbke & Moser 2007).

Comparing the results found in the literature with the results from the luminescent bacteria tests determined in the ring test, it seems that the order of toxicity (and, in the case of INC, also the EC_{50} values) is in the acceptable range (Farré et al. 2004). A relatively low sensitivity was observed in the tests with INC as well as tests with twelve German incineration ashes (Römbke & Moser 2007). The same result was found in a study run in the German Federal State of Baden-Wuerttemberg with three different ash types (Deventer et al. 2004). In the same study the umu-test was used for the evaluation of three different types of incineration ashes, concluding that there was - as in the ring test for the waste INC - no indication of water-extractable genotoxicity (Deventer et al. 2004). The same result was to be expected for the tests with SOI and WOO, as the (known) main contaminants of these waste samples are heavy metals and PAHs, which are not known to show strong genotoxicity (Brinkmann & Eisentraeger 2008).

As of now, very few waste tests have been performed with plants. Most often, ashes from waste to energy plants have been studied. For example, in tests with another *Brassica* species (*Brassica chinensis*) Wong & Wong (1989) determined that EC_{50} values were in the range of 10–25% fly ash when mixed with sandy control soil, depending on test duration and endpoint. In tests with twelve German ashes and both species (*A. sativa, B. rapa*), a wide range of EC_{50} values (3–30%) was found, which could not be correlated with the age of these ashes or their content of heavy metals (Römbke & Moser 2007). In the classification of the ashes, both species showed a similar sensitivity. In the ring test, *B. rapa* was more sensitive than *A. sativa*. However, Quilici et al. (2004) determined a broader range (EC_{50}s between 24 and >100%) when testing French ashes bottom.

The toxicity of PAH-contaminated soil was studied several times in plant tests; however, because of the influence of the respective soil parameters and the potential impact of other contaminants, the results can hardly be compared. It appears that plants do not usually react very sensitively to PAHs in soil (EC_{50} values determined in laboratory tests after spiking PAHs often >1,000 mg/kg for both species (Henner et al. 1999)). This impression is supported by the results of the ring test in which the SOI substrate was the least toxic. Results from plant tests with copper-contaminated wood substrate are not known. However, plant tests with copper-contaminated soils or soils spiked with copper salts have been performed. Based on these studies, it is likely that the heavy metal copper, while essential for plant physiology, can also be toxic at concentrations >100 mg/kg (Hock & Elstner 1984). Higher concentrations of copper occurred in the tests described here, assuming a WOO copper content of 2,000 mg/kg (see Chap. 3). Thus, comparing the results from the literature with the results from the ring test, it seems that the order of toxicity (and, in the case of INC, also the EC_{50} values) is in an acceptable range.

In the acute earthworm tests, the LC_{50} value determined for INC (45.5%) is in good agreement with the upper end of the range of LC_{50} values determined for 12 German incineration ashes (11.5 and 43.6%; Römbke & Moser 2007). Comparable incineration ashes from France with very low heavy metal contents showed a lower toxicity in the earthworm acute test (LC_{50} values 40–76%), but no details are known about the conditions of these tests (Quilici et al. 2004). In general, soils contaminated with PAHs usually have only a mild effect on earthworms. For example, after mixing a PAH-contaminated soil (sum of EPA-PAH: about 3,000 mg/kg) with artificial soil an LC_{50} of 32.9% was determined (Potter et al. 1999). This result seems to be comparable with the results of the ring test where the substrate SOI contained about 840 mg/kg PAH. Results from earthworm tests with copper-contaminated wood substrate are not known. On the other hand, the heavy metal copper is highly toxic for earthworms. Reproduction, not mortality, is the most sensitive endpoint (Rundgren & Van Gestel 1998). For example, in the field, significant effects on earthworm populations were found, starting at copper concentrations between 30 and 100 mg/kg soil (Belotti 1998). Assuming a copper concentration of 2,000 mg/kg in WOO it seems that the effects observed in the ring test occurred in an order of magnitude comparable to the one determined in soil studies.

1.3.5 Further Work Performed in the Ring Test

In addition to the tests with the basic and additional test sets, the participants in the ring test performed more tests, partly referring to the same data but evaluating them differently. For example, 20 batches of OECD artificial soil (OECD 1984) provided by the participating laboratories were characterized and compared to investigate how much its properties differed; this increased the variability of test results (Chap. 24). Although the collected soils were declared to be prepared strictly according to OECD guidelines, they were different even at the first look and the organic carbon content in the artificial soils varied from 1.4 to 6.0%. This indicates variability in organic carbon content of peat from different sources, producers and countries. Therefore, more guidance is needed to enable laboratories to prepare a more suitable control and dilution substrate.

The relation between toxicants, their bioavailability and effects in the waste substrates was also studied, focusing on the role of copper in the aquatic test systems (Chaps. 25 and 26). First of all, eluates prepared according to CEN 14735 (2005) were characterized in detail through a combination of a pH dependence (CEN/TS 14429 (2005)) and a percolation test (CEN/TS 14405 (2004)). These methods allow for chemical speciation modeling, which provides information on the "free" forms of inorganic and organic substances in solution as opposed to dissolved organic carbon (DOC) associated and colloid bound forms. The unbound form of the substances is considered to be bioavailable and thus responsible for the observed response. In addition, the repartitioning of substances between bound and unbound forms as a result of the dilution of the eluate with a medium to quantify the ecotoxicological response that can be modeled was shown not to be a simple dilution factor as the relative concentration of the "free" forms increased upon dilution. Therefore, the combination of more detailed leachate testing and ecotoxicological studies provides a basis for more detailed evaluation of ecotoxicological response than was possible earlier. For aquatic organisms, this fraction has been shown to relate rather well with measured ecotoxicity (in particular in the case of copper in waste wood). Methodological aspects of CEN 14735 (2005) may need modification in light of the suitability of the method for organic contaminants, the choice of liquid-solid separation in view of the level of dissolved organic carbon and colloids in solution and the relevant liquid to solid ratio for specific questions related to waste use, treatment and disposal.

Finally, given the scientific value of the ring test results, and their intended use as expert inputs strengthening the basis of European environmental policy, all results were published online (/http://ecotoxwasteringtest.uba.de/h14/), using a specific software named H14-Navigator. Due to the high complexity of the ecotoxicological domain, project and data structures, an approach of an entirely Topic Maps-based application has been chosen for this purpose: The H14-Navigator is a new kind of front end operating on a single topic map, which contains information about the H14 ring test and its results. It uses ontology for physical, chemical and biological properties, and knowledge models contained in the H14 topic map to generate interface structures, knowledge-oriented access to and navigation paths through a highly networked information space (Fig. 1.5). Where appropriate, the topic view is complemented by tables providing more detailed information. An intelligent search algorithm completes front end functionalities.

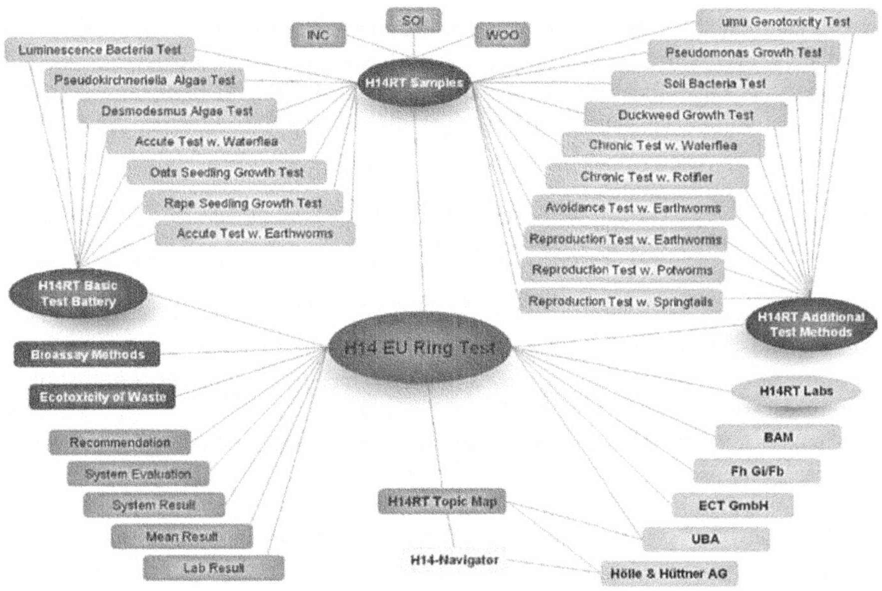

Fig. 1.5 A graph representing structures in the H14 ring test project, as well as in the H14 topic map; each node represents a topic, and each edge represents a typed relation between the involved topics, thus there is a one to one match between structural project components in reality and in the informational structures of the H14-Navigator backend

1.4 Discussion

1.4.1 Organization of the Ring Test

The results of this validation study show that the Standard CEN 14735 (CEN 2005) is basically suitable for the evaluation of the ecotoxicity of wastes under practical conditions. This statement is based on four facts:

- The high number of participating laboratories (60), delivering in total 1,048 data sets (basic and additional tests as well as reference tests) and, subsequently, the high number of valid data sets (74.1–92.6% in the basic tests).
- The low number of non-acceptable test results combined with an even lower number of statistical outliers or data outside of the range of the mean ± 2-fold SD, adding up to about 1,000 data sets (including reference tests).
- The important information gained about the performance of these tests in the reference tests conducted in parallel to the waste tests.
- The experiences gained in the additional tests can be used when alternatives and/or additions to the basic test battery are needed.

These statements are subject to the provisions set by CEN TC 292 (2005). In addition to the work related directly to this guideline, further insights were provided, concerning the understanding of the ecotoxicity of wastes (e.g. modeling the interactions between contaminants in eluates and effects on organisms).

1.4.2 Methodological Consequences

From the experience of this ring test and related activities it was concluded that waste wood containing leachable and legally admitted metal-containing wood preservatives, as well as incineration ash, are ideal substrates for reference materials in ecotoxicological testing. Apart from being readily available, stable and non-hazardous during processing, transport and application, it should even be possible to adjust their toxicological properties according to the level needed for a given purpose. Soil might be another option, however; it is recommended to select samples containing heavy metals above the geogenically typical range if a toxic effect is required and it is to remain stable over longer periods, e.g. if the material is meant for quality control purposes and is to be kept available on stock.

Open methodological questions refer to details of the preparation of the eluates (e.g. concerning the influence of the first separation step (centrifugation) or the establishment of new methods like the short-term column percolation test (ISO/DIS 21268-3 (2004e)). Some experiences (in particular the difference in reproducibility between aquatic and terrestrial tests) indicate that preparation and storage of eluates (e.g. volume collected after filtration, type of filtration membrane and so on) differed between participants, thus increasing the variability of aquatic test results. However, while no detailed recommendations can be given on how these conditions could be improved, it has to be remembered that the vast majority of ring test results indicate a high degree of robustness.

From the results of the ring test, the practicability and sensitivity of the aquatic tests using algae, daphnids and bacteria can be proved. In the case of the algae test, it was found that there are differences (non-significant) in sensitivity between both species, meaning that both species can be used. The high sensitivity of the *Daphnia* test found in the ring test was astonishing in the light of previous studies with waste eluates (e.g. Pandard et al. 2006). However, it should be kept in the basic test set as there is not enough experience available for other possible and comparable test methods. As regularly reported in the literature, freeze-dried bacteria are more sensitive towards chemical stress especially against heavy metals (Farré et al. 2004). In the waste tests, this correlation could only be confirmed for WOO, where freeze-dried bacteria were eight times more sensitive than freshly prepared or liquid dried bacteria (the latter two preparations differed only slightly in their EC50 values). Therefore, when recommending this test for waste eluates, freeze-dried bacteria should be used as the specific source for such tests.

Among the terrestrial tests, the plant test can be recommended, but further information is needed on whether both species are always necessary. In case just one species has to be selected, *B. rapa* is recommended due to its usually higher sensitivity. The acute earthworm test - despite its practicability - should be replaced because of its low sensitivity, caused probably by the fact that this test focuses on a lethal endpoint.

Some, mainly aquatic, tests can probably be automatised and/or miniaturized without a decline of practicability and sensitivity (actually, this step has already been done for the umu-genotoxicity test (Brinkmann & Eisentraeger 2008)). Microplates and other miniaturized systems using algae and *Daphnia* are already in use in most labs and they seem to be equally or slightly more sensitive than the classic set up. Based on these findings, it is recommended that Algaltoxkit and Daphtoxkit should be taken into consideration as valid alternatives to the standard algal growth inhibition assay with *P. subcapitata* and the acute *D. magna* test for the evaluation of the toxicity of solid waste eluates.

Further research is needed in the area of additional tests, for, with the exception of the Lemna test, the number of data sets for these tests was too small to perform a detailed evaluation. Their sensitivity is partly high, meaning that for example, the Collembola or earthworm reproduction test might replace the earthworm acute test. Special attention should be given to the earthworm avoidance test and, in particular, the Arthrobacter test, which may be included in the basic test set due to their combination of practicability and sensitivity. In any case it is to be investigated whether such methods can be used for the testing of specific waste material which cannot be tested with the basic test set.

Finally, the experiences from the many reference tests will influence the content of the currently used test guidelines. For example, the reference substance chloroacetamide used in the earthworm acute test should be replaced by a chemical like boric acid which is not toxic to humans (Römbke & Ahtiainen 2007). Boric acid can also be recommended as the standard reference substance in plant tests. In addition, details of the performance (in particular the validity criteria and the toxic effect range of reference results) of the algae and luminescent bacteria tests have to be clarified.

The results of the ring tests have clearly shown that biological test systems are an implementable and reliable instrument for the ecotoxicological characterization of wastes.

For waste types of unknown and/or complex composition, biological tests are a helpful instrument for the identification of hazardous waste, listed in mirror entries of the European waste list and for the risk assessment of waste, e.g. in reuse-scenarios. The development and application of the Extended Limit test design (see Chap. 29) is a good instrument to assess the ecotoxicological potential of waste with an economically justifiable and time adequate test effort. Furthermore, biological tests can be integrated in waste risk assessments enabling a more sustainable management of waste. The results and experiences from the ring test should be transferred to a harmonized approach for the implementation of the hazard criterion H14, as part of a future amendment of the European Waste List.

1.5 Recommendations

Summarizing the outcome of this ring test the following recommendations can be made:

- Despite some problems concerning the preparation of eluates (changes in the procedures described by CEN 14735 (2005) may have caused some of the variability observed in the aquatic tests) the experiences of the ring test support the proposals made in CEN 14735 (2005) concerning the performance of such tests.
- Special consideration should be given to the preparation of the terrestrial control and dilution medium (OECD artificial soil), as the variability of its properties may have contributed to the overall variability of test results.
- In general, the methods used do not need to be modified considerably for the testing of wastes (e.g., further guidance concerning substrate moisture handling should be given). Other methodological changes refer to test properties, which should be modified independently from the test item (e.g. the exchange of reference substances, and the use of ring test data for an improved range of toxicity criteria when testing reference substances).
- With small modifications (exchange of the acute earthworm test with a more sensitive soil invertebrate test) the basic test set as used in the ring test is considered to be a good starting point for the hazard and risk assessment of wastes. Available information points out that besides "traditional" test versions with daphnids and algae, miniaturized microplate methods can also be used.
- From the experiences with the additional test set, it is possible to identify alternatives to the current basic test set, especially when looking for terrestrial test methods with short durations (e.g. the earthworm avoidance test and, in particular, the Arthrobacter test, due to their combination of practicability and sensitivity).
- More importantly, the same experiences can be used to find test methods for those wastes which cannot be tested with the basic tests (e.g. the Lemna test for turbid eluates which cannot be tested in the algae test).
- The comparison of repeatability standard deviation s_r and reproducibility standard deviation s_R revealed that about 85% are within a range for the s_r/s_R ratio from 2 to 8, taking into account that this range already covers influence factors such as eluate preparation or the use of different test species. Considering also the differences between chemical/physicochemical and ecotoxicological ring tests, this variation is considered acceptable. Certainly, reproducibility and repeatability should be improved by constantly training the test performers and a well established test quality assessment.
- Independent of the aims of the ring test, the data collected can be used in combination with a detailed characterization of the eluates for the modeling of the bioavailable fraction of the respective contaminant, thus allowing the understanding of ecotoxicological effects observed in the tests. At the same time, such modeling can help to improve the methodological aspects of CEN guideline 14735 (2005).

- In any case, more experience with different types of waste materials is necessary for all the ecotoxicological test systems discussed so far, to define the range of materials which can be tested successfully. The same information is also necessary for the definition of limit values. In this context, the collection of all relevant data, as well as legal requirements in a data-base, together with appropriate software like the H14-Navigator, is recommended, to improve data evaluation.
- In line with the recommendations made in the literature (e.g. Wundram & Bahadir 1999; Pandard et al. 2006; Wilke et al. 2007) a combination of the results of a set of biological tests and chemical residue analysis is needed for an ecotoxicological characterization of wastes.

Acknowledgements We thank all the participants in the ring test for the dedicated work they carried out: Adam Scheffczyk, Adolf Eisentraeger, Andrea Ruf, Andreas Fangmeier, Andreas Haller, Anne van Cauwenberge, Berndt-Michael Wilke, Bernhard Foerster, Bona Griselli, Brigitte von Danwitz, Christiane Fahnenstich, Christine Bazin, Christoph Hafner, Claire van der Wielen, Corinna Firla, Detlef Dengler, Dirk Maletzki, E. Garcia John, Elisabetta Ciccarelli, Elsa Mendonça, Franz Rittenschober, Frank Riepert, Frederic Garrivier, Gabriela Sbrilli, Goeran Dave, Greet De Messemaeker, Gregoria Carbonell, Guido Persoone, Hansjuergen Krist, Hege Stubberud, Helga Neumann-Hensel, Henk te Winkel, Henner Hollert, Ines Fritz, Jaap Postma, Jakub Hofman, Juergen Zipperle, Kathleen O'Rourke, Kerstin Hund-Rinke, Kirit Wadhia, M.J Jourdain, Maike Schaefer, Maria Ana Cunha, Marit Kolb, Markus Barth, Martina Solenská, Mónica Amorim, Monika Pattard, Nadine Pounds, Pascal Pandard, Paulo Sousa, Pilar Andrés, Přemysl Soldán, Ralf Petto, Reinhilde Weltens, Roland Weiss, Rolf Altenburger, Roman Kuperman, Rune Berglind, Ruud Meij, Stefania Balzamo, Sylvia Waara, Thomas Junker, Thomas Moser, Tiago Natal da Luz, Toni Ratte, Tristano Leoni, Vít Matějů, Vladimír Kočí, Yves Barthel.

Chapter 2
Legal and Organisational Background of the Ring Test

H. Moser and H. Kessler

Abstract The European Waste List is a harmonized list of waste types, which classifies waste in hazardous and non-hazardous waste codes. For this classification, 15 hazard criteria have been derived from the Council Directive 91/689/EEC on hazardous waste. Some of the hazard criteria are based on the presence of dangerous substances. The criterion H14, "ecotoxic", lacks an assessment and testing strategy and no specific threshold values have been defined so far. Following a scientific workshop in September 2005, a European ring test on ecotoxicological characterization was organized to define suitable test methods for the biological assessment of waste and waste eluates. The selection of a basic and an additional test set was based on the information in the annexure of the European standard EN 14735. The ring test therefore was conducted as an evaluation study of EN 14735. After a high-quality sample preparation, three waste materials (incineration ash, contaminated soil, and waste wood) were tested by 60 labs. Besides the use of biological test methods, the waste eluates and the waste itself were chemically characterized. In addition to information regarding the quality of artificial soil and the statistical performance of the test systems, the ring test generated a lot of data and experience in ecotoxicological characterization of waste and its methodology. All results and documents will be accessible via a web based data bank application (see Chap. 27).

Keywords European waste list, Waste legislation, Test battery, Ring test structure

H. Moser (✉)
Federal Environment Agency, Woerlitzer Platz 1, 06844, Dessau, Germany

H. Moser and J. Römbke (eds.), *Ecotoxicological Characterization of Waste*,
DOI: 10.1007/978-0-387-88959-7_2, © Springer Science+Business Media, LLC 2009

2.1 The European Waste List

In 1994, a comprehensive list of all wastes was compiled, pursuant to the Waste Framework Directive (94/904/EC), and the so-called European Waste Catalogue (EWC 94/3/EC). In addition to the EWC, a List of Hazardous Waste (HWL) was created, based on the Hazardous Waste Directive (91/689/EEC), to provide a precise and uniform Europe-wide definition of hazardous waste and ensure the correct management and regulation of hazardous waste. By merging the EWC and the HWL, and updating and in some parts significantly extending the waste lists, the European Waste List (EWL 2000/532/EC and its amendments) was formed. It was made obligatory for Member States to implement the EWL by January 2002. So far, the list has been amended three times and can be revised when necessary, according to the Waste Framework Directive.

The European Waste List is a harmonized list of about 850 different waste types, thus forming a consistent waste classification system across the EU. It is intended to be a catalogue of all wastes, grouped according to generic industry, process and waste type. The waste types in the 20 main chapters are classified by a six-digit-code, which differentiates between hazardous and non-hazardous waste by marking hazardous waste with an asterisk (*). The EWL defines 405 waste types as hazardous and 200 waste types in so-called "mirror entries". Mirror entries summarize two or more codes for a group of waste materials, which can either be hazardous or non-hazardous, depending on the composition of the waste.

Table 2.1 Definition of the hazard criteria (Annex III, 91/689/EEC)

H1	"Explosive" - may explode when under effect of flame or sensitive to shocks
H2	"Oxidising" - exhibit highly exothermic reactions in contact with other substances
H3A	"Highly flammable" - Liquids with flash point <21°C, catch fire on contact with air, readily ignited, flammable gases, evolve highly flammable gas on contact with water
H3B	"Flammable" - Liquids having flashpoint between 21 and 55°C
H4	"Irritant" - Non corrosive substances which cause inflammation on contact with skin
H5	"Harmful" - if inhaled, ingested or penetrate the skin may involve limited health risks
H6	"Toxic" - may involve serious, acute or chronic health risks and even death
H7	"Carcinogenic" - may induce cancer or increase its incidence
H8	"Corrosive" - may destroy living tissue on contact
H9	"Infectious" - substances containing viable micro-organisms or their toxins which known or believed to cause disease in man or other living organisms
H10	"Toxic for reproduction" - affect the incidence of non-heritable adverse effects in the progeny and/or male or female reproductive functions or capacity
H11	"Mutagenic" - may induce hereditary genetic defects or increase their incidence
H12	Substances which release toxic gases in contact with water, air or an acid
H13	Wastes capable by any means after disposal of yielding another substance which possess any of the characteristics listed by this annex
H14	"Ecotoxic" - may present risks for one or more sectors of the environment

For the classification of waste as hazardous, hazard criteria (H1 to H15, see Table 2.1) are fundamentally defined in Annexure III of the Council Directive 91/689/EEC on hazardous waste, but for some of the listed H criteria, methods for implementation are still missing. The criteria, e.g. H3 to H8, H10, H11 (flammable, irritant, harmful, toxic, carcinogenic, corrosive, teratogenic, mutagenic) are based on the concentration of dangerous substances. The criterion H14 "ecotoxic" lacks an assessment and testing strategy. In addition, no specific threshold values have been defined so far.

2.2 Background of the Ring Test

In September 2005, the experts at an international workshop organized by the German Federal Environment Agency and the Joint Research Centre agreed that the ecotoxicological potential of waste can only be determined by biological test systems. A distinct need for a harmonized test battery was identified, to be developed and validated in the framework of the European Normalization Committee CEN TC 292, which works on the scientific standardization of methods for an assessment of solid waste and waste eluates (Gawlik & Moser 2005). As a result of the cooperation of the experts from the European Member States in the Normalization Committee, the guideline EN 14735 "Characterization of waste - Preparation of waste samples for ecotoxicity tests" was published, in which a collection of biotests applicable in waste is listed in EN 14735 Annexure B. A general agreement was reached by the workshop expert group that a suitable test battery will address the ecotoxicity of waste by using test organisms as representatives for various ecosystems or compartments and various trophic levels.

The validation of test procedures is an essential part in the standardization process of each ecotoxicological test. The framework of European standardization (CEN) demands an internationally conducted ring test with typical test substrates which provide valid information on the practicability and the reproducibility of the results for the test procedures employed. Therefore, the main aim of this ring test was the establishment of a test battery for the ecotoxicological characterization of wastes, using the EN 14735 standard "Characterization of waste - Preparation of waste samples for ecotoxicity tests" (CEN 2005) as the basis. In this context, the following issues were addressed:

- The validation of the preparation of the test substrates according to this standard
- The assessment of the suitability of the basic test battery (i.e. three aquatic and two terrestrial tests) in terms of practicability and sensitivity, including the question whether modifications of existing test methods are necessary or not
- The evaluation of the uncertainty level of the results for the various tests
- The identification of recommendations concerning a test battery for routine use

In addition, the suitability of additional test methods, not considered as part of the routine test battery, as well as basic issues such as the properties of control substrates, was studied.

Organizational Structure of the Ring Test

The ring test was organized as a project of the research program UFOPLAN of the German Federal Ministry of Environment, Nature Conservation and Nuclear Safety. The German Federal Environment Agency served as project executing organization.

The main activities of the ring test were centralized (Fig. 2.1)

1. Federal Environment Agency (UBA) Dessau-Rosslau, Germany:
   - General organization and communication between participants and third parties, as well as the organization of a demonstration workshop (June 21–22, 2006; Berlin, Germany), the final meeting (June 29, 2007; Berlin, Germany) and the final report. In this role, the UBA was supported by a Scientific Advisory Board, consisting of four scientists including a representative of CEN.
2. Federal Institute for Materials Research and Testing (BAM), Berlin, Germany:
   - Preparation, characterization and distribution of the three test substrates (differing greatly in their chemical and physical properties).

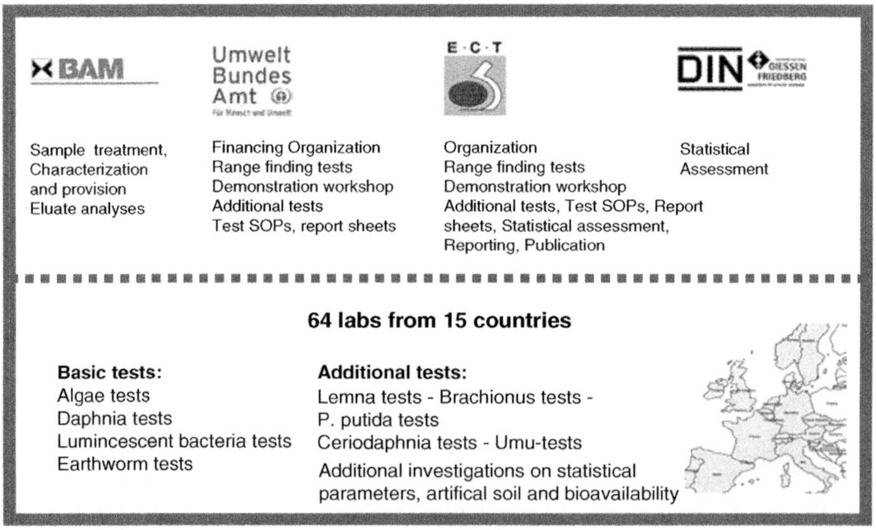

Fig. 2.1 Overview on the organization of the EU ring test

3. ECT Oekotoxikologie GmbH (ECT) Floersheim, Germany:

   - Scientific co-ordination of the ring test, including the preparation of Standard Operation Procedures (SOPs) and forms (paper, EXCEL-file), the organization of the demonstration workshop (December 15, 2005; Berlin, Germany) and the compilation of the final report.

4 University of Applied Sciences (FH) Giessen-Friedberg, Giessen, Germany:

   - Statistical assessment of the ring test results according to ISO 5725-2 (2002).

5. Hoelle & Huettner AG Tuebingen, Germany

   - Development and establishment of a data bank including all documents, raw data, exposure and statistical results derived from the ring test, accessible for further expert assessment.

A total of 60 laboratories returned data to the organizers. The participants were based in 15 countries: Austria (2), Belgium (5), Czech Republic (4), France (4), Germany (23), Ireland (1), Italy (5), The Netherlands (1), Norway (1), Portugal (4), Slovak Republic (1), Spain (4), Sweden (3), UK (1) and USA (1). The participants represented 16 universities, 16 public research institutions and 28 contract laboratories. The level of experience regarding ecotoxicity tests for waste from the laboratories was high and during the demonstration workshop the characteristics of the test battery and the specific requirements of the ring test waste materials were addressed.

2.3 Test Materials

The test materials were selected after consultation with CEN's committee CEN TC 292/WG7. Their processing, characterization and distribution were conducted by BAM (Berlin). The three materials covered a wide range of toxicity and different matrices of the main waste flows:

- Municipal waste incineration ash (INC), originating from a Dutch incineration plant for municipal waste ("bottom ash") and strongly contaminated by heavy metals.
- PAH contaminated soil (SOI), from a former gasworks site in Berlin (Germany).
- Waste wood (WOO), consisting of a mixture of treated and untreated wood samples from a commercial timber processing plant, which were treated with copper-based wood preservatives according to the regulations of different European countries.

A more detailed description of the samples and the sample treatment is given in Chap. 4.

2.4 Biological Test Systems

2.4.1 Basic Test Battery

The European standard EN 14735 provides a collection of appropriate test methods for the ecotoxicological characterization of waste and waste eluates. Based on this collection, a basic test set, containing three aquatic and two terrestrial procedures, was evaluated within the ring test. Table 2.2 gives a short description of the basic test set; more details on the test performance are given in Part IV.

Only tests that fulfilled the validity criteria given in the individual guidelines, which included a reference test (not in the case of plant tests, as no reference substance is formally required for this) and which were performed according to the Standard Operation Procedures (SOPs) provided by the organizers, were considered acceptable for assessment. In the case of the luminescent bacteria tests, sensitivity is mainly proved by performing limit tests; therefore, reference tests following a dose-response design were not used for the acceptance check of these tests.

A total of 634 data sets were produced in the basic tests. The results of 143 Algae tests, 161 Daphnia tests, 154 luminescent bacteria tests, 52 earthworm tests and 124 plant tests were submitted by the participants.

Table 2.2 Brief overview on the five tests belonging to the basic test battery

Name	Guideline	Species
Eluate (aquatic) waste tests		
Freshwater algal growth inhibition test with *Desmodesmus subspicatus* and *Pseudokirchneriella subcapitata*	ISO 8692 (2004)	*D. subspicatus* *P. subcapitata*
Determination of the inhibition of the mobility of *Daphnia magna* Straus (*Cladocera, Crustacea*) - Acute toxicity test	ISO 6341 (1996)	*Daphnia magna*
Determination of the inhibitory effect on the light emission of *Vibrio fischeri* (Luminescent bacteria test)	ISO 11348–1/2/3 (1999)	*Vibrio fischeri* (three sources)
Solid (terrestrial) waste tests		
Soil quality - Determination of the effects of pollutants on soil flora - Part 2: Effects of chemicals on the emergence and growth of higher plants	ISO 11268–1 (1997)	*Eisenia fetida, Eisenia andrei*
Soil quality - Effects of pollutants on earthworms - Part 1: Determination of acute toxicity using artificial soil substrate	ISO 11268–2 (2004b)	*Avena sativa, Brassica rapa*

2.4.2 Additional Test Systems

In addition to the basic test battery, ten additional test systems from the annexure of EN 14735 were performed in the ring test. These test methods are briefly described in Table 2.3. More details of the test method and the results are given in Part V.

As there was not much experience regarding these additional test methods in general, and as no SOPs were prepared for them, the acceptance of test data sets was handled on a case-by-case basis (i.e. in general no tests with a reference substance were required).

Table 2.3 Brief overview on the ten tests belonging to the additional test battery

Name	Guideline	Species
Eluate (aquatic) waste tests		
Water quality - Determination of the toxic effect of water constituents and waste water on duckweed (*Lemna minor*)-Duckweed growth inhibition test	ISO 20079 (01.11.2005)	*Lemna minor*
Determination of chronic toxicity to *Brachionus calyciflorus* in 48 h - population growth inhibition test	ISO/CD 20666 (2007)	*Brachionus calyciflorus*
Water quality - *Pseudomonas putida* growth inhibition test (*Pseudomonas* cell multiplication inhibition test)	ISO 10712 (1995)	*Pseudomonas putida*
Determination of chronic toxicity to *Ceriodaphnia dubia* in 7 days - Population growth inhibition test	AFNOR 90–376 (2000)	*Ceriodaphnia dubia*
Determination of the genotoxicity of water and waste water using the umu test	ISO 13829 (2000)	*Salmonella typhimurium*
Solid (terrestrial) waste tests		
Effects of pollutants on collembola (*Folsomia candida*) - Method for the determination of effects on reproduction	ISO 11267 (1999)	*Folsomia candida*
Effects of pollutants on earthworms (*Eisenia fetida*) - Part 2: Determination of effects on reproduction	ISO 11268–2 (1998)	*Eisenia fetida, Eisenia andrei*
Enchytraeid Reproduction Test	ISO 16387 (2004d)	*E. albidus, E. crypticus*
Soil quality - Avoidance test for testing the quality of soils and effects of chemicals on behaviour - Part 1: Test with earthworms (*Eisenia fetida* and *Eisenia andrei*)	ISO 17512–1 (2006)	*Eisenia fetida, Eisenia andrei*
Determination of the inhibition of dehydrogenase activity of *Arthrobacter globiformis* (Solid contact test using the redox dye resazurine)	ISO WD 10871	*Arthrobacter globiformis*

A total of 196 data sets was produced in the additional test battery: 51 *Lemna* tests, 10 *Brachionus* tests, 12 *Pseudomonas putida* bacteria tests, 10 *Ceriodaphnia* tests and 23 umu tests were submitted by the participants for waste eluates. The respective numbers for the solid waste tests are: 21 Collembola tests, 17 earthworm reproduction tests, 12 earthworm avoidance tests, 12 enchytraeid tests and 28 *Arthrobacter* tests.

2.4.3 Test Data Evaluation

The evaluation of results was done following ISO 5725-1 (1994) and ISO 5725-2 (2002), with the modification of the calculations being based on the logarithms of the EC_{50} values. All EC_{50} calculations were performed using the statistical program ToxRat (2006); i.e. in those cases where test participants did not use probit analysis themselves, the respective data sets were re-calculated to improve comparability of the results. As the test procedures in question are tedious and elaborate, compared with trace analytical investigations, replicate determinations within a given laboratory and within short intervals of time were not possible. Thus, repeatability and reproducibility were evaluated using the results on the variability in different laboratories.

In addition, results that were outside the range of geometric mean ± twofold standard deviation (warning limit approach; Environment Canada 2005) were taken into account. Another aspect considered was the factor between minimum and maximum values of each respective test. Details of the evaluation process are given in Chap. 4. Results regarding reproducibility and repeatability of the data are also given in Chap. 22.

2.4.4 Data Integration for Further Assessments

As the ring test is a good platform to gather additional experiences on the methodology and the performance of the test methods, all results and documents have been integrated in a web-based data bank called "H14-Navigator". The compilation of all ring test results, i.e. test specific experiences such as, e.g. reference tests, analytical information about the wastes and eluates, and the respective report sheets, forms a reliable structure for further data assessment or the integration of additional investigations. More details on the H14-navigator are given in Chap. 27.

Chapter 3
Selection and Characterization of Test Samples and Eluates

R. Becker, U. Kalbe, and A. Buchholz

Abstract This chapter comprises the origin of the starting materials and the procedures employed, such as drying, sieving where appropriate, homogenisation, bottling, and relevant chemical characterizations. Samples of municipal waste incineration ash, contaminated soil from a former gasworks site, and waste wood with a typically copper-based preservative were selected, to cover a range of toxicologically relevant properties and effects, different matrices, and some of the most important waste types in Europe. Individual units of each of the materials, containing between 100 g and 12 kg, with an upper particle size limit of 4 mm, were bottled. The homogeneity of selected chemical parameters (heavy metals, polycyclic aromatics, Cu) was tested in accordance with the respective guidelines. Sixty participants from 15 countries received between 100 g and 30 kg of each of the materials according to the extent of their specific test program. Ring test results on the determination of pH and selected metals in the eluates as well as the water holding capacity of the waste materials are presented and briefly discussed.

Keywords Municipial waste incineration ash, Gaswork contaminated soil, Copper-contaminated wood, Bulk material, Eluates

3.1 Introduction

Most matrix reference materials used for interlaboratory comparisons are selected because one or more specific property values are directly determined by the measurement method in question. Such chemical (often trace analytical) parameters are relatively easily explored with regard to composition and concentration before the

R. Becker (✉)
Federal Institute for Materials Research and Testing (BAM)
Berlin, Germany

H. Moser and J. Römbke (eds.), *Ecotoxicological Characterization of Waste*,
DOI: 10.1007/978-0-387-88959-7_3, © Springer Science+Business Media, LLC 2009

starting material is sampled. In the present case the ecotoxicological response had to be anticipated by educated judgement on the basis of general knowledge on the available substrates. The objective was to provide a range of matrices reflecting the major waste flows in Europe, different in their ecotoxicological response to the test under scrutiny. The starting materials selected were a compromise between the relevance of matrix and pollutant, the available technical facilities for their processing at BAM, the despatch to the participants (some based outside the EU) and the handling in their laboratories. Stench, strongly poisonous substrates and instable ecotoxicological properties had to be avoided. Incineration ash is currently one of the major waste types, with further increase in volume to be expected. Among the major organic substrates, waste wood is currently a subject of concern due to the leaching of Cu-based preservatives (Katz & Salem 2005). Polycyclic aromatic hydrocarbons (PAH) are among the most widely observed organic pollutants in industrially contaminated soils. Thus, incineration ash of municipal waste, industrial soil contaminated with aromatic hydrocarbons, and waste wood treated with currently permitted Cu-containing preservatives were the substrates of choice. The materials were adequately processed, homogenised and bottled, using specifically adapted procedures to ensure the best feasible homogeneity within each unit and the whole batch as detailed in the following sections.

3.2 Homogenization and Bottling

The techniques employed for the three substrates had to be adjusted according to the state of the starting materials upon delivery, the water content, particle size distribution and flow ability. The required upper limit for the particle size was 4 mm in all cases (in accordance with EN14375). The homogenized substrates INC and SOI were packed in PTFE/glass fabric foil (Armbrecht & Matthes GmbH, Gevelsberg, Germany) and bottled in 10 L tin-plate containers (230/217 × 281 mm, Dosen-Zentrale Zuechner GmbH, Hilden, Germany).

This container type is permitted for the transport of dangerous goods (Fig. 3.1). In the case of WOO, the containers (volume: 30 L, 328/312 × 410 mm, Dosen-Zentrale Zuechner GmbH) were laminated inside with self-adhesive PTFE/glass fabric (Armbrecht & Matthes GmbH). Small amounts (below 3 kg) of the materials were bottled in amber glass containers or plastic bags (WOO only). The individual demand of each participant for the three materials was established at the beginning of the project.

3.2.1 Incineration Ash (INC)

The starting material (719 kg) was obtained from a Dutch incineration plant for municipal waste and was recovered there as bottom ash. It was wet and composed of a variety of metal parts, stones and a mixture of caked slag-like materials displaying

Fig. 3.1 Tin-plate containers and large rotary tube sample divider

all the colours of the rainbow. As grinding was technically impossible and the ash portion was required for further processing, the material was dried in a gentle air stream in a 150-l Eirich intensive mixer (type Eirich R09 W, Eirich, Hardheim, Germany). then, the dried substrate was sieved in two steps (<10 mm, then <4 mm). The obtained material was light gray and dry. The major portion was significantly below 4 mm, with optically visible smaller portions up to 4 mm, composed of stones, glass, metals and slag-like material. The homogenisation was achieved by means of a large rotary tube sample divider (Fig. 3.1) modified to four discharge channels (Retsch, Haan, Germany). The bulk material was partitioned using the so-called "cross-riffling" procedure (Van der Veen & Nater, 1993; Berger et al. 2004) outlined in Fig. 3.2. A total 318 kg were bottled and distributed among the participants according to their specific demand.

The starting material was divided into four subsamples A-D, which were further sub-divided. These subdivided samples were unified in the indicated order to give I-IV, and were composed of equal portions of A-D. Further dividing of the sub-divided samples I-IV yielded the 16 subsamples from which the units for the participants were bottled by further division, using spinning-rifflers or sample dividers of appropriate size (Retsch RT 100 with or ten tubes).

3.2.2 Contaminated Soil (SOI)

The polluted sandy soil came from a former gasworks site in Berlin (Germany). The sample material was sieved using a vibrating sieve (type 930145, Braeuer, Dornbirn, Austria) and simultaneously dried to constant weight. The fraction <4 mm was homogenised and a total of 680 kg was bottled using the procedure described in Fig. 3.2. As a specific feature, the field-wet starting material appeared

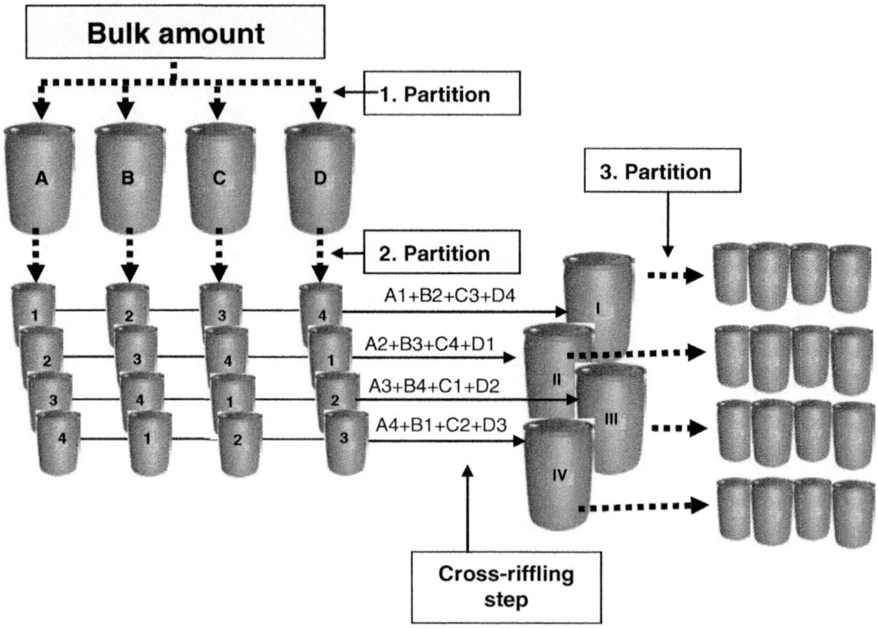

Fig. 3.2 Homogenisation and bottling of the materials

as normal sandy soil and only after drying and sieving could any further handling by the technical personnel be done, wearing specific full-body protection due to the tar dust it contained.

3.2.3 Waste Wood (WOO)

This substrate was a mixture of treated and untreated wood samples from a commercial timber processing plant, which were treated with copper-based wood preservatives according to the regulations of various European countries. The starting material was ground with a cutting mill (<4 mm, SM 2000, Retsch, Haan, Germany) by a contractor. 900 kg were received at BAM and homogenized using the procedure in Fig. 3.2. 617 kg were bottled in containers between 0.5 and 10 kg and despatched to the participants. This substrate demanded a complicated homogenization procedure due to its low bulk density and poor flowability. Thus, the bulk material had to be divided, back-mixed and finally bottled in small portions manually.

3.3 Homogeneity Testing

International guides and standards, developed to cover various aspects of reference material characterization and interlaboratory study designs such as ISO Guides 43 ISO 13528 and ISO 5725, suggest that a certain quantity of inhomogeneous material should be taken into calculation before the start of the interlaboratory comparison, to account for the overall variability of the results. In this case, however, a homogeneity study on the ecotoxicological materials was not feasible. Moreover, the preparation of the three batches was done with the best possible efficiency and within the set time frame, no repetition of the whole process for the sake of homogeneity would have been possible. Therefore, for each bottled material a parameter or a group of parameters, which were likely to contribute to the overall ecotoxic effect and could be efficiently determined using validated trace analytical methods, were selected. The sample intake for the respective analyses was smaller than that for the biological tests. Therefore, the homogeneity of the sample intake in the biotests were better than that of the analyses. Table 3.1 gives the parameters, methods and results of the homogeneity studies on INC, SOI, and WOO.

INC samples (165–190 g) from ten different bottled units and four different samples (43–45 g) from one unit were milled separately with a planetary ball mill (Type “pulverisette”, Fritsch, Idar-Oberstein, Germany) at 250 rounds/min for 4 × 4 min (all devices made of zirconium oxide). The product was sieved and the fraction >250 μm was milled with a Mixer Mill MM 301 (Retsch, Haan, Germany) and mixed with the sieving fraction <250 μm. Three grams of each milled sample were digested with aqua regia according to ISO 11466:1995 and the elements in Table 3.1 were determined under repeatability conditions. There is a strong difference between the variabilities of the contents of the metals. It is assumed that the high variability of the content of some metals - most conspicuous in the case of Cu and Pb - is due to a so-called “nugget-effect”. Presumably, some metals tend to occur

Table 3.1 Results of the homogeneity study on INC

Metal	Method[a]	Mean[b] (mg/kg)	Variability between bottled units (%)
Cd	ICP-OES	6.6	7.3
As	ICP-OES	7.4	6.9
Co	ICP-OES	19	7.2
Cr	ICP-OES	212	6.3
Cu	ICP-OES	6,500	48
Hg	CV-AAS	37	17
Mn	ICP-OES	800	5.7
Ni	ICP-OES	211	11.2
Pb	ICP-OES	1,623	18.7
V	ICP-OES	42	2.7
Zn	ICP-OES	2,639	7.4

[a]Inductive coupled plasma/optical emission spectrometry (ICP-OES) or cold vapour/atomic absorption spectrometry (CV-AAS)
[b]Grand mean of the homogeneity study

in compact particles, e.g., wire pieces in the ash, while others are more evenly distributed in the waste or during the incineration process. It should however be noted that such compact metal particles are hardly available in the test organisms or eluted in the aquatic tests, respectively. Therefore, the homogeneity of the portion of these metals which are available to the test organisms is probably comparable to that of the other metals.

The homogeneity of SOI was investigated routinely for soil reference materials. 5 g of soil were sampled four times from different sources of the ten bottled units. The 40 samples were analysed using pressurised fluid extraction (ASE 200, Dionex, Idstein, Germany) with methanol (two cycles) followed by HPLC-DAD (acetonitrile/water) determination of the 16 EPA-PAH according to ISO 13877:1998 (Table 3.2).

The similarity of the variabilities of the contents of individual PAHs within and between the bottled units indicated that this property of the material is evenly distributed over the batch. The wide particle size range of up to 4 mm leads to a greater variability compared to soil reference materials meant for chemical trace analysis. However, the relative standard deviations for the higher PAH approach the values typically observed for reference materials with a narrower particle size distribution (up to 0.25 mm). The relative low contamination with mineral oil hydrocarbons (TPH) is often observed on gasworks sites and is close to the background value of industrial areas (Table 3.3).

Table 3.2 Results of the homogeneity study on SOI

PAH congener	Mean (mg/kg)	SD_{within}[a] (mg/kg)	Rel. SD_{within}[a] (%)	$SD_{between}$[a] (mg/kg)	Rel. $SD_{between}$[a] (%)
Naphthalene	n.d.[b]	–	–	–	–
Acenaphthylene	n.d.	–	–	–	–
Acenaphthene	7.18	2.20	30.6	1.91	26.6
Fluorene	4.16	0.98	23.6	0.88	21.0
Phenanthrene	69.1	13.09	19.0	13.4	19.5
Anthracene	23.4	4.34	18.5	4.55	19.4
Fluoranthene	181.6	26.6	14.7	26.5	14.6
Pyrene	146.0	20.5	14.1	20.9	14.3
Benz[a]anthracene	87.2	11.6	13.3	11.2	12.9
Chrysene	69.4	9.14	13.2	8.74	12.6
Benzo[b]fluoranthene	78.6	9.92	12.6	8.04	10.2
Benzo[k]fluoranthene	31.0	3.71	11.9	3.57	11.5
Benzo[a]pyrene	59.0	7.36	12.5	6.15	10.4
Dibenz[ah]anthracene	9.37	1.15	12.2	0.84	8.92
Benzo[ghi]perylene	34.7	3.87	11.2	2.92	8.41
Indeno[1,2,3-cd]pyrene	35.2	3.66	10.4	3.11	8.83

[a](Relative) standard deviation within or between the units
[b]Not detected

Table 3.3 Characterization of INC, SOI, and WOO

Parameter	Method	Intake[a]	INC	SOI	WOO
pH; water, 26.5°C	ISO 10390:2005	–	10.42 ± 0.04	8.36 ± 0.02	5.41 ± 0.04
C:H:N analysis (%)	–	100 mg	C: 1.16 ± 0.989 N: 0.035 ± 0.002 H: 0.423 ± 0.002	C: 4.98 ± 2.47 N: 0.085 ± 0.028 H: 0.304 ± 0.028	C: 46.44 ± 0.4 N: 0.277 ± 0.091 H: 6.535 ± 0.044
Total carbon (%)	–	80 mg	–	2.73 ± 0.99	37.4 ± 1.12
Total inorg. C (%)	–	80 mg	–	0.099 ± 0.004	<0.2
Water content[b] (%)	ISO 11465:1993	5 g	2.08 ± 0.28	0.980 ± 0.014	8.77 ± 0.11
Water content (%)	*Karl-Fischer-titration*	50 mg	1.99 ± 0.04	1.29 ± 0.68	8.23 ± 0.10
TPH[c] (mg/kg)	ISO 16703:2004	5 g	–	152 ± 30	–

[a]Amount of substance for one determination
[b]105C (INC, SOI); 103°C (WOO) according to DIN 52183:1977
[c]Total petrol hydrocarbons (gas chromatography)

In the case of WOO, nine samples taken from different bottled units were analysed in triplicate (between-unit variability) and four samples taken from different spots of one unit were analysed to assess the within-unit variability. For each determination of copper, 2 g of the sample were combusted and ashed. The ash was digested with H_2SO_4/HNO_3. The solution was diluted and analysed using AAS. The Cu content of the units was 2.11 ± 0.06 g/kg and 2.09 ± 0.07 g/kg within the unit. This is an excellent homogenous distribution of the contained Cu.

3.4 Preparation of Wastes and Eluates in the Participating Laboratories

Participants were provided with standard operating procedures on how to apply the individual tests to the waste samples and the eluates. Each waste sample had to be diluted with an appropriate dilution material (e. g. OECD artificial soil in the invertebrate tests and LUFA standard soil 2.2 in the plant tests) that had to be used for the control as well. For all treatments (controls and dilution steps) water had to be added in such a way that the requirements of the respective test organisms were met.

Waste eluates had to be prepared using a ratio of one part of the waste sample and ten parts of water (*L/S* 10) according to EN 12457-2. Thus, 95 ± 5 g of the respective waste sample (calculated as dry mass) had to be filled into a 1 l flask. Water had to be added to ensure an *L/S* ration of 10 l water/kg dry mass within a variation of 2%. The water content of the received waste sample had to be considered as part of the total volume of leachant. The closed flask had to be agitated with 5–10 rpm for 24 ± 0.5 h at 20 ± 0.5°C. If a larger volume was needed, the eluates of several 1 l flasks could be added before being used in the tests. If the sedimentation of the suspended waste material was not achieved within 15 ± 5 min, the mixture had to be centrifuged for 30 min at 2,500 × *g* and filtered with a membrane filter of 0.45 µm pore size (e.g. PTFA or nylon filter). Rinsing the filter with water was definitely forbidden. Volume, pH, and conductivity had to be measured immediately after filtration. Additionally, the oxygen content as well as coloration and/or turbidity had to be determined to facilitate the performance and evaluation of the biotests. In order to perform tests following the dose-response design, the eluates had to be diluted using an appropriate dilution material (e.g. OECD reconstituted water). The same water had to be used for the control.

Details of the preparation for and design of the individual terrestrial and aquatic tests were laid down in the respective guidance documents given to the participants.

3.5 Characterization of Wastes and Eluates

Table 3.3 contains the results of a general characterization of the three materials. The strongly alkaline reaction of INC was expected and probably contributed to the ecotoxic effect in some of the tests.

Besides the biological tests, participants were encouraged to provide additional measurements on the test substrates, such as the water holding capacity, the pH of the eluates used for the aquatic tests, and the content of copper (a matter of concern in incineration ash as well (Lin et al. 2004)) and optionally, more metals in the eluates. It should however be noted that any interpretation of the measurement results, e.g. the copper data from the participants, were intrinsically distributed with a reproducibility standard deviation which may be significant when compared to the (unknown) reproducibility of the preparation of eluates (see Sect. 3.4).

This complicates their use in identifying reasons for the widely deviating results of the biotests. Therefore, a few laboratories were invited to send eluates to BAM to have them analysed under repeatability conditions. These results are summarised in Table 3.4. The minor amounts of chromium and arsenic in the "arsenic-free" wood substrate is likely to have originated from "chromated copper arsenate" which is an important wood preservative (Schoknecht et al. 1998) and might have made it into the starting material e.g. by contamination during impregnation.

The results for selected parameters measured by the participants are listed in Table 3.5. Besides the obvious swapping of data by some laboratories (no. 38: pH; 44, 60: water content) a conspicuous feature is the dependency of the variability of results reported for Cu in the eluates. The greater STD and the non-normality of the distribution of results in the case of SOI suggest problems during eluate preparation, especially with this substrate. The only available literature data on the interlaboratory variability of Cu content in the eluates from incineration ash is a reproducibility

Table 3.4 Eluates from selected laboratories analysed at BAM under repeatability conditions

No.	As (μg/L)	Cd (μg/L)	Cr (μg/L)	Cu (μg/L)	Hg (μg/L)	Mn (μg/L)	Ni (μg/L)	Pb (μg/L)	Zn (μg/L)
INC-02	<1	0.4	24.0	271	<0.1	26.4	4.2	76.2	169
INC-15	<1	0.2	10.9	89.8	<0.1	<2	1.9	12.3	3.7
INC-22	<1	<0.1	26.4	75.4	<0.1	<2	<1	11.8	<2
INC-23	<1	<0.1	33.5	157	<0.1	<2	<1	13.4	<2
INC-32	<1	0.2	17.0	136	<0.1	40.8	5.5	1,960	232
INC-59	<1	<0.1	19.6	169	<0.1	1.4	<1	19.2	7.3
INC-62	<1	<0.1	21.3	172	<0.1	<2	<1	18.6	4.8
SOI-02	<1	0.3	4.4	76.4	<0.1	39.8	2.3	29.9	79.4
SOI-15	<1	0.7	3.1	12.9	<0.1	3.4	3.6	13.2	5.2
SOI-23	3.0	0.3	7.4	37.9	<0.1	16.2	<1	14.4	25.7
SOI-32	<1	0.1	4.4	38.3	<0.1	29.6	2.3	149	146
SOI-59	4.3	0.2	10.7	52.4	<0.1	31.5	1.9	24.5	35.0
SOI-62	3.4	0.1	7.3	42.1	<0.1	20.5	2.3	16.6	17.7
WOO-02	74.3	1.2	246	30,100	<0.1	1,840	10.6	13.8	453
WOO-15	81.4	1.6	282	32,600	<0.1	1,990	10.7	9.5	418
WOO-22	74.7	0.9	262	26,200	<0.1	1,820	14.9	7.7	431
WOO-23	82.2	1.0	269	30,400	<0.1	1,930	13.4	18.7	426
WOO-32	61.7	0.9	222	20,400	<0.1	1,350	8.7	27.4	504
WOO-59	82.6	1.0	294	30,700	<0.1	1,810	9.5	8.3	407
WOO-62	79.1	1.1	272	30,800	<0.1	1,910	9.9	6.4	422

standard deviation of 26% which is in good agreement with the standard deviation reported in Table 3.5 (Kalbe et al. 2006). A similar picture is seen in the case of the water holding capacity (WHC; the data reported by 13–15 participants are not given). The WHC does not seem to have received much attention as a parameter in interlaboratory comparisons. The literature reveals that data on the variability of water retention capacity is found between laboratories in foodstuff only. In the case of dried fruits the between-laboratory variability was 10–20% (Robertson et al. 2000) and in the case of wheat flour below 5% (Gaines 2000). The results in the present ring test for the WHC of the substrates were (mean ± standard deviation): INC - 20.9 ± 12.2 mg/g; WOO - 304.4 ± 219.5 mg/g; SOI - 14.8 ± 15.1 mg/g. Considering the type of test substrates used in the ring test, the comparison with the literature data reveals that the WHC variability is acceptable.

3.6 Assessment and Recommendations

The homogeneity of the analysed parameters of the bottled substrates was satisfying with regard to the variability observed in the ring test. The particle size range of 0–4 mm as chosen here may work for many substrates, especially for tests with large sample intakes. In the case of ecotoxicity tests with sample intakes of a few g for a single determination, a reduction in particle size range for such substrates which

Table 3.5 Characterization of the substrates by the participants

	INC				WOO				SOI			
No.	pH	Dry matter (%)	Moisture (%)	Copper[a] (μg/L)	pH	Dry matter (%)	Moisture (%)	Copper[a] (mg/L)	pH	Dry matter (%)	Moisture (%)	Copper[a] (μg/L)
1	n.d.	97.8	2.2	166	n.d.	90.9	9.2	29.4	n.d.	99.2	0.8	12.4
2	10.7	97.8	2.2	271	5.3	91.4	8.6	30.1	8.4	99.4	0.6	76.4
5	10.8	97.8	2.2	220	5.0	91.1	n.d.	28.5	8.1	99.0	n.d.	66.0
6	10.6	97.7	2.3	170	4.8	90.6	9.4	19.2	8.0	99.2	0.9	8.3
7	11.8	97.5	2.5	193	4.4	90.7	9.3	31.3	7.6	99.2	0.8	17.4
8	10.8	97.3	2.7	<50	5.4	90.9	9.1	32.2	8.4	99.2	0.8	<50
9	9.3	97.7	2.3	n.d.	4.9	92.6	7.4	n.d.	8.0	99.2	0.8	n.d.
10	9.7	97.9	2.1	n.d.	4.9	93.1	6.9	n.d.	7.2	99.2	0.8	n.d.
12	n.d.	98.2	1.8	186	n.d.	91.5	9.3	32.3	n.d.	99.5	0.5	32.3
13	10.6	97.7	2.4	n.d.	5.0	90.8	10.1	n.d.	7.8	99.2	0.8	n.d.
14	n.d.	98.1	1.9	174	n.d.	91.7	8.3	30.5	n.d.	99.4	0.6	49.0
15	10.4	97.2	2.8	89.8	4.9	88.7	11.3	32.6	8.0	98.5	1.5	12.9
16	10.2	98.5	1.5	105	7.7	98.0	2.0	22.3	8.7	98.8	1.2	8.6
17	n.d.	97.4	2.6	n.d.	n.d.	89.3	12.0	n.d.	n.d.	99.1	0.9	n.d.
21	n.d.	99.6	n.d.	160	n.d.	98.6	n.d.	26.0	n.d.	99.2	n.d.	43.5
22	10.2	97.7	2.3	75.5	5.1	92.2	7.8	26.2	8.0	99.0	1.0	n.d.
23	n.d.	n.d.	0.8	157	n.d.	n.d.	1.1	30.4	n.d.	n.d.	0.3	37.9
24	9.7	98.2	1.8	182	4.9	91.7	8.3	36.8	7.4	99.2	0.8	14.0
27	n.d.	98.0	2.0	148	n.d.	92.0	8.0	39.0	n.d.	98.0	2.0	10.8
28	n.d.	98.4	1.6	166	n.d.	91.2	8.8	30.0	n.d.	99.7	0.3	13.8
29	n.d.	97.6	2.4	n.d.	n.d.	90.7	9.3	n.d.	n.d.	99.2	0.8	n.d.
30	n.d.	n.d.	n.d.	n.d.	n.d.	n.d.	n.d.	n.d.	n.d.	n.d.	n.d.	n.d.
32	n.d.	n.d.	n.d.	136	n.d.	n.d.	n.d.	20.4	n.d.	n.d.	n.d.	38.2
33	n.d.	97.5	2.5	202	n.d.	99.3	0.7	n.d.	n.d.	99.0	1.0	18.1
34	n.d.	97.3	2.7	135	n.d.	88.9	11.1	29.0	n.d.	98.8	1.2	42.0
35	n.d.	97.4	2.6	160	n.d.	90.9	10.1	43.0	n.d.	99.3	0.7	10.0
36a	n.d.	96.7	3.3	148	n.d.	91.4	8.6	31.1	n.d.	99.2	0.8	11.8
36b	n.d.	n.d.	n.d.	171	n.d.	n.d.	n.d.	33.7	n.d.	n.d.	n.d.	16.1

37	n.d.	97.2	2.8	154	n.d.	91.4	8.6	31.5	n.d.	99.3	0.7	<100
38	5.5	97.7	2.3	150								
39	n.d.	98.0	2.0	26	n.d.	91.0	9.0	23.2	n.d.	99.0	1.0	9.6
40	10.7	98.0	2.0	170	5.3	49.3	4.3	30.3	8.1	99.2	0.8	<70
41	10.5	97.5	2.5	140	5.3	89.9	9.1	30.0	7.3	99.0	0.7	12.6
44	10.9	97.7	2.3	191	5.5	99.2	0.8	31.6	7.7	91.0	9.0	47.0
46	9.7	97.7	2.3	n.d.	5.5	91.5	8.5	n.d.	7.9	99.1	0.9	n.d.
47	9.9	97.5	2.5	n.d.	5.3	90.0	10.1	n.d.	8.1	99.2	0.8	n.d.
49	10.6	98.1	2.9	240	5.4	91.6	8.4	81.6	8.7	99.3	0.7	32.2
50	n.d.	n.d.	n.d.	34	n.d.	n.d.	n.d.	27.6	n.d.	n.d.	n.d.	6.8
51	10.4	97.6	2.4	150	5.4	91.4	8.6	31.5	8.5	99.3	0.7	31.0
52	n.d.	n.d.	n.d.	175	n.d.	n.d.	n.d.	29.4	n.d.	n.d.	n.d.	27.0
53	10.7	98.9	1.2	181	5.4	92.8	7.8	31.2	7.9	99.8	0.2	40.7
56	n.d.	97.3	n.d.	197	n.d.	90.7	9.3	32.9	n.d.	98.9	1.1	20.6
58	10.7	97.8	2.2	180	5.2	90.5	9.5	30.0	8.4	99.0	1.0	<20
59	10.7	97.8	2.2	126/169	5.3	91.0	9.0	30.7	8.5	99.2	0.8	<10/52.4
60	n.d.	97.7	2.3	188	n.d.	99.2	0.8	35.1	n.d.	90.5	9.5	16.6
61	10.4	97.6	2.4	180	4.6	90.8	9.2	32.9	8.2	99.2	0.8	<50
62	n.d.	97.9	2.1	172	n.d.	91.1	8.9	30.8	n.d.	99.3	0.7	42.1
63	9.9	97.6	2.4	173	4.9	89.6	10.4	36.9	7.9	99.0	1.0	14.0
64	10.3	98.1	1.9	n.d.	5.2	94.9	5.1	n.d.	8.9	97.7	2.3	n.d.

n.d. Not determined

[a]Cu-value of laboratories 2,15,22,23,32,59, and 62 are taken from Table 3.4. Medians ± STD-Dev.: (INC) 170 ± 46.4 μg/L (one outlier removed); (WOO) 30.5 ± 4.20 mg/L (two outliers removed); (SOI) 17.5 ± 18.5 μg/L

tend to display inhomogenous distribution of relevant substances should nevertheless be considered. . It may easily be imagined that waste types with more heterogeneous composition (especially with particles of highly differing density and at the same time different toxicological properties) will undergo segregation after bottling. The crucial aspect of preparation of test batches with varying physical behaviour (e.g. flowability) is the total amount of substance to be processed. Nearly any kind of waste can be processed up to a total amount of a few kilograms. The effect of properties such as stench, complex matrix composition with a tendency to segregation, biohazard characteristics or even of the necessity to process, store and despatch a given material under cooling, tend to accelerate potentially with the total amount of material once the scale of several hundred kilograms is reached. This leads to the recommendation that a ring test provider should select candidate substrates carefully to avoid problems during processing and unclear health care situations during application of the material in the participating laboratories. Important considerations are transporting the bottled material safely to the laboratory and sometimes dealing with customs problems for destinations outside the European Union.

From the experience of this ring test and related activities, it can be concluded that waste wood containing elutable and legally admitted metal-containing wood preservatives, as well as incineration ash, are ideal substrates for reference materials in ecotoxicological testing. Apart from being readily available, stable and non-hazardous during processing, transport and application, their toxicological properties can be adjusted according to desired levels. Such a reference material can be achieved by blending two or more ashes (or wood materials, respectively) with different properties such that the desired level is obtained. Soil might be another proposition; however, it is recommended that samples containing heavy metals above the geogenically typical range be selected if a toxic effect is required and it is to remain stable over longer periods, e.g. because the material is meant for quality control purposes and therefore to be kept available on stock.

Chapter 4
Ring Test Data Evaluation

G. Donnevert, S. Uhlig, and T. Moser

Abstract The statistical evaluation of the ring test data (only standard test battery) was performed in a step-wise process. First, the individual test results (EC/LC_{50} values) were recalculated (using probit analysis and the ToxRat program). Secondly, they had to fulfil several acceptance criteria. Data sets which passed these acceptance criteria were evaluated in two different ways in parallel: according to the approach usually used for the validation of chemical and physicochemical methods in environmental analysis (ISO 5725-2 (2002)) and according to the warning limit approach following Environment Canada (2005). However, for the evaluation of the results from ecotoxicological tests, the ISO approach had to be modified. Assuming that EC/LC_{50} values are log-normally distributed, the log-transformed EC/LC_{50} values were used instead of the original EC/LC_{50} values, to calculate means and standard deviations. Hence the re-transformed total means are not arithmetic but geometric means. Since the ecotoxicological test procedures in question are tedious and elaborate compared with trace analytical investigations, replicate determinations within a given laboratory and within short intervals of time were impossible. Thus, repeatability and reproducibility were evaluated using the results of the different laboratories together with the confidence intervals of the results, which were each calculated by application of the same algorithm. In addition, the results of the individual tests were used to calculate warning limits according to Environment Canada (2005). The test results that were identified as statistical outliers (ISO approach) or those outside the warning limits (Environment Canada) were excluded from the calculation of the final geometrical mean EC/LC_{50}. As a measure of robustness, the ratio between reproducibility and repeatability standard deviation, the standard deviations calculated according to the warning limit approach and the factor between minimum and maximum EC/LC_{50} value are presented. Finally, recommendations to improve the statistical evaluation of data from ecotoxicological ring tests are given.

Keywords Acceptance criteria, Performance data, Repeatability, Reproducibility, Statistical evaluation, Validation, Warning limits

---

G. Donnevert (✉)
University of applied sciences (FH) Giessen-Friedberg, Wiesenstr. 12, 35390, Giessen, Germany

H. Moser and J. Römbke (eds.), *Ecotoxicological Characterization of Waste*,
DOI: 10.1007/978-0-387-88959-7_4,

4.1 Introduction

ISO 5725-2 "Basic method for the determination of repeatability and reproducibility of a standard measurement method" is the relevant standard for method validation of interlaboratory experiments. This standard is internationally accepted for the evaluation of the so-called validation ring tests, especially in chemical and physicochemical analytics. In the field of water analysis, performance data according to ISO 5725-2 are published in national and international standards. With the increasing request for measurement uncertainty information, these performance data have gained even more importance, because they can be used for a preliminary estimation of the measurement uncertainty of an analytical method. For better comparability of performance data, the test results of ring tests for biological methods are evaluated likewise according to ISO 5725-2, where appropriate. However, for evaluation of the results from the toxicity tests under investigation, the procedure of ISO 5725-2 needed some modification, as EC/LC_{50} values are assumed to be log-normally distributed. But before the test results were subjected to statistical evaluation they had to fulfil several acceptance criteria.

4.2 Acceptance Criteria for Test Results

All test results were evaluated in a tiered process as follows:

(a) Compliance with validity criteria given in the guideline?
Example: Mortality of earthworms <10% in the control
Note: This check was always performed.

(b) Test performance according to the SOP?
Example: Number of replicates correct? Test conditions measured?
Note: Clear non-compliance with the SOPs led to rejection of the test results.

(c) Result of reference test within the required range?
Example: *Daphnia* EC_{50} of Potassium dichromate: 0.6–2.1 mg/L
Note: This acceptance criterion was strictly used for earthworm and *Daphnia* tests, partly used for bacteria and Algae tests and not used for plants.

All three criteria were combined to determine whether a specific test result was acceptable and therefore, to be included in the statistical and graphical assessment.

In complex situations (e.g. slightly missed ranges or minor violation of SOPs), a case-by-case decision was made. After consultation with representatives from DIN/ISO for the algae test, the ranges for the outcome of reference testing were extended. This decision was based on the experiences in the ring test, as the ranges mentioned in the guideline and the SOP were too narrow.

4.3 Recalculation of Test Results

The compilation of test results showed that participants had used many different statistical methods to evaluate their test results. Furthermore, some laboratories had calculated the wrong results, e.g. EC/LC_{50} for 48 h test duration instead of 24 h. Therefore, all participants were asked to send their raw data for recalculation. Not all laboratories provided raw data, but the available data were used to recalculate the test results with the software ToxRat® (2006) using the same algorithm. For this purpose probit analysis using linear maximum likelihood regression was applied. This does not imply that probit analysis is the best algorithm for all cases, but it seemed to be suitable to get comparable results. The recalculation yielded not only EC/LC_{50} values but also the 95% confidence limits for EC/LC_{50} in most cases. This confidence limit was important for the calculation of repeatability and reproducibility standard deviation using the adapted statistical evaluation (see Sect. 4.4.2).

4.4 Evaluation Following ISO 5725-2

4.4.1 Requirements for a Statistical Evaluation of Ring Tests Acc. to ISO 5725-2

A statistical evaluation of a ring test according to ISO 5725-2 applies to a balanced uniform-level experiment. Therefore the requirements given in Table 4.1 must be met.

4.4.2 Adaptation of the Statistical Evaluation for the Ecotoxicological Ring Test

Statistical evaluation of the results of the toxicity tests applied in the ring test for the ecotoxicological characterization of waste in absolute accordance with ISO 5725-2 was not possible. The first reason for this was that EC/LC_{50} values are not normally but approximately log-normally distributed, and the second reason was that for most of the participating laboratories it was not possible to perform more than one test with the same test system and test substrate in parallel (replicate determinations). Thus, in order to achieve conformity with the requirements of ISO 5725-2 and to get sound performance data, some adaptations of the statistical evaluation procedure were necessary.

According to ISO 5725-2 every laboratory should report the same number of test results ($n \geq 2$) per test and test sample. These results, which must have been obtained under repeatability conditions, are needed for calculating the repeatability standard deviation s_r (see glossary). As most of the participants reported just single

Table 4.1 Requirements set by ISO 5725-2

Requirement	Comment
Every participant uses the same method, i.e. every laboratory should strictly follow the SOP	This requirement was not strictly fulfilled (for details see Chap 7M)
The test materials must be homogenous	This requirement was met as far as possible (for details see Chap. 3)
Statistical evaluation is possible for quantifiable data only, test results greater than or less than a value cannot be considered	Therefore some of the test results could not be considered
The analytical method must yield results on a continuous scale	Therefore EC_{50} values had to be reported, not dilution steps (in earthworm tests: LC_{50})
Evaluation according to ISO 5725-2 implies that the test results for one parameter and substrate are approximately normally distributed	This requirement is met by the decadic logarithm of EC/LC_{50} only
For statistical evaluation a minimum number of eight valid data sets is required	This requirement was easily met by the basic test battery, but for the additional tests mostly there were less than 8 and often less than 5 accepted test results
The test results must be uniform	To improve the consistency of test results acceptance criteria were defined and most of the results were recalculated
For calculation of the repeatability standard deviation, every laboratory should report the same number of test results ($n \geq 2$) per test and test sample	Most of the participating laboratories reported single test results only, because the performance of the tests is tedious and elaborate. Even those who had reported more than one result might not meet the demand "within short intervals of time"

test results, repeatability could not be calculated in the way it is described in ISO 5725-2. However, as the results of toxicity tests are calculated from a series of exposure concentrations, confidence limits can be calculated. The 95% confidence limits of the EC/LC_{50} estimate the internal variation of the test. Therefore, they contain precise information that corresponds approximately to repeatability conditions. Thus it was possible to use the upper and lower confidence limits of the individual test results to calculate the repeatability standard deviation. As already mentioned, the EC/LC_{50} values are log-normally distributed and therefore, for further evaluation, the EC/LC_{50} values and the respective 95% confidence limits are log-transformed. If the EC/LC_{50} values are log-transformed, the 95% confidence limits are equal to the individual test result (EC/LC_{50}) ±2SD. The standard deviation of the individual laboratory test results can be derived from the upper and lower confidence limit using the formula

$$\text{STD} = \frac{\log \text{uCL} - \log \text{lCL}}{4},$$

STD is the logarithmic within-laboratory standard deviation, uCL is the upper confidence limit of the 95% confidence interval of the test result, lCL is the lower confidence limit of the 95% confidence interval of the test result.

Example: Test result from the algae test for INC:

$EC_{50} = 10.24\% \Rightarrow \log EC_{50} = 1.01$
uCL = 12.00%
lCL = 8.72%

$$\mathrm{STD} = \frac{\log 12 - \log 8.72}{4} = 0.035.$$

For the statistical evaluation of the ring test, the software ProLab® (2006) was adapted; i.e., the calculation of the repeatability standard deviation was performed integrating the respectively calculated logarithmic within-laboratory standard deviation (STD).

4.4.3 Statistical Outliers

Statistical tests are performed to indicate two types of outliers:

Type B outliers are test results which deviate so much from the other test results that they are considered irreconcilable with the other data, i.e. they show a large systematic error. The statistical test for type B outliers is Grubbs' test on a significance level of 1%.

Type C outliers are test results whose within-laboratory variances deviate so much from the other test results that they are considered irreconcilable with the other data. The statistical test for type C outliers is Cochran's test on a significance level of 1%. However, ISO 5725-2 says that those test results may be considered outliers. The decision to exclude these test results from evaluation is the responsibility of the group of experts. For calculation of s_R and s_r in this ring test both types of statistical outliers were excluded.

4.4.4 Presentation of the Results of the Statistical Evaluation Following ISO 5725-2

Statistical evaluation was executed using the adapted software ProLab® (2006) (see Sect. 4.4.2). Figure 4.1 gives an example of the presentation of statistical evaluation results. The ordinate represents the concentration in the logarithmic scale; at the left side, the concentrations for EC/LC_{50} are given in %. The abscissa is labelled with the laboratory codes, and the test results are arranged in ascending order. The horizontal line marked "mean" represents the mean of the log EC/LC_{50} values, i.e. the geometric mean of the test results. The different bars show the ± log standard deviation range calculated from the confidence intervals of the test results

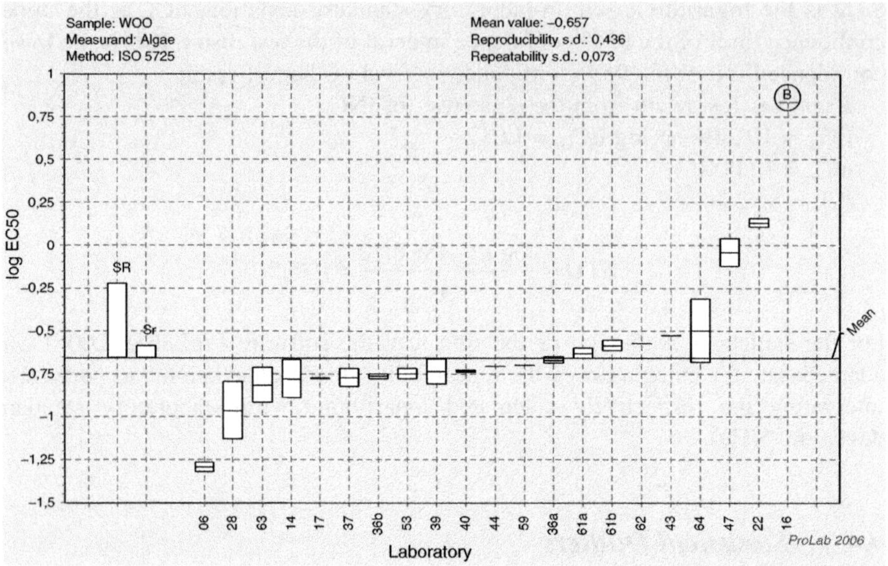

Fig. 4.1 Example of the graphical representation of the statistical evaluation (ISO 5725-2 2002)

(see Sect. 4.4.2), and the middle line represents the logarithm of the respective EC/LC$_{50}$ values. If there is only a thin line the confidence interval cannot be calculated. A statistical outlier type B (see Sect. 4.4.3) is marked with a circle. The two bars at the left represent reproducibility standard deviation (s_R) and repeatability standard deviation (s_r).

The higher the ratio between s_R and s_r, the higher is the contribution of the laboratory bias. With reference to ring tests for chemical and physicochemical methods which were evaluated according to ISO 5725-2, the test method can be considered fairly robust as long as this ratio is below 4.

4.5 Alternative Methods for the Statistical Evaluation of Ring Tests

The precision of ring test results can be evaluated by various methods. In addition to the approach following ISO 5725-2, Environment Canada (1999) proposed the calculation of the factor between the lowest and the highest EC/LC$_{50}$ values. Further, the test results can be used to calculate warning limits and to plot warning charts (Environment Canada 2005).

4.5.1 Factor Between the Lowest and the Highest EC/LC_{50} Values

This approach for data assessment was developed for tests with chemicals spiked into standard media (Chapman 1995; Environment Canada 1999). For these tests a factor of four is considered acceptable. But complex substrates such as the waste samples used in this ring test may lead to higher factors. Outliers were not omitted for the calculation of the factor between the lowest and the highest EC/LC_{50} values (min-max factor).

4.5.2 Calculation of Warning Limits and Plotting of Warning Charts

Warning limits and the corresponding warning charts are normally used for the interpretation of the results of tests with reference toxicants in one laboratory. They are intended to assess changes in the sensitivity of organisms and precision within the laboratory. In routine application, a laboratory's historic results are used to calculate the upper and lower warning limits. These are calculated as ±2 standard deviations from the geometrical mean of the historical values. The upper and lower warning limits and the geometrical mean are plotted on a warning chart. If a new reference test is performed and the respective EC/LC_{50} is within the warning limits, the result is considered satisfactory (see Environment Canada 2005).

For the assessment of the results of the toxicity tests in this ring test, the warning limit approach was adapted. In contrast to the usual application, in this ring test, warning limits were calculated by the EC/LC_{50} of all test results from the single laboratories for the same test system (or even species) and the same test substrate. Based on these calculations the warning charts were plotted. An example is given in Fig. 4.2.

In these warning charts the ordinate represents the concentration in a logarithmic scale, and the abscissa gives the lab codes in ascending order. The black dots show the EC/LC_{50} values of the single laboratories. Test results outside the warning limits which were interpreted as outliers for the warning limit approach are marked with circles. These test results were not always identified as outliers in the statistical evaluation according to Sect 4.4.3; often they were marked as stragglers only, not as significant outliers (ISO 5725-2, 2002). A straggler is a test value which is suspected to be an outlier, as its test statistic lies between 95 and 99% probability. However Grubbs' test for type B outliers is performed on a significance level of 1% only.

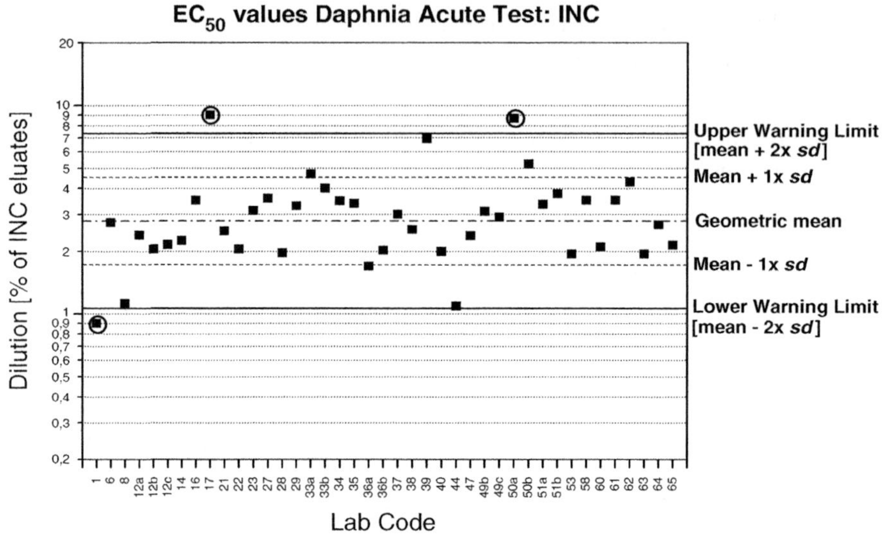

Fig. 4.2 Example of the graphical representation of the warning chart approach (Environment Canada 2005) Circled dots = results interpreted as outliers

4.6 Assessment of the Results of Statistical Evaluation

In Table 4.2, the overall performance of the test evaluation is presented: First the total number of tests that fulfilled the acceptance criteria was determined. Second, all data sets in which no EC/LC_{50} could be determined (i.e. where they had to be given as < or > a tested dilution) were excluded from the statistical evaluation. As this was the case in most tests with the substrate SOI (which often showed no toxicity even in 100% test substrate) a statistical evaluation of the ring test data with that substrate was regarded as not applicable. The remaining data sets were checked for passing the outlier tests (i.e. following the ISO approach) and being outside the Warning Limits (i.e. following the approach presented by Environment Canada). In the tests performed according to ISO 5725-2, six EC/LC_{50} values were identified as outliers, while according to the warning limit approach nineteen EC/LC_{50} values were outside the given range. For the ISO approach statistical outliers were excluded from the evaluation, i.e. from calculation of geometrical mean, as well as s_R and s_r, because these test results would have a high impact on the performance data. In contrast to this, mean and standard deviation for the warning limit approach were calculated for all accepted test results (i.e. including outliers). The final EC/LC_{50} values for each test system and substrate were calculated excluding all outliers, irrespective of the evaluation method used (Table 4.2).

Table 4.2 Total number of tests, number of accepted tests, number of tests where EC/LC_{50} values could be determined, number of tests without outliers (ISO 5725-2 2002) and number of tests with results outside the warning limits (Environment Canada 2005)

Test system	Subgroup	Test substrate	Total number of tests	Number of accepted tests	Number of tests with EC/ LC_{50} values	Number of tests with outlier (ISO)	Number of tests outside warning limits (EC)
Algae	All	INC	48	35	32	0	0
	D. subspicatus	INC	21	14	13	0	0
	P. subcapitata	INC	27	21	19	0	0
	All	SOI	48	35	n.a.	n.a.	n.a.
	All	WOO	47	36	35	0	2
	D. subspicatus	WOO	21	15	14	0	0
	P. subcapitata	WOO	26	21	21	1	1
Daphnia		INC	54	47	41	0	3
		SOI	54	51	2	0	0
		WOO	53	51	50	0	3
Lumin Bac.	All	INC	49	45	22	1	0
	Freeze	INC	25	23	19	1	1
	All	SOI	48	42	16	0	1
	Freeze	SOI	25	21	n.a.	n.a.	n.a.
	All	WOO	57	53	53	0	1
	Freeze	WOO	24	20	20	0	2
	Fresh/liquid	WOO	33	33	33	1	1
Earth-worm		INC	18	14	8	0	0
		SOI	17	15	1	0	0
		WOO	17	15	14	0	1
Plant *Avena*		INC	22	19	18	1	1
		SOI	21	18	n.a.	n.a.	n.a.
		WOO	21	18	17	0	1
Plant *Brassica*		INC	21	18	18	1	1
		SOI	20	17	10	0	0
		WOO	19	17	16	0	0

All data are given for the different test systems (including subgroups) and three waste types. *n.a.* Not applicable because in too many tests no definite LC/EC_{50} value could be determined

The highest number of results was provided for the *Daphnia*, luminescence bacteria and algae tests. The acceptance criteria were passed by 71.4–100% (Algae, *Desmodesmus subspicatus*, WOO and Luminescent Bacteria, fresh/liquid, WOO respectively) of the total number of tests. Only in the case of the algae *D. subspicatus* tests with the INC eluate, the acceptance rate was considerably lower (60.9%). In general, almost all accepted tests were included in the statistical evaluation of the data according to ISO 5725-2 and Environment Canada (2005). Because of the low toxicity of INC to luminescence bacteria and earthworms, only 48.9 and 61.5% of the accepted tests were included in the statistical evaluation.

4.7 Discussion

ISO 5725-2 is not applicable without adaptation for the evaluation of ring tests for toxicity tests. On the one hand test results have to be evaluated logarithmically, and on the other hand, e.g. for plant tests, a multiple performance of the test, which is demanded to determine repeatability, is too elaborate. The use of confidence limits of test results for the estimation of repeatability is a new approach which has not been used before in the evaluation of ring tests. It is assumed that the confidence limits represent the within-laboratory variation caused by sample preparation and performance of the test. Although with the available data it is not possible to supply evidence that the confidence limits of the test results were the appropriate measure of laboratory precision, the results of the statistical evaluation prove the applicability of the adapted evaluation process. During the ring test some laboratories repeated the tests with a particular test system on the different waste substrates several times but often the time interval between two tests was two weeks or even one month and more. Therefore, it is doubtful whether these time intervals meet the requirement for repeatability of ISO 5725-2, which demands repeated test performance "within short intervals of time". However, it is worthwhile to mention that despite inhomogenity of the test substrate and a greater time interval between repeated tests the EC/LC_{50} values generated with a particular test system for a particular test substrate did not differ considerably (see Chap. 22).

The first possibility for the assessment of variation is the comparison of repeatability standard deviation s_r and reproducibility standard deviation s_R (Table 4.3). In chemical and physicochemical ring tests a s_R/s_r ratio of less than 4 is an indication of robustness. In the ecotoxicological ring test with wastes, this ratio covers a range between 2.16 and 15.0. About 42% of all ratios are within the "chemical" range while 83% are located between 2 and 8. Taking into account the differences between chemical/physicochemical and ecotoxicological ring tests this variation is considered to be acceptable.

Secondly, in former guidelines Environment Canada has offered advice that variation in repeated tests of a reference toxicant (e.g. for the construction of a warning chart) would be considered reasonable if the coefficient of variation CV (i.e. the standard deviation divided by the mean, expressed as a percentage) is less than 30%, and preferably ≤20%. These values apply for the arithmetic mean and the respective standard deviation. However, Environment Canada (2005) detected that the arithmetic mean and its standard deviation are subject to bias in the case of toxicity tests. Therefore, the guideline for the interpretation of warning limits was converted to a logarithmic basis, in an approximate way (Environment Canada 2005). The new guideline value for the assessment of the warning limits is a standard deviation of 0.132. This corresponds to the "reasonable" CV of 30%, while a standard deviation of 0.0338 represents the "preferred" variation (CV = 20%) in a set of test results. These values apply to any set of results, because they were derived from ratios on a logarithmic scale. The calculated standard deviations for any set of logarithmic EC/LC_{50} may be compared to those guideline values; i.e. in this case s_R or the onefold standard deviation of the mean in the Warning Chart Approach (since they are in most cases the same, only the former is given in Table 4.3). However, as ring test results are determined from a number of laboratories under slightly different test conditions, the variation can be expected to be bigger than in tests with a reference toxicant performed in one laboratory over a period of time, as it is the case for the original use of warning charts. When comparing the s_R values/standard deviations from the Warning Chart Approach (see Sect. 4.5.2) with these values, it can be determined that three values are smaller than 0.132, three are above 0.132 but smaller than 0.2, six are above 0.2 but smaller than 0.3 and nine are greater than 0.3 (Table 4.3).

Recommendations for reasonable variation of ring test results have not yet been specified, and a universally valid value does not seem to be reasonable for the different toxicity tests.

Table 4.3 Compilation of the main results of the ring test: Reproducibility standard deviation s_R (= warning limit standard deviation), repeatability standard deviation s_r, ratio between s_R and s_r, minimum/maximum EC/LC$_{50}$ factor, number of tests used for the calculation of the final means (see also Table 4.1) and final geometrical means

Test system	Subgroup	Test substrate	s_R	s_r	Ratio s_R/s_r	Min/Max EC/LC$_{50}$ factor	No. of tests for final mean calculation	Final mean
Algae	All	INC	0.482	0.070	6.89	59.0	32	4.08
	D. subspicatus	INC	0.437	0.080	5.46	14.9	13	8.80
	P. subcapitata	INC	0.372	0.063	5.90	20.4	19	2.42
	All	SOI	0.143	0.035	4.09	n.a.	n.a.	56.5
	All	WOO	0.493	0.066	7.47	135.2	33	0.50
	D. subspicatus	WOO	0.125	0.058	2.16	2.3	14	1.34
	P. subcapitata	WOO	0.436	0.073	5.97	135.2	19	0.22
Daphnia	–	INC	0.210	0.056	3.75	10.0	38	2.71
	–	WOO	0.369	0.050	7.38	30.6	47	0.38
Lumin Bac	All	INC	0.254	0.054	4.70	7.8	21	35.4
	Freeze	INC	0.220	0.057	3.86	7.8	17	28.9
	All	SOI	0.094	0.029	3.24	2.1	15	65.5
	Freeze	SOI	0.072	0.031	2.32	n.a.	n.a.	65.8
	All	WOO	0.508	0.044	11.55	113.9	52	2.56
	Freeze	WOO	0.332	0.052	6.38	19.7	18	0.59
	Fresh/liquid	WOO	0.181	0.036	5.03	10.6	32	5.60
Earthworm	–	INC	0.086	0.038	2.26	1.9	8	45.5
	–	WOO	0.172	0.063	2.73	3.2	13	20.1
Plant *Avena*	–	INC	0.278	0.087	3.20	17.3	17	29.4
	–	SOI	0.090	0.006	15.00	n.a.	n.a.	n.a.
	–	WOO	0.297	0.093	3.19	10.5	16	10.0
Plant *Brassica*	–	INC	0.319	0.033	9.67	21.7	17	23.9
	–	SOI	0.155	0.014	11.07	2.7	10	63.0
	–	WOO	0.298	0.115	2.59	11.3	16	2.64

All data are given for the different test systems (incl. subgroups) and three waste types. s_R and s_r are calculated excluding outliers type B and type C; *n.a.* Not applicable

The minimum/maximum EC/LC$_{50}$ factor calculated for all tests with the different test systems (and subgroups) used with the different waste types ranged from 1.9 (earthworm, INC) to 135.2 (algae, all and *Pseudokirchneriella subcapitata*, WOO). However, high values are often determined in those cases where subgroups were pooled (e.g. algae, INC, algae, WOO, luminescence bacteria, WOO). The only exception is the algae, *P. subcapitata*, in WOO. Excluding these values, the factor between minimum and maximum EC/LC$_{50}$ ranged from 1.9 to 30.6 and the vast majority of the values were below 20 (about 50% even less than 10). Taking into account the heterogeneity of the test substrates, this can be regarded as acceptable. Summarising this discussion, the test results, in particular the final means, are considered acceptable. They will be discussed in detail in the following chapters.

4.8 Recommendations

For future ring tests with toxicity tests, the following aspects should be kept in mind to receive test results which are well fit for method validation:

- Variability of the SOPs should be limited. If different modifications of the method are allowed, there should be enough participants for every variant to enable a separate evaluation of test results for every modification.
- The optimal algorithms for calculation of test results should be specified in advance and applied either by the participating test laboratories or during recalculation.
- An acceptance procedure for data evaluation should be drafted in advance, e.g. as a flow chart, and strictly followed.
- Laboratories should be asked to provide more than one result per test and test sample (for aquatic tests only) to enable calculation of repeatability according to ISO 5725-2.

It is recommended to investigate in detail whether the confidence interval of terrestrial and aquatic (if only one test run is performed) test results is appropriate to obtain reliable data for the calculation of repeatability standard deviations. Furthermore, a team of experts should define guideline values for "reasonable" reproducibility standard deviation for each test method. Concerning ISO 5725-2, the ISO committee responsible for the revision of this International Standard should be requested to integrate the special requirements of toxicity tests (e.g. lognormal distribution of test results) into the next edition.

Glossary

Confidence limits

(Environment Canada 2005) on an EC$_{50}$ or LC$_{50}$ represent upper and lower concentrations, within which the true endpoint is thought to lie, for a stated level of probability.

The 95% confidence limits represent a statement that there is a 19 out of 20 chance that the true endpoint falls within those specified limits.

EC_{50}

(Environment Canada 2005) is the median effective concentration. It is the concentration of material in water (e.g., mg/L) or soil or sediment (e.g., mg/kg) that is estimated to cause a specific toxic effect to 50% of the test organisms. In most instances the EC50 and its 95% confidence limits are statistically derived by analyzing the percentages of organisms showing the specific effect at various test concentrations, after a fixed period of exposure. The duration of exposure must be specified (e.g., 72-h EC50).

Geometric mean

(Environment Canada 2005) is a measure of central tendency for a set of observations. It can be useful because it is less influenced by extreme values than is the more familiar arithmetic mean. For n values in a set, the geometric mean is the nth root of the product of all the values (i.e., multiplied). It can also be calculated as the antilogarithm of the arithmetic mean of the logarithms of the values.

LC_{50}

(Environment Canada 2005) is the median lethal concentration, i.e., the concentration of material in water, soil or sediment that is estimated to be lethal to 50% of the test organisms. The LC_{50} and its 95% confidence limits are usually derived by statistical analysis of percent mortalities in several test concentrations, after a fixed period of exposure. The duration of exposure must be specified (e.g., 48-h LC_{50}).

Normal distribution

(Environment Canada 2005) is a symmetric bell-shaped array of observations. The array relates frequency of occurrence to the magnitude of the item being measured. In a normal distribution, most observations will cluster near the mean value, with progressively fewer observations toward the extremes of a range of values. The shape is determined by the mean and standard deviation, with 68.3%, 95.4%, and 99.7% of the observations included within ±1, ±2, and ±3SD of the mean, respectively.

Outlier

(Environment Canada 2005) is an extreme observation, a measurement that does not seem to fit the other values from a test.

Probit

(Environment Canada 2005) is a unit of divergence from the mean of a normal distribution, expressed in terms of a standard deviation of the distribution. The practical use of probits, in estimating an LC_{50} or EC_{50}, is to straighten the sigmoid curve of the accumulated normal distribution, which shows percent effect as a function of log concentration.

Repeatability standard deviation s_r

(ISO 5725-1) is a measure of dispersion of the distribution under repeatability conditions, i.e. conditions where independent test results are obtained with the same method on identical test items in the same laboratory by the same operator using the same equipment within short intervals of time.

Reproducibility standard deviation s_R

(ISO 5725-1) is a measure of dispersion of the distribution under reproducibility conditions, i.e. conditions where test results are obtained with the same method on identical test items in different laboratories with different operators using different equipment.

Warning chart

(Environment Canada 2005) is a graph used to follow changes over time, in the endpoints which measure toxicity of a reference toxicant. The date of the test is on the horizontal axis and the effect-concentration is plotted on the vertical logarithmic scale.

Warning limits

(Environment Canada 2005) allow an investigator to evaluate the variation in toxicity tests with a reference toxicant. The limits are ±2SD, calculated logarithmically, from the historic geometric mean of the test endpoints.

Chapter 5
Range of Reference Tests in Aquatic Tests

M. Pattard, J. Römbke, and T. Moser

Abstract In the EU Waste Ring test 2006/2007 a very high number of reference tests were performed for three standard (Algae, *Daphnia* and Luminescent bacteria) as well as various additional (*Lemna*, *Brachyonus*, *Ceriodaphnia*, and *Pseudomonas*) test systems. These results were required by the respective test guidelines to prove the sensitivity of the test organisms. In addition, tests with two reference substances provided results confirming the validity of the umu tests performed. Besides the required reference substances, in some tests additional chemicals were tested. The number of the required reference tests, mainly potassium dichromate in the standard tests, was so high that these results could be used for the re-evaluation of the data ranges listed in the test guidelines. The range given in the current Algae test guideline, based on old ring test data, should be changed and so a new range is proposed here. In the additional tests, the number of data sets provided, often with different reference substances, was much lower. Since potassium dichromate did not cause problems in the various tests, it is recommended that this substance be used as the standard reference substance in aquatic tests.

Keywords Algae, Daphnia, Luminescent bacteria, Lemna, Ceriodaphnia, Brachyonus, Pseudomonas, Potassium dichromate

5.1 Introduction

In the EU Waste Ring test 2006/2007 a very high number of reference tests were performed for three standard (Algae, *Daphnia* and Luminescent bacteria) as well as various additional (*Lemna*, *Brachyonus*, *Ceriodaphnia*, and *Pseudomonas*) test systems (CEN 2005). These results were required by the respective test guidelines to prove the sensitivity of the test organisms. In this context, it is very important to

---

J. Römbke (✉)
ECT Oekotoxikologie GmbH, Boettgerstr. 2-14, 65439, Floersheim, Germany

H. Moser and J. Römbke (eds.), *Ecotoxicological Characterization of Waste*,
DOI: 10.1007/978-0-387-88959-7_5, © Springer Science+Business Media, LLC 2009

test a reference substance at regular intervals to ensure the constant sensitivity of the test systems (Gourmelon & Ahtiainen 2007). While in the case of standard tests this requirement was fulfilled in almost all cases, the number of reference tests performed for the additional tests was much lower. However, even for these tests, the reference data gained form a valuable source of information as the published experience for most of these tests is very limited.

In this chapter the results obtained by the participants from reference tests are compiled for each test system separately. In addition, these results are compared with the range of every reference substance used, as required in the respective guideline or found in the literature (only in the case of the umu genotoxicity test, such a comparison does not make sense as here only a specific minimum induction rate had to be met (ISO 2000). In Table 5.1 the reference substances recommended by the respective guidelines and the expected effect ranges are summarised.

In the following sections, the results of the reference tests are compiled, starting with the tests of the standard battery. In case more than six test results were available, the data set was assessed using the Warning Limit approach published by Environment Canada (2005). Tests with a lower number of data are shown in graphs and the minimum - maximum factor is calculated (Environment Canada 1999). Afterwards, the mean LC/EC_{50} (determined via probit analysis and almost always using the statistical program ToxRat (2006)) of each test system is compared with guideline requirements. Finally, recommendations are given concerning the further development of reference testing in aquatic ecotoxicology.

Table 5.1 Ecotoxicological tests and reference substances

Organism group	Organism	Guidelines	Reference substance, range
Standard test battery			
Algae	*Desmodesmus subspicatus, Pseudokirchneriella subcapitata*	ISO 8692 (2004)	Potassium dichromate 0.39–1.43 mg/L 0.38–2.60 mg/L
Water flea	*Daphnia magna*	ISO 6341 (1996)	Potassium dichromate 0.6–2.1 mg/L
Luminescent bacteria	*Vibrio fischeri* (three sources)	ISO 11348-1/2 (2005)	Effect ranges (20–80%) for three substances, differentiated for two sources of bacteria
Additional test battery			
Macrophytes	*Lemna minor*	ISO 20079 (2004b)	3,5-Dichlorophenol 1.8–3.6 mg/L
Rotifers	*Brachionus calyciflorus*	ISO 20666 (2007)	Copper sulphate (as Cu^{2+}) 35.8–71.2 µg/L
Water flea	*Ceriodaphnia dubia*	ISO 20665 (2006)	Sodium pentachlorophenolate 216–344 µg/L
Sludge bacteria	*Pseudomonas putida*	ISO 10712 (1995)	3,5-Dichlorophenole 10–30 mg/L µg/L
Genotoxicity	*Salmonella typhimurium*	ISO 13829 (2000)	4-Nitroquinoline-*N*-oxide; aminoanthracene; IR > 2

5.2 Results of the standard test battery

5.2.1 Algal Growth Inhibition Tests

In total, 35 participants performed Algae growth inhibition tests with the waste materials, about half of them either with the species *Desmodesmus subspicatus* (40%) or *Pseudokirchneriella subcapitata* (60%). They provided 46 valid reference test data (45 with potassium dichromate and one with 3,5-dichlorophenol (3,5-DCP). The latter was accepted as the measured EC_{50} of 3.94 mg/L confirmed the sensitivity of the test system (required range: 1–5 mg (a.i.)/L), but it will not be discussed further. All test results, and differentiated for the two species, are summarized in Figs. 5.1–5.3.

According to the SOP of the EU Waste Ring test 2006/2007, the EC_{50} (growth rate, 72 h) value for potassium dichromate should be in the range of 0.84 mg/L ± 0.12 mg/L (SD) for *D. subspicatus* and 1.19 mg/L ± 0.27 mg/L (SD) for *P. subcapitata* (ISO 2004a). However, using these values an unexpected high number of tests had to be considered as non-acceptable, as the EC_{50} values were outside these ranges. So members of the ISO Technical Committee 147, responsible for aquatic tests, were contacted to clarify whether these values are really state-of-the-art. According to the TC 147 members, the basis of these values could be considered preliminary, as they were based on relatively few raw data. Therefore, for the evaluation of the Waste Ring test data a broader range was defined, using the mean and SD of all valid data sets provided in the ring test: 0.39–1.43 mg/L and 0.38–2.61

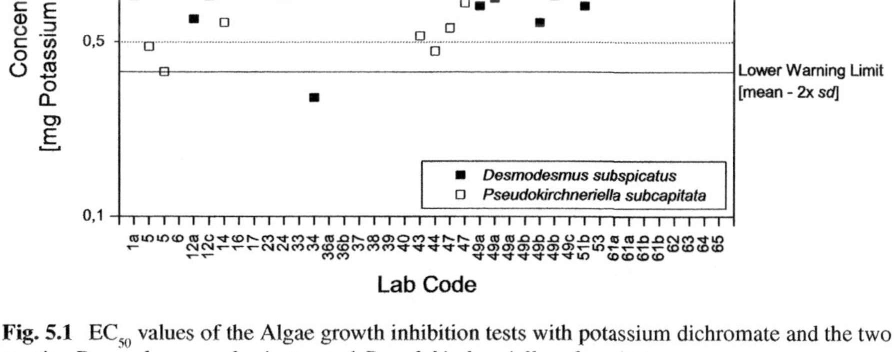

Fig. 5.1 EC_{50} values of the Algae growth inhibition tests with potassium dichromate and the two species *Desmodesmus subspicatus* and *Pseudokirchneriella subcapitata*

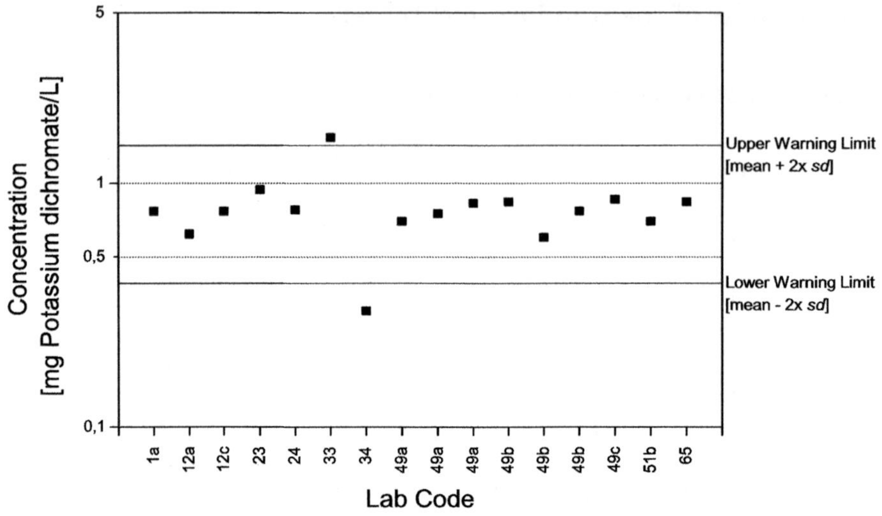

Fig. 5.2 EC_{50} values of the Algae growth inhibition tests with potassium dichromate and the species *Desmodesmus subspicatus*

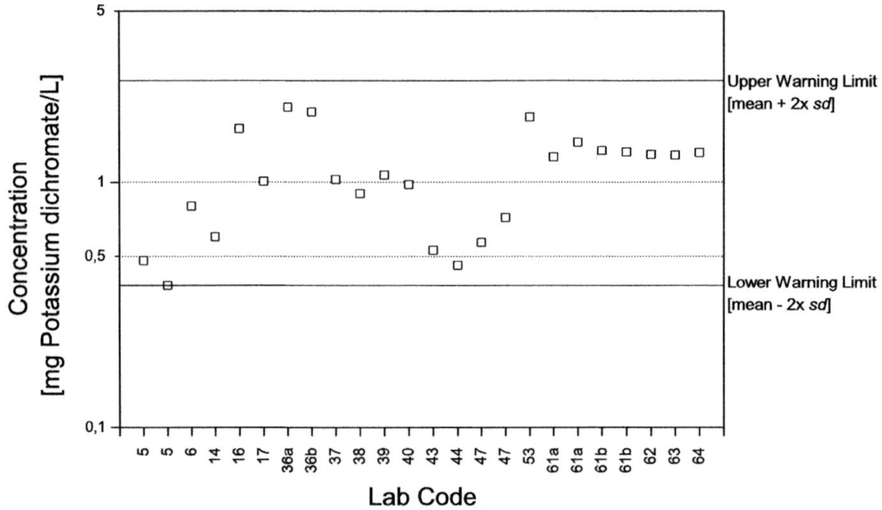

Fig. 5.3 EC_{50} values of the Algae growth inhibition tests with potassium dichromate and the species *Pseudokirchneriella subcapitata*

mg/L for *D. subspicatus* and *P. subcapitata*, respectively. According to these criteria, seven tests were considered not acceptable (see also Chap. 7).

Later on, after further and more detailed data evaluation, additional reference tests were included (mostly those where the waste tests had to be classified as invalid for other reasons), leading to a refinement of the recommended range for reference tests with Potassium dichromate and the two Algae species separately and together:

- *D. subspicatus*: 0.79 ± 0.25 mg/L
- *P. subcapitata*: 1.10 ± 0.48 mg/L
- Both species together: 0.97 ± 0.43 mg/L

It is recommended that these values from the waste ring test and other experiences (Chao & Chen 2000) be used for an updating of the ISO guideline. It is worth mentioning that despite a tendency indicating a higher sensitivity of *D. subspicatu,s* there is no significant difference between the two species regarding their reaction to Potassium dichromate.

5.2.2 Acute Mobility Test with Daphnia

The acute mobility test with *Daphnia magna* using potassium dichromate was performed by 46 participants. All of the determined EC_{50} values were within the range of 0.6–2.1 mg/L required by the guideline (ISO 1996) with the exception of two tests where the EC_{50} values were either too low or too high, respectively. In one case a reference test with zinc-sulphate was performed, which was considered to be valid as the resulting EC_{50} value of 1.81 mg/L is within the range of 1.3–2.8 mg/L expected for this substance (Picado et al. 2007). All test results are summarized in Fig. 5.4. Only in two cases the result of the reference tests was outside the recommended

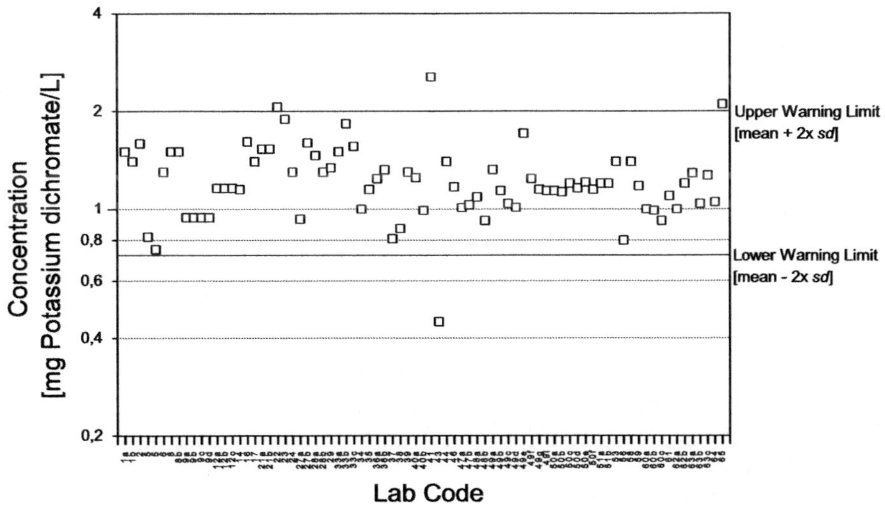

Fig. 5.4 EC_{50} values of the *Daphnia* acute mobility tests with potassium dichromate

range (see Chap. 8 for details). In general, the mean EC_{50} value was determined as 1.24 ± 0.32 mg/L. The results found in this ring test prove that potassium dichromate is a suitable reference substance, confirming the requirement of a standardized test protocol, especially for important influence factors like temperature or cultivation (Lewis & Wang 1985). In addition, these results confirm that the range required in the guideline is correct; thus no action concerning the revision of the guideline is needed.

5.2.3 *Luminescent Bacteria Test*

According to the ISO guideline and the respective SOP, an EC_{50} value for a specific reference substance is not required. Instead, certain concentrations specified in the guideline have to cause an effect in the range of 20–80% of the control. The required concentrations for the three potential reference substances, differentiated for the two main sources of luminescent bacteria, are specified in Table 5.2. No test result had to be rejected due to a failure in these limit tests; i.e. all provided test results using the recommended concentrations in the range of 20–80% (see Chapter 3).

Reference tests with luminescent bacteria were performed by 19 participants. However, two participants used the substance Phenol as reference (mean EC_{50}: 17.44 mg/L), both with freeze bacteria. As no information concerning an expected range is available, these results will not be discussed further here. All other participants used either 3,5 Dichlorophenol (five tests), Zincsulfate heptahydrate (six tests) or Potassium dichromate (14 tests) for this purpose, because these substances are also used for the quality check of the bacteria batches used in the test. In total 25 EC_{50} values were delivered, eight for liquid/fresh bacteria and 17 for liquid bacteria. Therefore, the number of tests for any combination of the two bacteria sources (liquid/fresh versus freeze) and the three reference substances is too small for any elaborated evaluation (for the same reason, no figure of these data points is provided). However, when comparing the test results determined in the Ring test (Table 5.3) with the effect concentrations required in the ISO guideline (Table 5.2) there is no indication of a difference. It should be noted that this comparison can only be a qualitative one as the required concentrations refer to a broad effect range (20–80%) and because the number of tests per group is very small. Thus, these results in toto support the selection of these concentration ranges. However, it should be discussed whether the current requirements could be changed in favor of

Table 5.2 Concentrations of three reference substances required to cause an effect of 20–80% after a contact time of 30 min according to ISO guideline 11348 (2005)

Reference substance	Liquid-dried or freshly prepared bacteria	Freeze-dried bacteria
3,5- Dichlorphenol	6.0 mg/L	3.4 mg/L
Zinc sulfate heptahydrate	109.9 mg/L	9.67 mg/L
Potassium dichromate	11.3 mg/L	52.9 mg/L

Table 5.3 EC_{50} values determined for the three reference substances in the ring test with the two sources of luminescent bacteria

	Liquid-dried/freshly prepared bacteria		Freeze-dried bacteria	
Reference substance	*n*	EC_{50}(mg/L)	*n*	EC 50 (mg/L)
3,5-Dichlorphenol	1	4.13	4	4.03
Zinc sulfate heptahydrate	1	29.70	5	5.76
Potassium dichromate	6	4.18	8	27.0

n Number of tests performed for each combination of bacteria source and reference substance

asking for dose-response reference tests, meaning that as in other test systems EC_{50} values will be required in the future. A future revision of the ISO guideline should also integrate the results of the Luminescent bacteria tests from other ring tests (Farré et al. 2006; LANUV-NRW 2006).

5.3 Results of the additional Test Set

The results of four additional tests are summarised in Table 5.4. In the case of the umu genotoxicity test, all reference tests performed revealed higher induction rates than required (ISO 2000); therefore these results are not listed here in detail. In the other four tests, the number of reference tests performed was small, ranging from seven data sets in the *Lemna* test to three in the *Pseudomonas* test. In one case, individual participants performed reference tests with substances other than those recommended in the guidelines, meaning that the resulting EC_{50} values could not be evaluated as no required range was available. For these reasons, it was decided to present the individual valid data in the table along with the respective mean values. The situation is briefly discussed separately for each test system.

5.3.1 Duckweed Growth Inhibition Test (Lemna minor)

Seven participants performed a reference test using 3,5-Dichlorophenol as given in the guideline. All EC_{50} values determined were within the required range of 1.8–3.6 mg/L. Actually, with the exception of one test (3.50 mg/L) all results were very close to each other. The determination of the *Lemna* species, the use of a standardized medium and the application of growth related endpoints such as frond area or frond number (see Chap. 12) were assumed in order to achieve reliable reference test results (Wang 1991). Therefore, no changes concerning the selection of the reference substance or the required range are recommended.

Table 5.4 Results of the reference tests performed in four additional aquatic tests

Lab code	Reference substance	LC/EC50	Mean values (and SDs, if appropriate
Duckweed growth inhibition test (Lemna minor)			
2	3,5-Dichlorophenol	2.66 mg/L	2.77 ± 0.44 mg/L
5	3,5-Dichlorophenol	2.70 mg/L	
8	3,5-Dichlorophenol	3.50 mg/L	
12	3,5-Dichlorophenol	2.20 mg/L	
14	3,5-Dichlorophenol	2.54 mg/L	
16	3,5-Dichlorophenol	3.21 mg/L	
37	3,5-Dichlorophenol	2.60 mg/L	
Chronic toxicity to Brachionus calyciflorus			
36	Copper sulphate (as Cu^{2+})	23.80 µg/L	28.40 ± 13.05 µg/L
38	Copper sulphate (as Cu^{2+})	12.00 µg/L	
61	Copper sulphate (as Cu^{2+})	38.80 µg/L	
63	Copper sulphate (as Cu^{2+})	39.00 µg/L	
47	Potassium dichromate	5.98 mg/L	Required range not known
62	Potassium dichromate	3.70 mg/L	
Chronic toxicity to Ceriodaphnia dubia			
38	Sodium pentachlorophenolate	314 µg/L	330 ± 60 µg/L
61	Sodium pentachlorophenolate	253 µg/L	
63	Sodium pentachlorophenolate	400 µg/L	
64	Sodium pentachlorophenolate	350 µg/L	
Pseudomonas putida growth inhibition test			
1	3,5-Dichlorophenol	22.50 mg/L	
16	3,5-Dichlorophenol	7.36 mg/L	
23	3,5-Dichlorophenol	23.60 mg/L	

5.3.2 Chronic Toxicity to *Brachionus calyciflorus*

Four participants performed this chronic test with the reference substance Copper sulphate as required in the guideline, while two other laboratories provided data with the test substance Potassium dichromate. Two of the former tests provided results which are within the recommended range of 35.8–71.2 µg/L Copper sulphate (as Cu^{2+}), but close to the lower end of that range. The other two EC_{50} values were clearly lower. No obvious reason could be identified for the organisms tested in the ring test reacting more sensitively than those in the French Ring test, which was the basis of the guideline recommendations. The two EC_{50} values in the tests with Potassium dichromate differed by about a factor of two. Since no information was available about the "normal" effect range for this compound, it was not possible to evaluate these results. It may be useful to find if potassium dichromate could be an alternative to the currently recommended copper sulphate.

5.3.3 *Chronic Toxicity to Ceriodaphnia dubia*

Four participants performed this test with the recommended reference substance Sodium pentachlorophenolate. Despite the fact that the resulting EC_{50} values were very similar, less than a factor of two apart, one value (400 µg/L) was outside the required range as indicated in the guideline (216–344 µg/L). However, in this case, it may be that the range found in the Ring test on which this requirement was based, was too narrow and other experiences from the methodological development and validation of *Ceriodaphnia* test should be taken into account (Bailey et al. 2000).

5.3.4 *Pseudomonas putida Growth Inhibition Test*

Three participants performed reference tests with the recommended reference test substance 3,5-Dichlorophenol. While one test result was outside the required range of 10–30 mg/L, the other two fitted very well. In this case of the failed reference test also, other validity criteria (i.e. lack of sufficient multiplication) were not fulfilled, which explains the invalid results for the reference substance. These results do not indicate the need for a change of the reference substance or modifications to the required range of EC_{50} values.

5.4 Recommendations

The performance of reference tests confirmed in most cases the sensitivity of the respective test systems in terms of the type of reference substance and the required range. This is true especially of the *Daphnia*, *Lemna* and *Pseudomonas* tests. An adaptation of the required range of EC_{50} values is obvious in the Algae tests where, based on the high number of results gained in this Ring test, detailed recommendations are given on how this range could be modified. Less obvious, but possibly the range of required EC_{50} values in the *Brachionus* and *Ceriodaphnia* tests should be checked as it seems that the currently required ranges are based on a quite small number of data sets.

A very special case is the luminescent bacteria test, as so far no dose-response testing is required for it, but only a limit test approach. It should be decided whether the ISO guideline should be modified in such a way that it will be in line with the other aquatic tests. One suggestion would be to use 3,5-Dichlorphenol or potassium dichromate (or both?) for the determination of EC_{50} values. The data gained through this ring test could be a starting point for such an approach but the number of data available is certainly too small to give recommendations for detailed numbers, partly because all tests have to be performed in duplicate for the two main sources of luminescent bacteria.

In this context it may also be discussed whether the number of reference substances currently used in aquatic tests can be used, partly because some of those used now (e.g. Sodium pentachlorophenolate) cause health problems. Potassium dichromate and 3,5-Dichlorophenol are the most obvious choices for such a move.

Chapter 6
Range of Reference Tests in Terrestrial Tests

A. Haller, M. Goth, and M. Pattard

Abstract Standardized test methods are necessary to evaluate the adverse influences of chemical substances, contaminated soils, or waste materials. Therefore, the participants of the EU Waste Ring test 2006/2007 performed a number of reference tests for two standard (earthworm acute, plant growth) and three additional (earthworm, enchytraeid, collembolan reproduction) test systems. No data were provided for two other additional tests, the earthworm avoidance and the *Arthrobacter* contact test. The number of the required (chloroacetamide in the earthworm test with *Eisenia fetida* or *Eisenia andrei*) and proposed (boric acid in the plant tests with *Avena sativa* and *Brassica rapa*) reference tests was high (14 and 9, respectively) enough to confirm the suitability of the substances and required/proposed ranges of EC_{50} values. In the three additional tests the number of data sets provided, often with different reference substances, was much lower. For reasons of availability (pesticides) and human toxicology (chloroacetamide) boric acid is recommended as reference substance for all terrestrial test systems. Therefore, and despite some data already provided (in particular for the plant growth test), more data from tests with boric acid are needed to identify the most suitable range of effects. The information gained in this ring test forms a good starting point for this purpose.

Keywords Earthworms, Enchytraeids, Collembolans, Mites, Chloroacetamide, Boric acid, Dimethoate, Phenmedipham

6.1 Introduction

Standardized test methods are necessary to evaluate the adverse influences of chemical substances, contaminated soils, or waste materials. In this context, it is very important to test a reference substance at regular intervals to assure a constant sensitivity of the

A. Haller (✉)
ECT Oekotoxikologie GmbH, Böttgerstr. 2-14, 65439, Flörsheim, Germany

H. Moser and J. Römbke (eds.), *Ecotoxicological Characterization of Waste*,
DOI: 10.1007/978-0-387-88959-7_6,

test systems (Gourmelon & Ahtiainen 2007). Therefore, the participants of the EU ring test for the ecotoxicological characterization of waste materials were asked to perform such tests in parallel to the waste tests (EN 2005). While in the case of basic tests this requirement was fulfilled in almost all cases, the number of reference tests performed for the additional tests is much lower. However, even for these tests the reference data gained form a valuable source of information, as the published experience for most of these tests is very limited. The terrestrial basic test battery consisted of the earthworm acute test and the plant emergence and growth test. Reproduction tests with earthworms, springtails, enchytraeids, and the earthworm avoidance tests, as well as the *Arthrobacter* contact test, were performed as additional tests.

In this chapter, the results obtained by the participants from reference tests are compiled for each test system separately. In addition, these results are compared with the range of every reference substance found in the respective guideline or in the literature. Except for the contact test, each of these tests lists a reference substance and a specific range of toxicity which should be met, when conducting a reference test under the required conditions. However, in the plant test guideline, no specific reference testing is required. Therefore, in this case the ring test SOP for plant testing recommends boric acid as the reference substance, as this compound has recently been proposed in the literature (Römbke & Ahtiainen 2007). In Table 6.1 the reference substances recommended by the respective guidelines as well as the expected effect ranges are summarized.

In the following sections, the results of the reference tests are compiled, starting with the tests of the standard set. If more than six test results were available, the data set was assessed using the warning limit approach published by Environment Canada (EC 2005). Tests with a lower number of data are shown in graphs and the minimum-maximum factor is calculated (EC 1999). Afterwards, the mean LC/EC_{50} (determined via probit analysis and almost always using the statistical program

Table 6.1 Ecotoxicological tests and reference substances

Test + group	Organism	Guidelines	Reference substance, range
Basic test battery			
Mortality (earthworm)	*Eisenia fetida, Eisenia andrei*	OECD 11268-1	Chloroacetamide, 20–80 mg/kg
Emergence, growth (plants)	*Brassica rapa, Avena sativa*	ISO 11269-2	Boric acid, 100–400 mg/kg
Additional test battery			
Reproduction (earthworm)	*E. fetida, E. andrei*	ISO 11268-2	Carbendazim, 1–5 mg (a.i.)/kg
Reproduction (springtail)	*Folsomia candida*	ISO 11267	Phenmedipham, 15–30 mg (a.i.)/kg
Reproduction (enchytraid)	*Enchytraeus crypticus*	ISO 16387	Carbendazim, 1.2 ± 0.8 mg (a.i.)/kg
Avoidance (earthworm)	*E. fetida, E. andrei*	ISO 17512-1	Boric acid 750 mg/kg
Contact test (bacteria)	*Arthrobacter globiformis*	DIN 38412-48	Benzalkonium chloride (BAC) 600 mg/kg

ToxRat (2006)) of each test system is compared with the guideline requirements. Finally, recommendations are given concerning the further development of reference testing in terrestrial ecotoxicology.

6.2 Results of the Basic Test Battery

6.2.1 Earthworm Acute Tests

In total, 19 participants performed earthworm acute tests with the waste materials. They provided 16 valid reference test data (15 with chloroacetamide and one with carbendazim). The latter is usually required in the earthworm reproduction test but was accepted as the measured EC_{50} of 0.9 mg/kg (OECD 2004b) confirmed the sensitivity of the test system (required range: 1–5 mg (a.i.)/kg). All test results are summarized in Table 6.2.

According to the SOP of the EU Waste Ring test 2006/2007, the LC_{50} value for chloroacetamide should be in the range of 10 to 30 mg/kg, but in the guideline the range of LC_{50} value is much wider (20–80 mg/kg; ISO 1993). The value in the SOP was based on results reported in the literature (e.g. Römbke et al. 1992; Kula 1998), indicating that the original range based on a small ring test when the acute earthworm test was developed (Edwards 1984), is starting with a value too high. However, to avoid confusion, it was decided to use on results of the reference tests for the drawning of warning limits, which are shown together with both ranges in Fig. 6.1. With the exception of one high (59.9 mg/kg) and one low (13.1 mg/kg), all LC_{50} values were within a range of 20–40 mg/kg. Even the two extreme values were inside the borders set by the SOP and the ISO guideline (10–80 mg/kg) and only the very low value was slightly outside the warning limits (calculated as 15.5 and 63.9 mg/kg). In summary, the results of the reference tests in the earthworm acute test confirm the sensitivity of the earthworms used. However, chloroacetamide shows developmental toxicity, meaning that its regular use should not be continued (ScienceLab 2005).

6.2.2 Plant Seedling Emergence and Growth Tests (Avena sativa, Brassica rapa)

The seedling emergence and growth tests with *A. sativa* and *B. rapa* were performed by 23 and 21 participants, respectively. Four (*A. sativa*) and five (*B. rapa*)

Table 6.2 LC_{50} values (mg/kg dry weight) of reference tests with chloroacetamide in the earthworm acute test performed by the participating laboratories (lab code)

	Laboratory code														
No.	6	9	10	11	16	22	24	26	38	51a	51b	59	61	63	64
LC_{50}	34.8	40.5	38.1	32.0	31.1	31.7	13.1	21.9	27.4	36.0	59.9	25.0	27.1	36.5	41.2

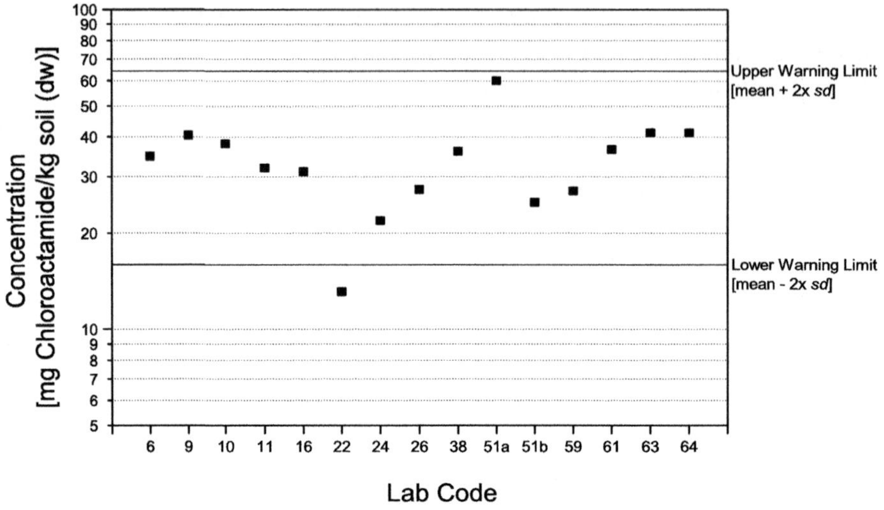

Fig. 6.1 LC_{50} values of the earthworm acute reference tests with chloroacetamide

Table 6.3 EC_{50} values for boric acid determined in the plant growth tests with *Avena sativa* and *Brassica rapa* (mg/kg dry weight of field or artificial soil

	Laboratory code								
	Avena sativa				*Brassica rapa*				
No.	6	38	61	63	6	13	38	61	63
EC_{50}	236.9	196.0	324.0	330.4	207.4	78.5	240.0	234.0	178.3

participants performed a valid reference test with boric acid, but the majority did not, mainly because the reference test was not explicitly required by the guideline. Five EC_{50} values could not be determined: two were limit tests, two others failed for unknown reasons and for the last one the EC_{50} could not be calculated. All test results are summarized in Table 6.3.

The required range for the EC_{50} values in tests with boric acid (100–400 mg/kg) is based on literature data (Stephenson et al. 1997) and was stated in the Ring test SOP. The results of four participants and with both species were within this range. No result was above the upper limit, while in one test with *B. rapa* the EC_{50} value was lower than the lower limit recommended by the SOP (Fig. 6.2). Despite the low number of data the test results presented here do not contradict the proposal to use boric acid as obligatory reference for plant growth tests.

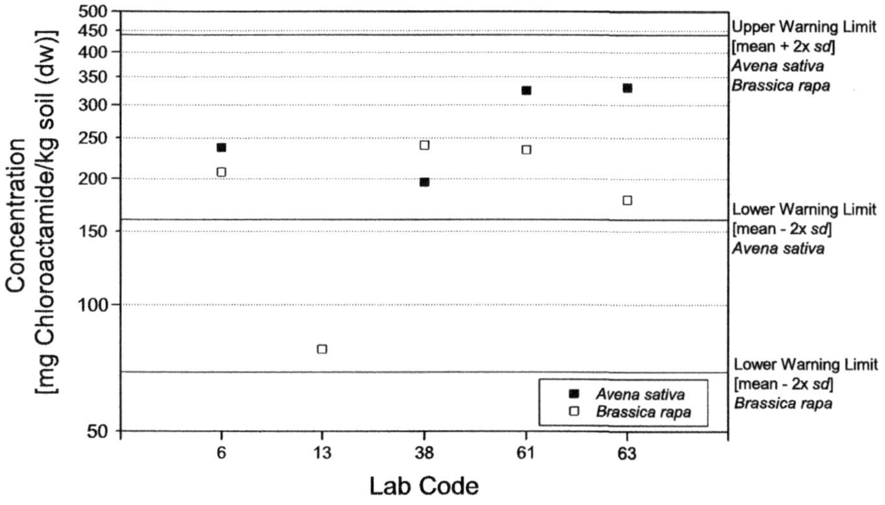

Fig. 6.2 Results (LC_{50} values) of the reference plant tests with *A. sativa and B. rapa*

6.3 Results of the Additional Test Battery

The results of three additional tests are summarised in Table 6.4 (no reference data were provided in the earthworm avoidance test and in the *Arthrobacter* contact tests). In addition to the EC_{50} values, the NOECs (No-observed-effect-concentration) are given, as in some tests only they were determined.

6.3.1 Earthworm Reproduction Tests

Three of the participants performing earthworm reproduction tests with *Eisenia fetida/Eisenia Andrei* delivered results of reference tests, one of them three times (Table 6.4). In the case of the tests with carbendazim one result was below, one well within and one above the range given in OECD Guideline No. 222 (1–5 mg (a.i.)/kg). In the literature, EC_{50} values between 0.6 and 2.7 mg/kg artificial soils are reported (Garcia 2004). The other two participants performed the test with potassium dichromate, fulfilling the internal validity criteria of 340–730 mg/kg. This is very close to information from the literature (336.3 mg/kg; Stantec 2004).

6.3.2 Enchytraid Reproduction Tests

Three participants performed a reproduction test with *Enchytraeus crypticus* with a reference substance test: two of them with carbendazim and one with boric acid

Table 6.4 Results of the reference tests performed in five additional terrestrial tests. All data given in mg/kg soil dry weight

Lab code	Reference substance	LC/EC_{50} (mg/kg)	NOEC (mg/kg)
Earthworm reproduction tests (*E. fetida/E. andrei*)			
22	Carbendazim	–	<0.1
48	Carbendazim	–	<12.6
51	Carbendazim	2.7	
51a	Potassium dichromate	436.0	–
51b	Potassium dichromate	339.0	–
Enchytraeid reproduction tests (*E. crypticus*)			
3	Boric acid	46.0	–
16	Carbendazim	22.4	–
22	Carbendazim	–	LOEC < 40
Collembolan reproduction tests (*F. candida*)			
11	Phenmedipham	109.5	–
16	Dimethoate	1.0	0.4
18	Phenmedipham	72.4	
22	Phenmedipham	–	<23.6

(Table 6.4). The EC_{50} values determined here in the two Carbendazim tests were slightly lower than the only value found in the literature: 44 mg/kg (Kuperman et al. 2006). However, the EC_{50} values in the tests with carbendazim are much higher than those values given in the ISO guideline which, actually, were determined for the larger species *E. albidus*. Therefore, there is a need to specify the range of reference test results for the species *E. crypticus* which seems to be less sensitive to this fungicide. While no information on the toxicity of boric acid to this species could be found in the literature, preliminary EC_{50} values (R. Kuperman, pers. Comm.) of the ring test gained with a field soil (43–66 mg/kg) indicate a similar result to those found here with artificial soil (EC_{50}: 46 mg/kg).

6.3.3 Collembolan Reproduction Tests

Four participants performed an additional reproduction test with *F. candida* with the waste materials (Table 6.4). Three of them used phenmedipham as reference substance, as proposed in the ISO guideline (ISO 1999) but in one laboratory the insecticide dimethoate was tested. According to the guideline, an effect of phenmedipham should occur in the range of 15–30 mg/kg active substance, while in the literature the EC_{50} for *F. candida* in OECD artificial soil is given as 39.2 mg/kg artificial soil (Amorim et al. 2005b). The EC_{50} of dimethoate for collembolan reproduction in artificial soil was determined here as 1.0 mg/kg, which is close to the value 3.8 mg/kg found in the literature (Martikainen 1996). However, he used animals twice the age as used here and as recommended in the ISO guideline, which may explain the difference. These results do not give a clear picture, as the values reported here are about twice as high as expected from literature (two out of three phenmedipham tests), while the third value with this reference substance and the only value with dimethoate are lower than expected.

6.3.4 Arthrobacter Bacterial Contact Tests

According to the ISO draft guideline 10871 (ISO 2008), originally developed for sediment testing but almost identical to the soil method still to be standardized by ISO, the chemical Benzalkonium chloride (BAC) has to be tested in a limit test using a concentration of 600 mg/kg. In a Ring test mentioned in the draft guideline, this concentration caused an inhibition of 58%. Accordingly, the guideline asks for an effect range of 30–80% when testing BAC at 600 mg/kg; otherwise the test would be unacceptable. All six participants in this Ring test achieved this range. It would be worthwhile to check whether boric acid would also be a suitable reference substance in this test.

6.4 Recommendations

The performance of reference tests confirmed the sensitivity of the respective test systems in terms of the type of reference substance and the required range. The exception is the earthworm acute test, where the current reference substance chloroacetamide should be replaced because of its developmental toxicity (in the meantime, based on the results of this ring test, the required range could be changed from 20–80 to 10–50 mg/kg). However, for reasons of availability of the pesticides used as reference substances it is proposed to use boric acid for all terrestrial test systems in future. Despite some data already provided (in particular for the plant growth test), more data from tests with boric acid are needed to identify the most suitable range. The information gained in this ring test forms a good starting point for this purpose.

Part II
Presentation of the Basic Text Battery

Chapter 7
Algae Tests

R. Weltens

Abstract Unicellular Algae are an important food source in the aquatic ecosystem, thus representing a relevant trophic level in the aquatic ecotoxicity test battery. The algal growth inhibition test is described in ISO 8692 (2004) and OECD 201 (2006). After the classical setup, the ISO protocol also describes a miniaturized setup to be used for screening purposes. Both protocols were performed within this ring test, using very different test volumes. Also two algal species were used: *Desmodesmus subspicatus* (44%) and *Pseudokirchneriella subcapitata* (56%). In total 143 test results were presented, of which 106 were accepted. The invalid tests were rejected mainly because reference data were lacking or out of range, and in a few cases because of low growth rate in controls. As almost no practical problems were reported, and the mean EC_{50} value of the reference substance showed low variation despite the differences in setup, it can be concluded that the algal test is robust and reliable. The interlaboratory variation to the EC_{50} values was more considerable for the complex waste samples. The samples clearly differed in toxicity WOO > INC > SOI. Mean LC_{50} values were respectively 1.3%, 8.8% and >highest test concentration for *Desmodesmus*, and 0.22%, 2.4% and >highest test concentration for *Pseudokirchneriella*. The differences between the species were not statistically significant. The sensitivity of the algal growth inhibition test for these waste samples was comparable to the results of the *Daphnia acute immobility test*. The algal test is recommended as part of the waste test set because of its ecological relevance and robustness. No preference for species or setup was detected.

Keywords Freshwater, Microalgae, *Desmodesmus*, *Pseudokirchneriella*, Growth inhibition test, Ring test, Waste, Extract, Miniaturized and classic bioassays

---

R. Weltens (✉)
VITO, Environmental Toxicology, Boeretang 200, B 2400 MOL, Belgium

H. Moser and J. Römbke (eds.), *Ecotoxicological Characterization of Waste*,
DOI: 10.1007/978-0-387-88959-7_7,

7.1 Introduction

Algae are primary producers and as such have an important role in the aquatic ecosystem. They are a major food source for higher plankton feeding organisms. The objective of the growth inhibition test on Algae is to determine the effect of a substance or a sample on the growth parameters of freshwater microalgae. Exponentially growing test organisms are exposed for 72 h to (a dilution series of) the test substance/sample. The growth rate during exposure is compared to the growth rate of unexposed controls. Growth and growth rate inhibition are quantified from measurements of the algal biomass (or surrogate parameter) as a function of time. Biomass is estimated from surrogate parameters like cell counts, fluorescence and optical density.

The test is described by OECD 201 (2006) and ISO 8692 (2004) and is widely used for the ecotoxicological evaluation of chemicals and waste water. Many chemicals are already documented concerning their EC_{50} for Algae. The classical protocol prescribes the use of 250 mL recipients during the test, but many miniaturised alternatives are in use, and accepted (ISO 8692 Annex A (2004)).

7.2 Method

The performance of the 72 h test with micro Algae (*Pseudokirchneriella subcapitata or Desmodesmus subspicatus*) was described in a SOP that was based on the ISO guideline No. 8692 (2004). The test was performed with a suitable medium (ISO 8692 (2004)) as control and a dilution series of waste eluate fractions in this artificial medium. All dilutions were sterilized by filtration. Tests were performed using the following conditions:

- *Species. P. subcapitata* and *D. subspicatus*
- *Test vessels.* The SOP recommended 250–300 mL erlenmeyer flasks, but in many/most labs a miniaturized version was used with very different volumes (800 μL up to 500 mL)
- *Temperature* 23 ± 2°C
- *Continuous light.* 60–120 $\mu E^{*}/m^{2}/s^{1}$, 6,000–10,000 lux
- pH in controls was 8.1 ± 2 at the beginning of the test
- Inoculum from an exponentially growing preculture of cells (the preculture is started 2–4 days before the beginning of the test with a sufficiently low cell density (5–10E + 3 cells/mL). The health condition of the Algae is checked immediately before the start of the experiment: no cell aggregations or contamination by other unicellular organisms should occur), with a maximum of 10E + 4 cells/mL in the final volume
- Vessels/plates were shaken either continuously or periodically, to prevent settling and sticking of the algal cells to the walls
- *Test concentrations.* Based upon preliminary experiments the following concentrations were recommended:

(% extract)	Dilution 1	Dilution 2	Dilution 3	Dilution 4	Dilution 5
SOI	100	50			
INC	20	10	5	2.5	1.25
WOO	1.67	1.25	0.883	0.625	0.417

7.2.1 Measurements

- *pH*. At the beginning and at the end of the experiment in one vessel of controls and dilutions.
- *Temperature*. Continuously recorded, preferably in a separate vessel (or else in the incubator).
- *Biomass*. At the beginning in the control replicates every 24 h for 72 h. Different surrogate parameters were used by different labs to measure biomass (cell number determined by coulter counter, fluorescence signal…).

7.2.2 Calculations

The main assessment endpoint was inhibition of the growth rate, determined as EC_{50} values (50% inhibiting median concentration). These values were determined by the probit analysis method, using the statistical program ToxRat® (2006). The overall assessment of the test results from the different participants was performed in a two-step procedure (Chap. 4). First, the total number of tests that fulfilled the acceptance criteria was determined. Secondly, all data sets in which no LC_{50} could be determined (i.e. where they had to be given as < or > a tested dilution) were excluded from the statistical evaluation. The remaining data sets were checked for passing the outlier tests (i.e. following the ISO approach (ISO 1994)) and being outside the warning limits (i.e. following the approach presented by Environment Canada (2005)). The 'final mean (geometric) LC_{50} values' for each substrate were calculated excluding all outliers, irrespective of the evaluation method used.

7.2.3 Validity Criteria

- The growth factor in the controls was at least 67 times from $t = 0$ up to 72 h (or 1.4 specific growth rate per day).
- pH should not shift by more than 1.5 unit in the controls during the test.
- The results of a reference test with the chemical potassium dichromate should be in line with predicted values (see acceptance criteria).
- Variation coefficient for control growth rates: <5%.

7.2.4 Acceptance Criteria

Test results were *not* accepted in case:

1. No reference test was provided.
2. The EC_{50} for potassium dichromate (PDC) was outside the range: mean of all PDC tests ±2SDs:

   – *D. subspicatus*: mean = 0.75 mg/L (0.39 - 1.43 mg/L)
   – *P. subcapitata*: mean = 0.99 mg/L (0.38–2.61 mg/L)

3. Validity criteria were not fulfilled (except the VC < 5%; this was not used as an acceptance criterion).
4. There were considerable deviations from the SOP.
5. If (biological) outlier occurred: EC_{50}</> by a factor of 10 from the mean.

7.3 Results

A total of 48 participating laboratories provided 143 test data sets of which 37 tests were not valid (25.9%). Reasons for rejecting are listed in Table 7.1.

The results of the Algae tests for the individual samples of INC, SOI and WOO are presented below.

7.3.1 Incineration ash (INC)

35 tests were accepted (Table 7.1).

The mean EC_{50} value was 4.4%, overall reproducibility was 0.482 and repeatability was 0.070 (Table 7.2)

Table 7.1 Summary of the acceptance of Algae tests: number of labs (individual lab numbers)

Criteria	INC	SOI	WOO	Total
Number all tests	48	48	47	143
1. No reference test	6	7	7	123
2. Outside range of reference	2	3	2	116
3. Not valid	3	2	1	110
4. Diff. to SOP	1	0	0	109
5. Biol. Outlier	1	1	1	106
Accepted number	*35*	*35*	*36*	*106*
Range data (>/<)	−3	−29	−1	–
Stat. Outlier (ISO 1994)	−1	?	0	–
W-L Outlier (EC 2005)	0	?	−1	–

? = Not known since EC_{50} values could not be determined (e.g. >100%)

Figures 7.1–7.3 show the results of the statistical evaluation of the results. The EC_{50} logs from the different labs are shown in increasing order (mean and confidence intervals (boxes)). s_R is the reproducibility standard deviation and s_r is the repeatability standard deviation (first two boxes on each Figure). Figure 7.1 shows the results from all the labs (two species) and Figs. 7.2 and 7.3 show the results for the individual algal species. These figures show that test performance was comparable for both species, although mean values are different.

Warning charts for INC are shown in Figs. 7.4–7.6 for both algal species *Desmodesmus* and for *Pseudokirchneriella* respectively. All EC_{50} values were within the twofold warning limits (Fig. 7.4) (geometrical mean EC_{50} ± 2SD; Environment Canada 2005).

Table 7.2 Summary of the results for individual algal species and total values for the sample INC

Parameter	*D. subspicatus*	*P. subcapitata*	Total
n	13	22	35
Mean EC_{50} (% eluate)	8.80	2.42	4.08
Reproducibiltiy	0.44	0.37	0.48
Repeatability	0.08	0.06	0.07

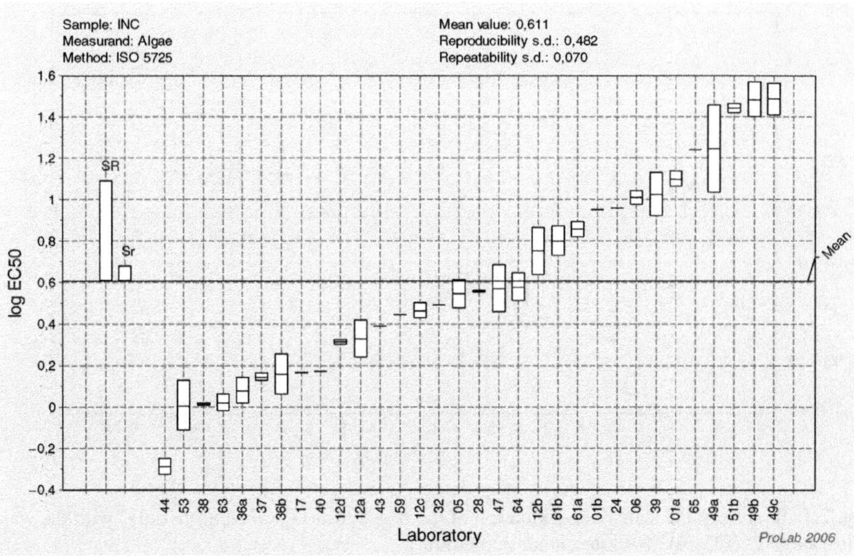

Fig. 7.1 Results of the statistical evaluation of the Algae tests (both species) with the waste substrate INC. (EC_{50} as % waste eluate in medium)

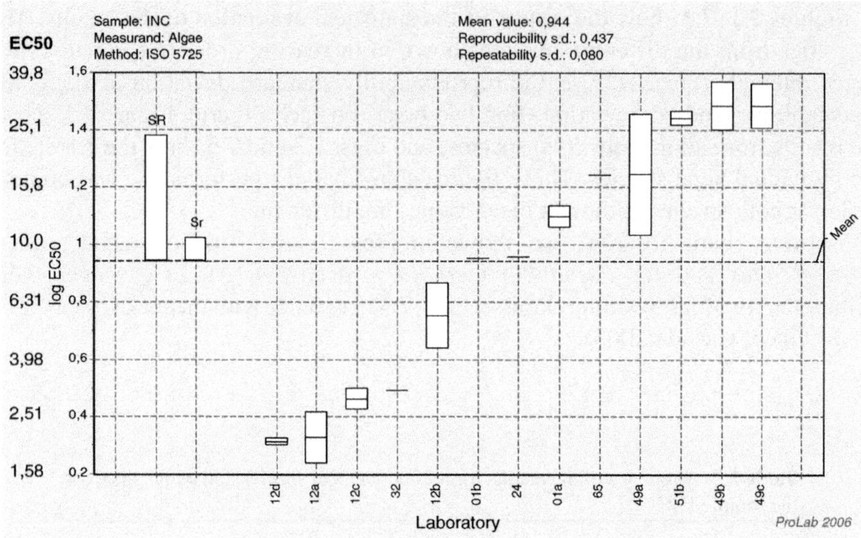

Fig. 7.2 Results of the statistical evaluation of the Algae tests (*D. subspicatus* only) with the waste substrate INC. (EC_{50} as % waste eluate in medium)

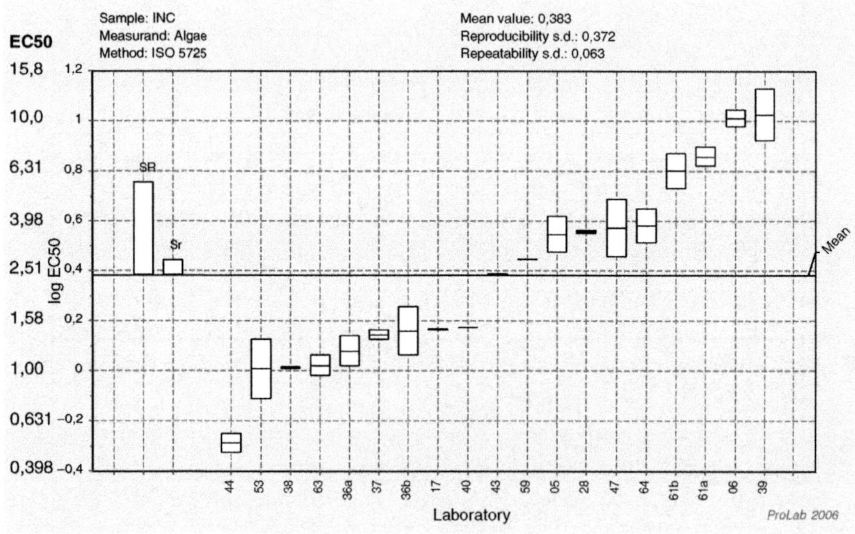

Fig. 7.3 Results of the statistical evaluation of the Algae tests (*P. subcapitata* only) with the waste substrate INC (EC_{50} as % waste eluate in medium)

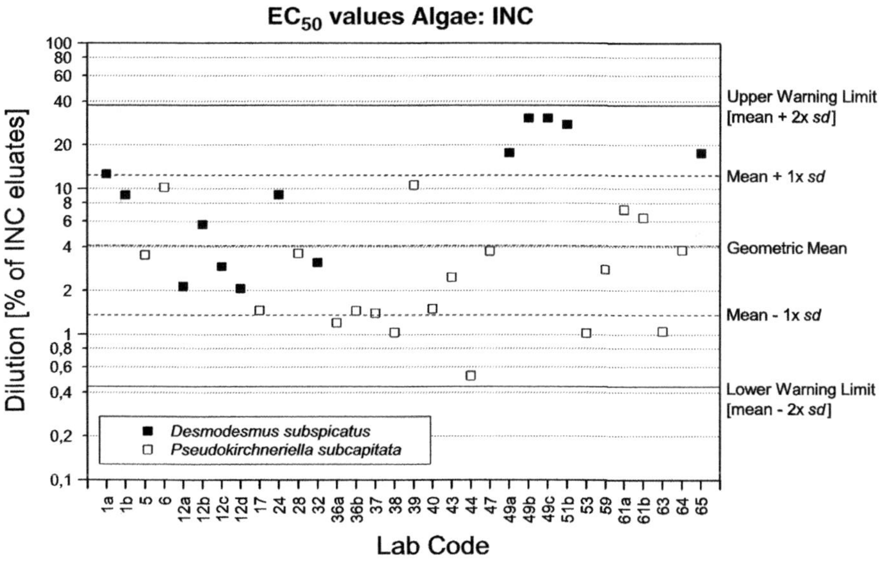

Fig. 7.4 Results (EC$_{50}$ as % waste eluate in medium) of the Algae tests (both species) with the waste substrate INC, in relation to the warning limits ±1 and 2xSD

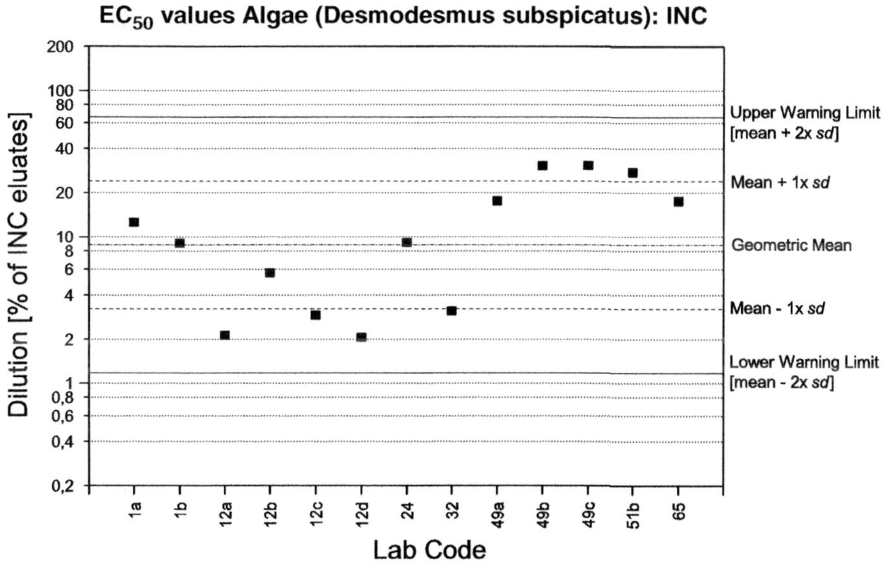

Fig. 7.5 Results (EC$_{50}$ as % waste eluate in medium) of the Algae tests (*D. subspicatus* only) with the waste substrate INC, in relation to the warning limits ± 1 and 2xSD

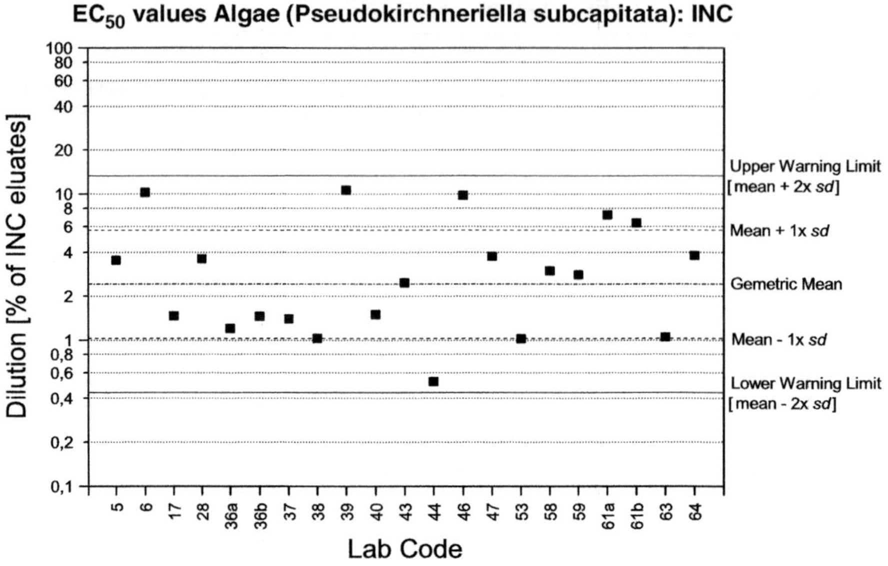

Fig. 7.6 Results (EC_{50} as % waste eluate in medium) of the Algae tests (*P. subcapitata* only) with the waste substrate INC, in relation to the warning limits +/1 and 2xSD

7.3.2 Gasworks Soil (SOI)

35 soil tests were accepted (Table 7.1).

In most cases the EC_{50} value was higher than the highest test concentration (50, 80, 90 or 100% eluate). In a few cases however, effects were seen (1 response in *Desmodesmus*, 5 in *Pseudokirchneriella*). Results are summarized in Fig. 7.7. The mean EC_{50} value for those Algae tests that responded to exposure to the SOI sample was 56.5% eluate ($n = 6$, Fig. 7.8). Figure 7.8 shows the results of the statistical analyses of these limited tests. Only 4 test results could be used to calculate confidence intervals.

7.3.3 Contaminated Wood (WOO)

36 tests were accepted (Table 7.1).

Table 7.3 summarizes the results for the WOO sample. The EC_{50} values differed by a factor of five between both species (Table 7.3; Fig. 7.9). The mean EC_{50} value was 0.5% eluate. According to the ISO 5725 protocol, the general reproducibility was 0.487 and repeatability, 0.116.

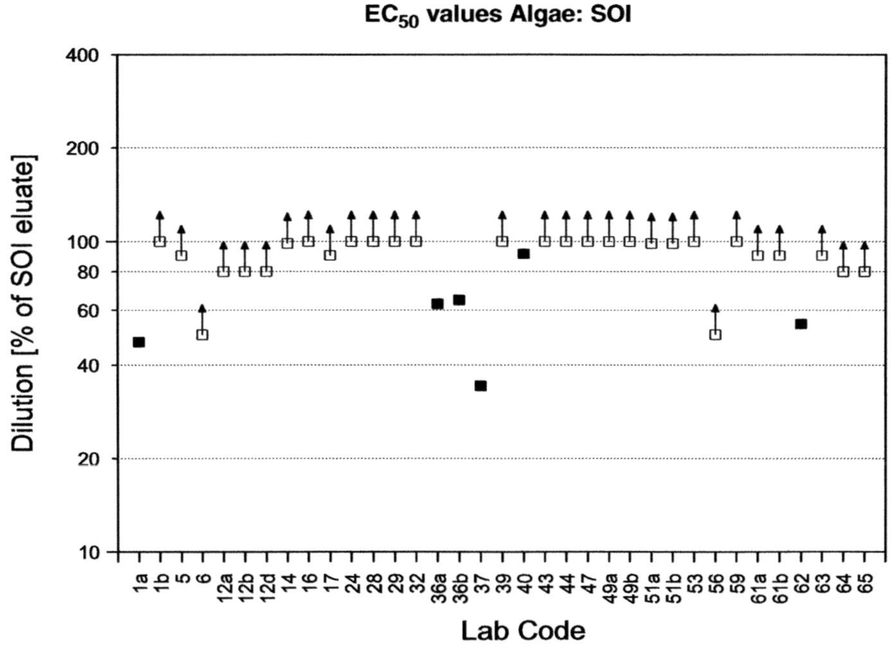

Fig. 7.7 Results (EC_{50} as % waste eluate in medium) of the Algae tests with the waste substrate SOI (both species)

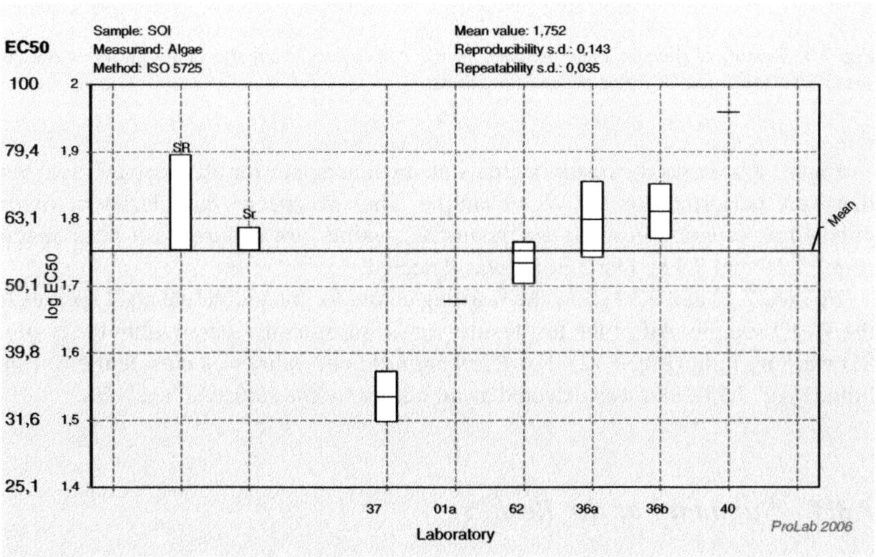

Fig. 7.8 Results of the statistical evaluation) of those Algae tests that showed effects to the waste substrate SOI (both species). (EC_{50} as % waste eluate in medium)

Table 7.3 Summary of the results for individual species and total values for the WOO sample

Parameter	*D. subspicatus*	*P. subcapitata*	Total
	14	22	36
Mean EC_{50} (%)	1.34	0.22	0.50
Reproducibiltiy	0.13	0.44	0.49
Repeatability	0.06	0.07	0.12

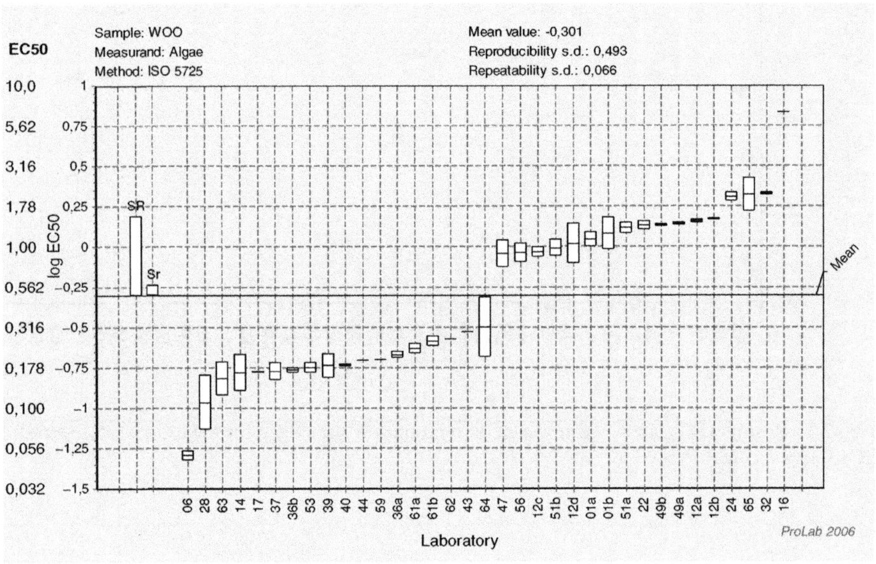

Fig. 7.9 Results of the statistical analyses of the Algae tests (both species) with the waste substrate WOO (EC_{50} as % waste eluate in medium)

Figure 7.9 shows the results of the statistical analyses for all accepted algal tests that were performed on the WOO sample. The two species can clearly be distinguished as separate groups; the mean EC_{50} values are different for both species (Figs. 7.10 and 7.11). One outlier was detected.

Figures 7.12 and 7.13 show the warning charts for the individual algal species for the WOO sample. All of the test results for *D. subspicatus* were within the twofold SD warning limit (Fig. 7.12). For *P. subcapitata* one value was outside the warning limits (Fig. 7.13), and was detected as an outlier by the statistical analyses.

7.3.4 Summary of the Results

Table 7.4 summarises the test results for all three samples of the algal growth inhibition test

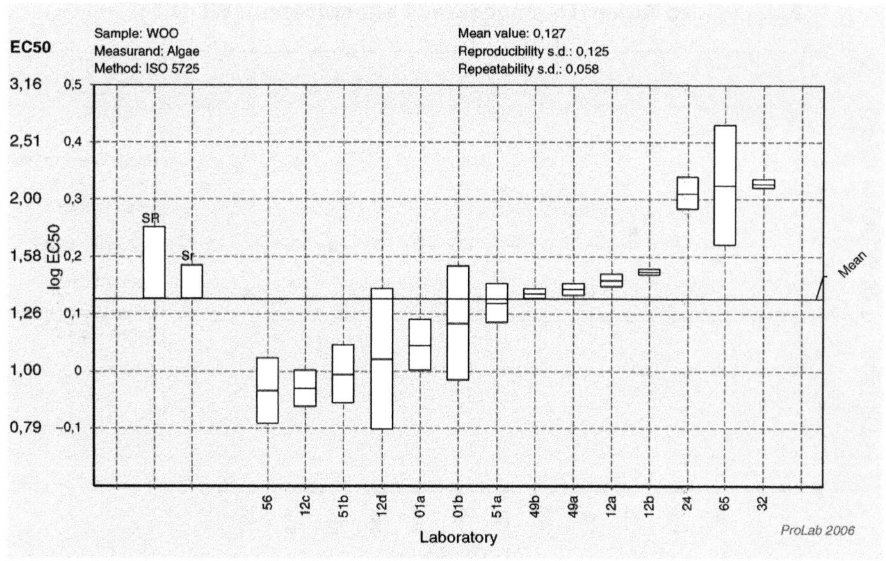

Fig. 7.10 Results of the statistical analyses of the Algae tests (*D. subspicatus* only) with the waste substrate WOO (EC_{50} as % waste eluate in medium)

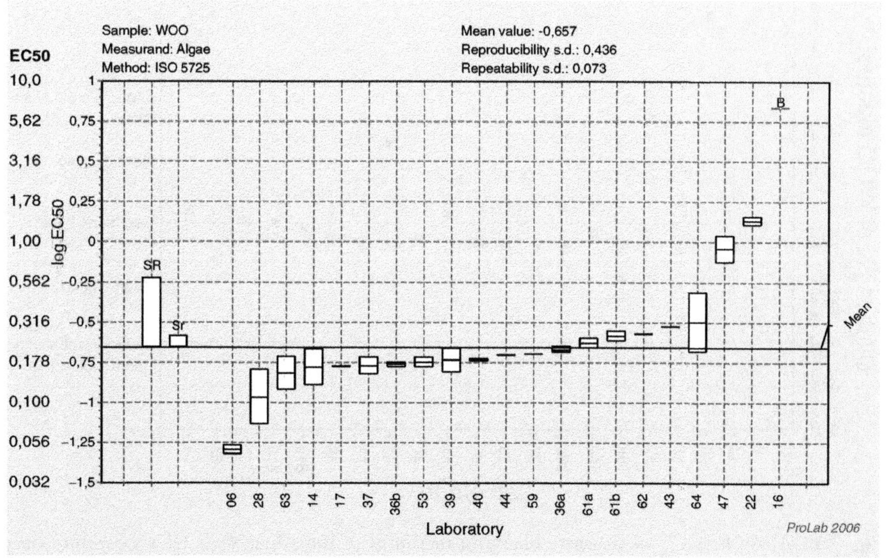

Fig. 7.11 Results of the statistical analyses of the Algae tests (*P. subcapitata* only) with the waste substrate WOO (EC_{50} as % waste eluate in medium)

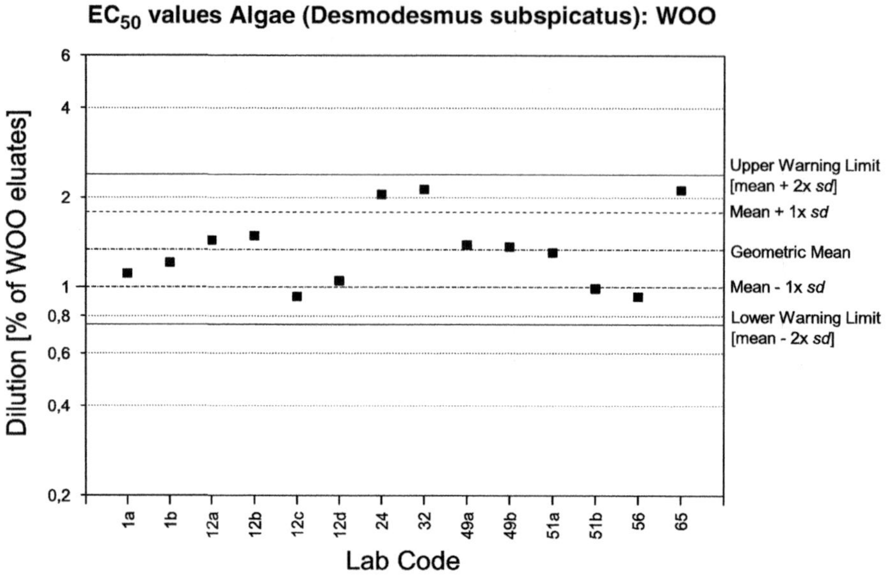

Fig. 7.12 Results (EC_{50} as % waste eluate in medium) of the Algae tests (*D. subspicatus* only) with the waste substrate WOO, in relation to the warning limits +/−1 and 2xSD

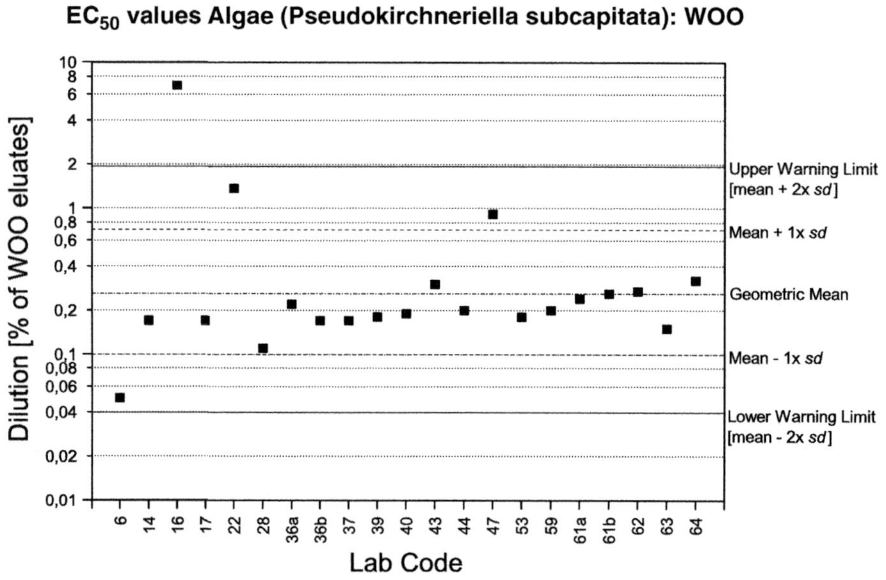

Fig. 7.13 Results (EC_{50} as % waste eluate in medium) of the Algae tests (*P. subcapitata* only) with the waste substrate WOO, in relation to the warning limits +/−1 and 2xSD

Table 7.4 Summary of the results of the algal acute tests with three waste types (data expressed as % waste eluate in artificial medium) and the reference substance (potassium dichromate; PDC)

Substrate	Mean EC50 Min-Max Max/Min	*D. subspicatus*	*P. subcapitata*
INC	4.08	8.80	2.42
	(0.52–30.7)	(2.06–30.67)	(0.52–10.61)
	59	14.9	20.4
SOI	56.5[a]	–	–
WOO	0.50	1.34	0.22
	(0.05–6.9)	(0.93–2.13)	(0.05–6.9)
	138	2.3	138
PDC	0.99	0.79	1.15
	(0.46–2.03)	(0.62–1.02)	(0.46–2.03)
	4.4	1.65	4.4

Given are the mean EC_{50} values, the lowest and the highest EC_{50} values measured and the ratio of highest and lowest value

[a] Only 6 out of 35 tests showed toxicity. 56.5% is the mean of these values. Most tests showed no response at the highest test concentration.

7.4 Discussion

7.4.1 Performance of the Test

Almost no practical problems were reported. 106 out of 143 test results were accepted (74%). The results on the reference substance show very high reproducibility (see Table 7.4: max/min for PDC = 4.4). These observations demonstrate that the algal test - despite the very different setups and different species used - is easily performed and provides reliable results. The reproducibility is lower for the waste samples than for the PDC tests, especially for the WOO sample where the *Pseudokirchneriella* results show a 138-fold difference between minimum and maximum. An increased variation is often seen in biotests when testing complex environmental samples because of the multiple interactions between different components and different biological pathways.

The EC_{50} values for the samples varied but were nevertheless almost all within the twofold SD range. Although not statistically different, the mean EC_{50} values differed for the two species, but were not consistent. The clearest separations between the responses of both species were seen in WOO.

7.4.2 Methodology

Only a few labs used the method as described in the SOP. Most labs used deviated volumes (from μL scale (microplates) up to 500 mL).

Tables 7.5 and 7.6 list the accepted tests for *P. subcapitata and D. desmodesmus* for the INC sample, for different classes of test volumes. No good comparison between different test volumes and the standard volume was possible as the number of test results for the classic setup was too low. The limited set of data did not permit final conclusions on the possible effects of the volume and setup on the test results. No major differences were seen between the different groups, but the minaturized methods seem to be more sensitive (Tables 7.5 and 7.6). The toxkit results are discussed elsewhere (Chap. 13).

The much higher variation of EC_{50} values in the waste samples compared to the results with PDC (see Tables 7.4–7.6) has to be attributed to differences in methods, interlaboratory differences in the eluate composition (see chapter homogeneity) and also to the complexity of the samples.

Table 7.5 Accepted test results for *P. subcapitata* for the INC sample

			Reference test (see Sect. 4.1.2)	
Vessel - plate	Species	EC_{50} (% dilut.)	Substance	EC_{50} (mg/L)
100 mL	*P. subcapitata*	1.02	PDC	1.85
250 mL	*P. subcapitata*	1.47	PDC	1.01
50 mL	*P. subcapitata*	10.2	PDC	0.80
		4.24	Mean	1.22
		5.20	SD	0.56
MicroPlate	*P. subcapitata*	3.61	PDC	1.71
MicroPlate	*P. subcapitata*	1.20	PDC	2.03
MicroPlate	*P. subcapitata*	1.45	PDC	1.94
MicroPlate	*P. subcapitata*	1.40	PDC	1.03
MicroPlate	*P. subcapitata*	1.03	PDC	0.90
MicroPlate	*P. subcapitata*	10.6	PDC	1.07
MicroPlate	*P. subcapitata*	2.46	PDC	0.53
MicroPlate	*P. subcapitata*	2.80	PDC	~0.6
MicroPlate	*P. subcapitata*	6.33	PDC	1.35
MicroPlate	*P. subcapitata*	3.81	PDC	1.32
		3.47	Mean	1.34
		2.99	SD	0.02
25 mL	*P. subcapitata*	1.49	PDC	0.98
25 mL	*P. subcapitata*	1.05	PDC	1.29
Toxkit	*P. subcapitata*	3.52	PDC	0.48
Toxkit	*P. subcapitata*	0.52	PDC	–
Toxkit	*P. subcapitata*	3.74	PDC	0.57
		2.06	Mean	0.53
		1.47	SD	0.06

Table 7.6 Accepted test results for *D. subspicatus* for the INC sample

			Reference test (see Sect. 7.2)	
Vessel - plate	Species	EC_{50} (% dilut.)	Substance	EC_{50} (mg/L)
100 mL	*D. subspicatus*	27.7	PDC	0.70
250 mL	*D. subspicatus*	12.6	PDC	0.77
250 mL	*D. subspicatus*	9.12	PDC	0.78
250 mL	*D. subspicatus*	3.12	PDC	1.02
250 mL	*D. subspicatus*	17.6	PDC	0.70
250 mL	*D. subspicatus*	30.6	PDC	0.84
250 mL	*D. subspicatus*	30.7	PDC	0.86
300 mL	*D. subspicatus*	17.7	PDC	0.84
		18.62	Mean	0.81
		10.3	SD	0.10
25 mL	*D. subspicatus*	2.92	PDC	0.77
25 mL	*D. subspicatus*	2.06	PDC	0.77
MicroPlate	*D. subspicatus*	9.02	PDC	0.77
MicroPlate	*D. subspicatus*	2.13	PDC	0.62
MicroPlate	*D. subspicatus*	5.67	PDC	0.62
		4.36	Mean	0.71
		2.99	SD	0.09

7.4.3 Different Species

A comparison of both species showed that *D. subspicatus* was slightly less sensitive to the waste samples INC and WOO, but slightly more sensitive to PDC than *P. subcapitata*. The range of EC_{50} values however overlapped and no statistically significant differences were seen between both species. The reproducibility and repeatability was higher for *D. subspicatus*. As results are comparable no preference in species can be detected.

7.4.4 Sensitivity in Comparison to Other Test Systems

The sensitivity of the Algae test is high and comparable to the *Daphnia* acute test results for these samples. Lapa et al. (2002b) and Ferrari & Férard (1999) performed ecotoxicity tests on waste material (leachates from bottom ashes and fly ashes) and reported that in these cases the Algae were more sensitive than Microtox or Daphnia. Deventer (2006) showed that Algae were often the most sensitive aquatic test organisms for different types of waste (eluate fractions).

Algae have been incorporated in previous projects on the ecotoxicity of waste (Gowlik and Moses 2006). Devillers et al. (2006) could not statistically verify the need for Algae within the test set, but other studies included algae as a separate trophic level (Donat 2006; Deventer 2006).

7.4.5 Recommendations

It is recommended that the Algae test should be part of the final set for waste testing (ISO 2005) as Algae represent a very important trophic level in the aquatic ecosystem. Algal tests have already been used for the ecotoxicity evaluation of chemicals (obligatory for classification and labelling of chemicals EC 67/548/EC and amendments), waste waters (review VMM 1996), sediment pore waters (Aminal 2000) and waste materials (Bernard et al. 1996; Kaneko 1996; Lapa et al. 2002a,b; Devillers et al. 2006; Donat 2006) for many years now and have proved to be sensitive to different types of pollution. In this ring test, again, they proved to be sensitive to the pollution present in waste eluate fractions. Microplates and other miniaturized systems are already in use in most labs and they seem to be equally or slightly more sensitive than the classic setup (see also chapter on Toxkits). Although differences may exist, no preference between different volumes could be specified. Both algal species can be used. There were some differences in sensitivity between both, but they were not consistent. No preference in species could therefore be detected.

Chapter 8
Daphnia Tests

P. Pandard

Abstract The proposed test strategy combined the direct and indirect approaches, integrating organisms representing the terrestrial and aquatic ecosystems. For the latter, several tests were selected; the *Daphnia magna* mobility test, the algal growth inhibition test and the inhibition of light emission of *Vibrio fischeri* were considered as a basic test set. From a methodological point of view, no significant problems were reported during the ring test. A total of 149 tests out of the 161 tests conducted were considered acceptable. The overall results showed that the *D. magna* mobility test was highly sensitive to incineration ash and contaminated wood eluates. On the other hand, high variability in the results was observed mainly for the contaminated wood eluate (INC, min-max: 10; WOO, min-max: 31). The CVs were 62 and 103% for INC and WOO respectively. In contrast to the terrestrial tests, the variability of the results shows not only the variability of the ecotoxicity test itself, but also integrates the variability of the overall procedure: sampling of the test portion, eluate preparation (leaching procedure, separation of the solid and liquid phases) and the ecotoxicity test. It may partly explain the high variability of the results of the waste tests compared with those observed for tests with the reference chemical.

Keywords *Daphnia magna*, Mobility, Eluate, Ring test, Waste

8.1 Introduction

Daphnids are freshwater organisms that have been used intensively throughout the last three decades for assessing the effects of chemicals in regulatory testing or for measuring the toxicity of water samples. These planktonic crustaceans are widely distributed in ponds and lakes and are representative of many ecologically filter feeding zooplankton (Liber & Solomon 1994). Both scientific and practical reasons led to the use of daphnids in aquatic toxicity testing. These organisms are (a) an important

P. Pandard (✉)
INERIS, Parc Technologique ALATA, BP 2, 60550, Verneuil en Halatte, France

H. Moser and J. Römbke (eds.), *Ecotoxicological Characterization of Waste*,
DOI: 10.1007/978-0-387-88959-7_8, © Springer Science+Business Media, LLC 2009

link in many aquatic food chains, (b) daphnids have a relatively short life cycle and are relatively easy to culture in the laboratory, (c) they are sensitive to a broad range of aquatic contaminants, and (d) their small size requires only small volumes of test water, limiting the sampling and transport of water samples (EC 1990).

Several organisations such as the ISO, OECD, EPA, and ASTM have standardized acute and chronic test methods such as *Daphnia* sp. acute immobilisation test (ISO 1996; OECD 2004a) and *Daphnia magna* reproduction test (OECD 1998). More recently, *D. magna* has been identified as a possible candidate for assessing endocrine disruption effects (Olmstead & LeBlanc 2000; Oda et al. 2005). Therefore, Japan proposed the enhancement of OECD Test Guideline NO. 211: *D. magna* Reproduction Test (OECD 1998) as a relevant screening and risk assessment method for endocrine disrupting chemicals in crustaceans (NIES 2005).

D. magna immobilisation tests have been widely used for measuring the acute toxicity of waste water (Costan et al. 1993; Clément et al. 1997; Rojickova-Padrtova et al. 1998; Vindimian et al. 1999; Manusadzianas et al. 2003), in addition to classification, labelling and risk assessment of chemicals. The applicability of such a bioassay for testing waste eluates has also recently been demonstrated (LFU 2004; Pandard et al. 2006).

8.2 Method

The *D. magna* immobilisation test was performed according to the ISO standard 6341 (ISO 1996) and was described in detail in a SOP. In brief, young daphnids, aged less than 24 hours at the start of the test, were exposed to dilutions of the waste eluates obtained according to EN 14735. *D. magna* immobilisation was recorded after 24 h and compared with control values. The results were expressed as EC_{50} (% V/V), determined by the probit analysis method, using the statistical program TOXRAT® (2006).

The ISO 6341 synthetic water was used as control and dilution water. The tests were performed under controlled conditions: temperature 20 ± 2°C, complete darkness or diffuse light, and no pH adjustment of the test solutions.

The health of the culture of *D. magna* and the sensitivity of the test organisms were checked using potassium dichromate by all the laboratories except one, which used zinc sulphate (External: Chap. 5). Test results were considered to be valid when (a) the percentage of mobility in control vessels was less than 10% at the end of the test; (b) the dissolved oxygen concentration at the end of the test was ≤2 mg/L in control and test vessels; and (c) the EC_{50} for the potassium dichromate was within the range of 0.6–2.1 mg/L (ISO 1998).

The overall assessment of the test results from the participants was performed in several steps (for details see External: Chap. 4). First, it was determined how many of the total number of tests fulfilled the acceptance criteria. Secondly, all data sets in which no EC_{50} could be determined (i.e. where they had to be given as < or > a tested dilution) were excluded from the statistical evaluation. The remaining data sets were checked for passing the outlier tests (i.e. following the ISO approach) and being outside the warning limits (i.e. following the approach presented by Environment Canada). The 'final mean (geometric) EC_{50} values' for each substrate were calculated excluding all outliers, irrespective of the evaluation method used.

8.3 Results

A total of 46 participating laboratories provided 161 data sets for the three eluates (Table 8.1). Only 12 tests (7.5%) were not acceptable. One test was not valid as the SOP was not fulfilled. The EC_{50} for the reference chemical was outside the recommended range (0.6–2.1 mg/L; ISO 1998) for two laboratories, excluding two tests for each waste. Five results were considered biological outliers. In six tests with INC, 49 tests with SOI and in one test with WOO, no EC_{50} could be determined. Therefore, the final statistical assessment was based on 93 tests.

The results of the *Daphnia* mobility tests are presented following the same scheme as in the other chapters of Part III, according to ISO (1994) and Environment Canada approach (EC 2005) (See Fs. 8.1–8.5; Table 8.2).

Among the 47 test results obtained for the eluate of municipal waste incineration ash, six did not allow the calculation of EC_{50} values. So the final statistical evaluation was performed on 41 test results. The maximum and minimum EC_{50} values differed by a factor of 10 (min: 0.9%; max: 9.0%). 80% of the values were in the range EC_{50} ± 1SD; only 7% (= three tests) of the EC_{50} were outside the warning limits (EC_{50} ± 2SD).

For the soil eluate, almost no inhibitory effects were recorded on the mobility of *D. magna*. In just two tests an EC_{50} could be determined: 86.5 and 95.5%. These results were consistent with those obtained with the other tests of the basic set.

Regarding the contaminated wood eluate, one result did not allow the calculation of EC_{50} values. So the final statistical evaluation was performed on 50 test results. For this eluate, a high variability of the results was observed, leading to a CV higher than 100% (Table 8.3). The maximum and minimum EC_{50} values differed by a factor of 30.7 (min: 0.06%; max: 1.84%). 66% of the values were in the range of EC_{50} ± 1SD (0.16–0.79%). 4% (= two tests) EC_{50} values were outside the warning limits (EC_{50} ± 2SD).

Table 8.1 Summary of the acceptance of *Daphnia* mobility tests

Substrate	No. Labs	No. Tests	Invalid	No Reference	Accepted
INC	46	54	5	2	47
SOI	45	54	1	2	51
WOO	45	53	0	2	51
Sum	n.a.	161	6	6	149

n.a. Not applicable

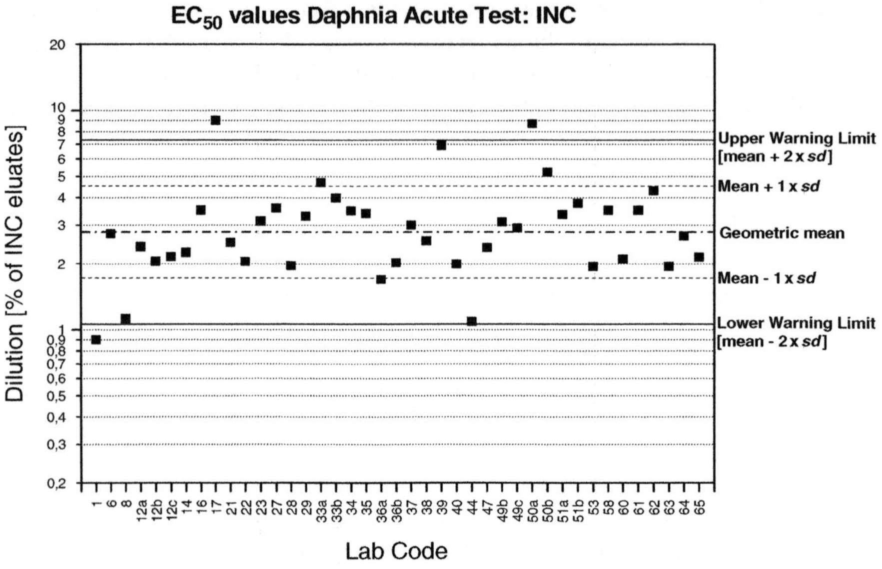

Fig. 8.1 Results (EC_{50} as % waste eluate) of the *Daphnia* mobility tests with the waste substrate INC (presentation according to Environment Canada 2005)

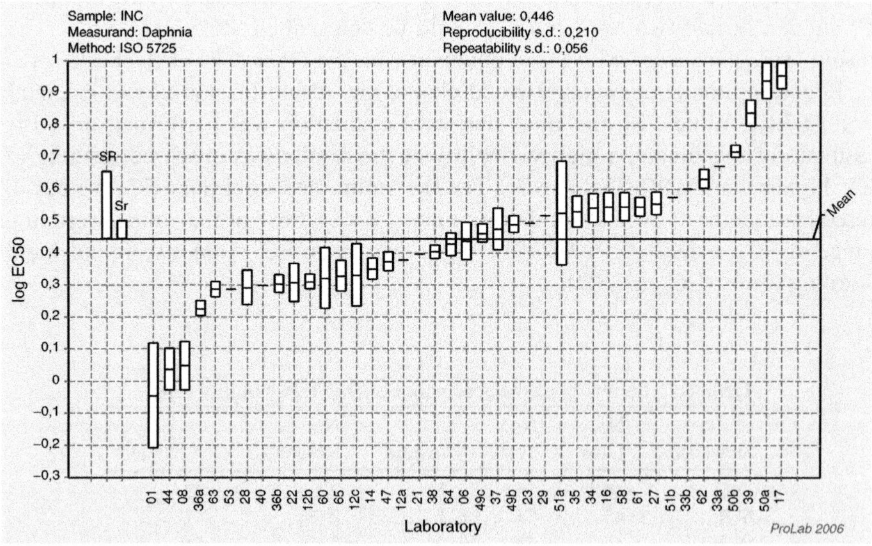

Fig. 8.2 Results (log EC_{50} as % waste eluate) of the *Daphnia* mobility tests with the waste substrate INC (presentation according to ISO 1994)

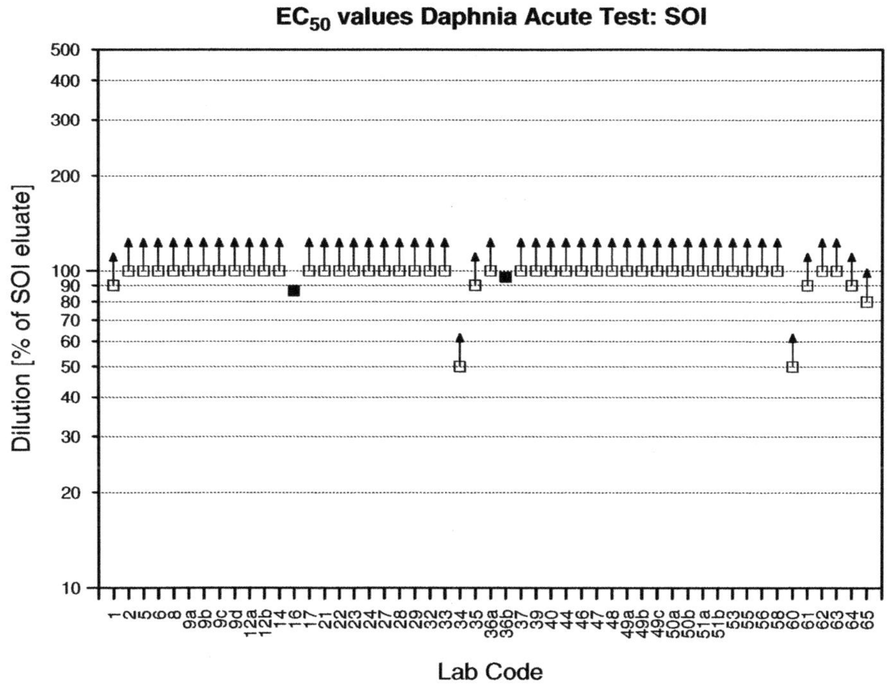

Fig. 8.3 Results (EC_{50} as % waste eluate) of the *Daphnia* mobility tests with the waste substrate SOI (presentation according to Environment Canada 2005)

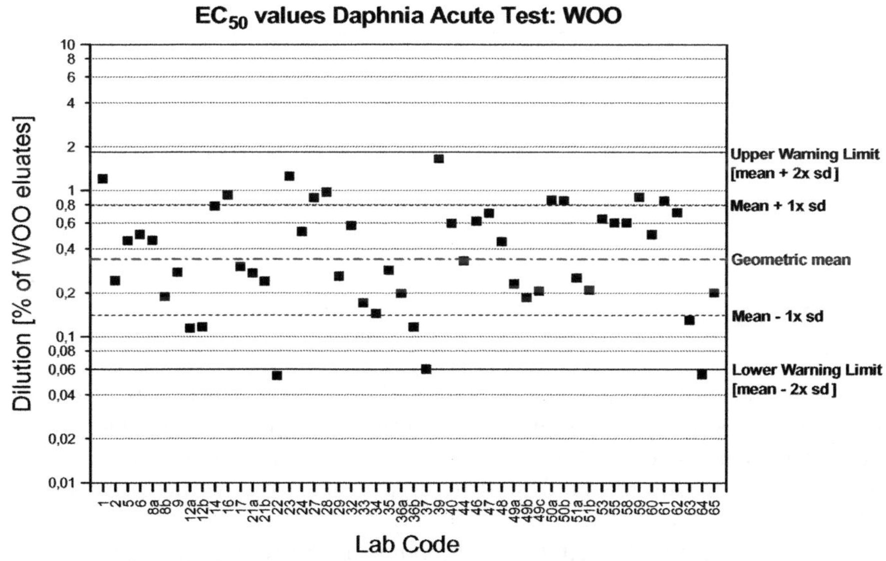

Fig. 8.4 Results (EC_{50} as % waste eluate) of the *Daphnia* mobility tests with the waste substrate WOO

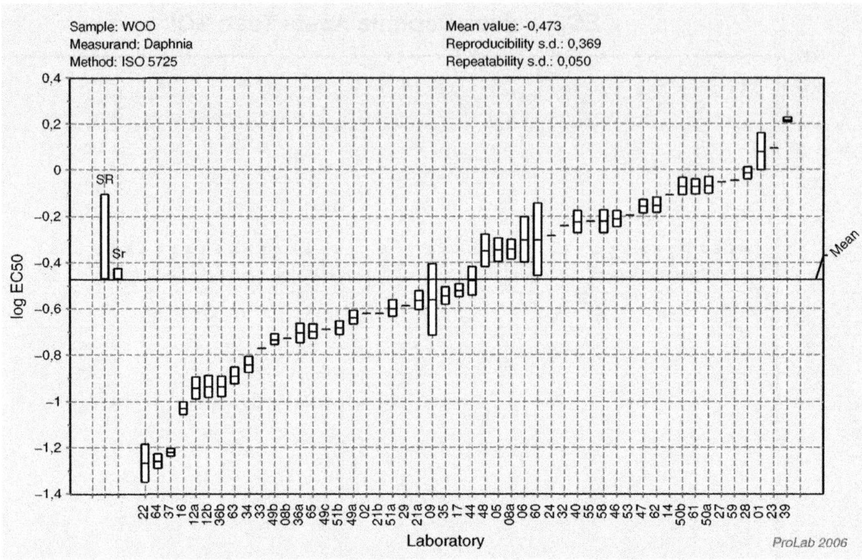

Fig. 8.5 Results (log EC_{50} as % waste eluate) of the *Daphnia* mobility test with the waste substrate WOO

Table 8.2 Summary of the results of the *Daphnia* mobility tests and the three waste types (data expressed as % waste eluate)

Substrate	EC_{50}	Ratio s_R/s_r	Lower WL	Upper WL	Factor Min-Max
INC	2.71	3.8	1.1	7.3	10.0
SOI	n.a.	n.a.	n.a.	n.a.	n.a.
WOO	0.38	7.4	0.06	1.84	30.7

Final (geometrical) mean EC_{50} values, s_R/s_r ratio, lower and upper 95% confidence limits and the factor between minimum and maximum EC_{50} values

8.4 Discussion

From a methodological point of view, no significant problems were reported during the ring test. Despite extreme pH values for the incineration ash and the contaminated wood eluates (INC: 10–11; WOO: ≈5), the pH values of the eluate dilutions were compatible with the survival of the test organisms in the range where toxic effects were observed.

The overall results showed that the *D. magna* mobility test was highly sensitive to the incineration ash and contaminated wood eluates. For INC, the *Daphnia* acute immobilisation test was the most sensitive compared to chronic toxicity tests such as algal growth inhibition test, *Ceriodaphnia dubia* reproduction test and *Brachionus calyciflorus* reproduction test. Regarding WOO, *D. magna* was more sensitive than Algae but slightly less than the two other organisms. Such results are relatively uncommon compared to those usually observed for chemicals, waste water (Costan et al. 1993; Vindimian et al. 1999) or even for waste eluates (40 wastes belonging to 27 waste types, Pandard et al. 2006). It can probably be linked to the specificity of the contaminants, mainly heavy metals, extracted from the waste by leaching.

The ratio of 92.6% of data acceptance is considered satisfactory. Nevertheless, a high variability of the results was observed, mainly for the contaminated wood eluate (Tables 8.2 and 8.3). However, referring to the s_R/s_r ratio, the variability of the tests with INC can be considered as acceptable as the ratio of 3.8 is lower than four. According to the results of the ring tests with chemical and physico-chemical methods, s_R/s_r ratios lower than four indicate a robust method (see External: Chap. 4 for details). However, a s_R/s_r ratio of 7.4 and a min-max factor of about 30 as determined for WOO clearly indicates a very high variability of the test results.

In contrast to the terrestrial tests, the variability of the results presented in this chapter integrates not only the variability of the ecotoxicity test itself, but also the variability of the entire procedure including sampling of the test portion, eluate preparation (leaching procedure, separation of the solid and liquid phases) and ecotoxicity test. A preliminary analysis of the raw data provided by the participating laboratories clearly showed that some of them did not strictly follow EN 14735 for eluate preparation (e.g. amount of waste for the leaching test, volume of eluate collected) and storage time of eluates before performing aquatic tests. It may explain partly the high variability of the results (Tables 8.2 and 8.3), e.g. when compared with the one observed in tests with the reference chemical potassium dichromate (min-max ratio: 2.6; see Chap. 5).

Some laboratories extended the duration of the *D. magna* mobility test to 48 h. These results showed that the toxicity of INC and WOO eluates increased significantly with increasing exposure time: from 45 to 91% according to the type of waste or the laboratory (Table 8.3). Therefore, it is recommended to extend the exposure time to 48 h, as for chemical testing.

Table 8.3 *Daphnia* mobility test-comparison of the EC_{50} 24 h and 48 h

Substrate		EC_{50} 24 h	EC_{50} 48 h	Ratio EC_{50} 24 h/EC_{50} 48 h
INC	Lab 61[a]	4.29	2.37	1.81
	Lab 63[b]	2.41	1.62	1.49
	Lab 64[a]	3.47	2.32	1.50
WOO	Lab 61[a]	0.42	0.29	1.45
	Lab 63[b]	0.21	0.11	1.91
	Lab 64[a]	0.075	0.05	1.50

[a]Mean value of five tests
[b]Mean value of ten tests

8.5 Recommendations

The *D. magna* mobility test was highly sensitive to incineration ash and contaminated wood eluates. On the basis of these results, this test could be part of the final test set. Nevertheless, the number of wastes selected in the ring test is low and cannot be considered representative of the different waste types of the European Waste List. Broader studies (LFU, 2004: 24 wastes; Pandard et al. 2006: 40 wastes) that included *D. magna* in their test strategy did not retain this acute test in their minimum set. Furthermore, Rojickova-Padrtova et al. (1998), did not select the *D. magna* mobility test from the toxicity results of 50 environmental samples for their optimal test set.

Chapter 9
Luminescent Bacteria Test

M. Pattard and H. Moser

Abstract The luminescent bacteria test has become a basic test for ecotoxicological testing of chemicals, waste water and eluates from soil and sediment. It has been selected for the basic test set as a representative method to assess the ecotoxicological hazard potential of waste eluates to terrestrial invertebrates. The luminescent bacteria test can be carried out by using freshly prepared bacteria, as well as liquid-dried or freeze-dried bacterial preparations. The methodology of the luminescent bacteria test did not cause any problems during the ring test and the total ratio of acceptance (90.9%) is considered to be good. The freeze-dried bacteria proved to be more sensitive towards chemical stress, especially heavy metals, than freshly-prepared or liquid-dried bacteria.

The sensitivity of the bacteria test was moderate compared to the other standard aquatic tests. WOO was by far the most toxic waste, while INC and in particular SOI were clearly less toxic, whereas the Min-max-Factor was relatively high in the WOO tests and reasonable for the other two wastes, INC and SOI. The reproducibility and robustness of the test results were high: out of 140 tests that were methodologically acceptable only three tests were identified as statistical outliers. For the characterization of waste it is recommended to keep the luminescent bacteria test as part of the basic test set, but it is necessary to give clear definitions for the preparation of luminescent bacteria.

Keywords Luminescent bacteria, *Vibrio fischeri*, Ring test, Waste eluate

9.1 Introduction

Bacteria are important members of the aquatic ecosystem. They fulfil ecosystem functions such as degradation of the organic matter. Bacteria can be easily cultured and bred in the laboratory. For these reasons, they are perfectly applicable as test organisms for the assessment of the toxicity of chemicals and other samples in aqueous

M. Pattard (✉)
Federal Environment Agency, Schichauweg 58, 12307, Berlin, Germany

H. Moser and J. Römbke (eds.), *Ecotoxicological Characterization of Waste*,
DOI: 10.1007/978-0-387-88959-7_9, © Springer Science+Business Media, LLC 2009

solution. The number and characteristics of ecotoxicological test methods with bacteria are large. For nearly 25 years the luminescent bacteria test with the marine bacterium *Vibrio fischeri* (former *Photobacterium phosphoreum*) has become a basic test for ecotoxicological testing of chemicals, waste water and eluates from soil and sediment. The test system gives a rapid method for determining the toxicity of an aqueous solution by determining the inhibition of the luminescence emitted by *V. fischeri* after a given exposure time, usually 30 min. For these reasons the luminescent bacteria test has been selected as a representative method to assess the ecotoxicological hazard potential of waste eluates to terrestrial invertebrates (CEN 2005).

9.2 Method

The test measuring the inhibition of the light emission of *V. fischeri* according to the ISO-Guideline No. 11348-Part 1–3 (1998) as part of the basic test battery was described in a Standard Operation Procedure (SOP). Measurements can be carried out using freshly prepared bacteria (Part 1), as well as liquid-dried (Part 2) or freeze-dried (Part 3) bacterial preparations. When presenting the test results, three preparations are differentiated: tests with freeze-dried bacteria vs. those in which freshly prepared or liquid bacteria were used (partly, the latter two groups are combined in the graphs). In special cases these different techniques may deliver different results for the sensitivity of the test organisms, especially in their reaction regarding heavy metals. In short, this acute test was performed with waste eluates in various dilution steps. Specified volumes of the diluted test sample (waste eluate) are combined with the luminescent bacteria suspension in a test tube. The test criterion is the luminescence, measured after a contact time of 30 min taking into account a correction factor (fkt), which is derived from the intensity changes of control samples during the exposure time.

The inhibitory effect of the waste eluates on the light emission of *V. fischeri* can be determined as an EC_{50} value. These values were determined by the probit analysis method, using the statistical program ToxRat® (2006). Test results were considered valid when the mean correction factor in the control was in the range of 0.6–1.8 and the parallel determination deviates not more than 3%. The ISO guideline 11348 (1998) recommends testing at least one of the three reference substances (3,5-dichlorophenol, potassium dichromate and zinc sulfate heptahydrate) parallel to each testing. For each reference substance and the different bacteria, a defined concentration is given, which has to cause 20–80% inhibition after 30 min contact time. The laboratory has to use all three reference substances to check the bacteria batch before testing.

The overall assessment of the test results from the different participants was performed in several steps (for details see Chap. 4).First, the total number of tests that fulfilled the acceptance criteria was determined. Secondly, all data sets in which no EC_{50} could be determined (i.e. where they had to be given as < or > a tested dilution) were excluded from the statistical evaluation. The remaining data sets were checked for passing the outlier tests (i.e. following the ISO approach) and being outside the warning limits (Environment Canada 2005). The 'final mean (geometric) EC_{50} values' for each substrate were calculated excluding all outliers, irrespective of the evaluation method used. Graphs were prepared for all the results together as well as separately for the two or three groups of preparations.

9.3 Results

A total of 41 participating laboratories provided 154 test data sets (Table 9.1). The results of fourteen tests were invalid (9% of the data sets), as either the validity criteria were not met (e.g. the mean correction factor was outside of the range (0.6–1.8), the parallel determination deviated more than 3% or the SOP was not followed (0.6% of the participants). One statistical outlier was found, but no biological outliers were identified, meaning that the statistical assessment was based on 140 tests, and the total range of acceptance was 90.9%. All laboratories presented results from the limit test with one reference substance. No limit test was outside the range of 20–80% effect.

In the following sections, the results of the bacteria tests with INC, SOI and WOO are presented, following the same scheme as in the other chapters of Part II.

The number of acceptable tests performed with INC was 45 but an EC_{50} value could be determined only for 22 tests (Figs. 9.1 and 9.2). In the tests with INC, only one EC_{50} value for *Vibrio fischeri* was statistically identified as an outlier (Fig. 9.1). Accordingly, the factor between reproducibility and repeatability was small (4.7). All EC_{50} values were within the warning limits (Fig. 9.2). Six test results were outside the range of the mean ±1SD. No reason can be given for the one outlier.

In SOI, the effects on the bacteria were clearly lower. The factor between reproducibility and repeatability was 3.2 (Fig. 9.3). The number of acceptable tests performed with SOI was 37 (Fig. 9.4) but the EC_{50} could be determined only for 14 tests. With one exception, all results were within a range of the mean ±2SD. There was no statistical outlier determined (Fig. 9.5).

The high content of water soluble heavy metals, in particular copper, resulted in the transfer of these contaminants into the water fraction during the leaching process of the waste material WOO. The waste eluate caused clear effects in the aquatic test systems. The luminescent bacteria test showed different results for the different

Table 9.1 Summary of the acceptance of bacteria tests with Vibrio fischeri

Substrate	No. Labs	No. tests	Invalid tests	Accepted tests
INC	41	49	4	45
SOI	40	48	6	42
WOO	40	57	4	53
Sum	n.a.	154	14	140

n.a. Not applicable

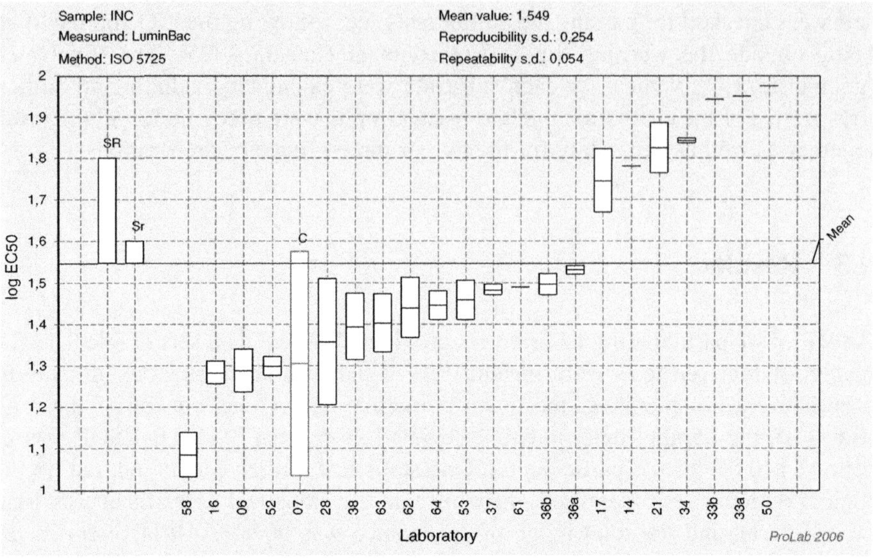

Fig. 9.1 Log EC_{50} (% waste eluate) for the waste substrate INC with *Vibrio fischeri* (presentation according to ISO 5725–1 (1994)). C: statistical outlier

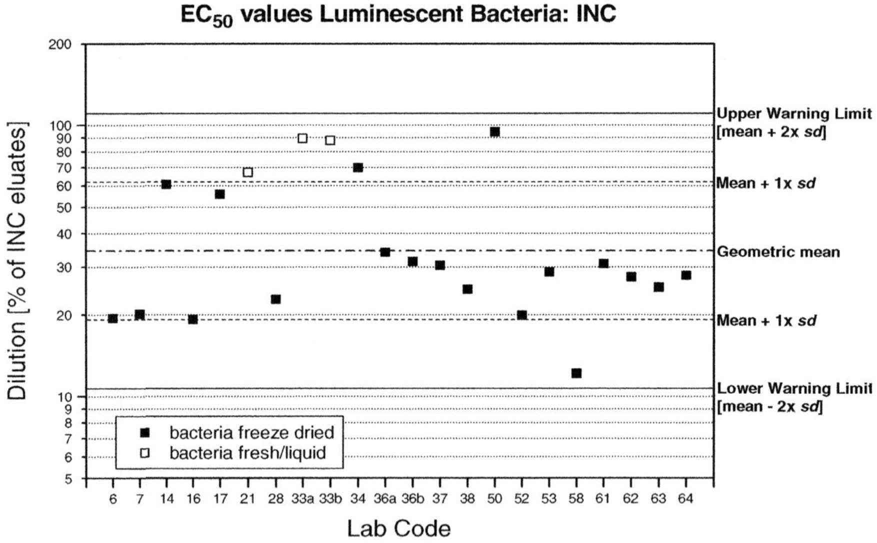

Fig. 9.2 EC_{50} values (% waste eluate) for the waste substrate INC with *Vibrio fischeri*

preparations of the bacteria. The factor between reproducibility and repeatability was 11.5 for all preparations of bacteria. According to Fig. 9.6 the test results can easily be separated into two groups, based on the preparation of bacteria. Only one

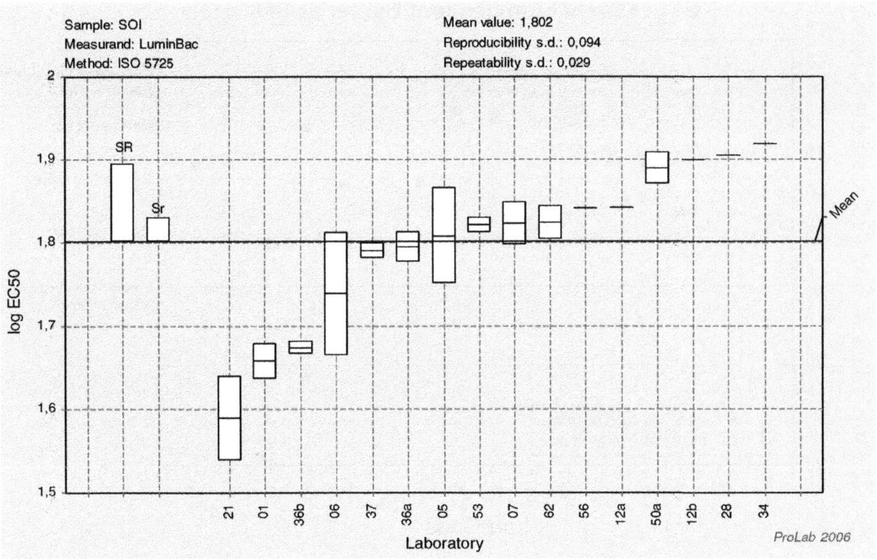

Fig. 9.3 Results (log EC_{50} as % waste eluate) for the waste substrate SOI with *V. fischeri* (presentation according to ISO 5725–1(1994))

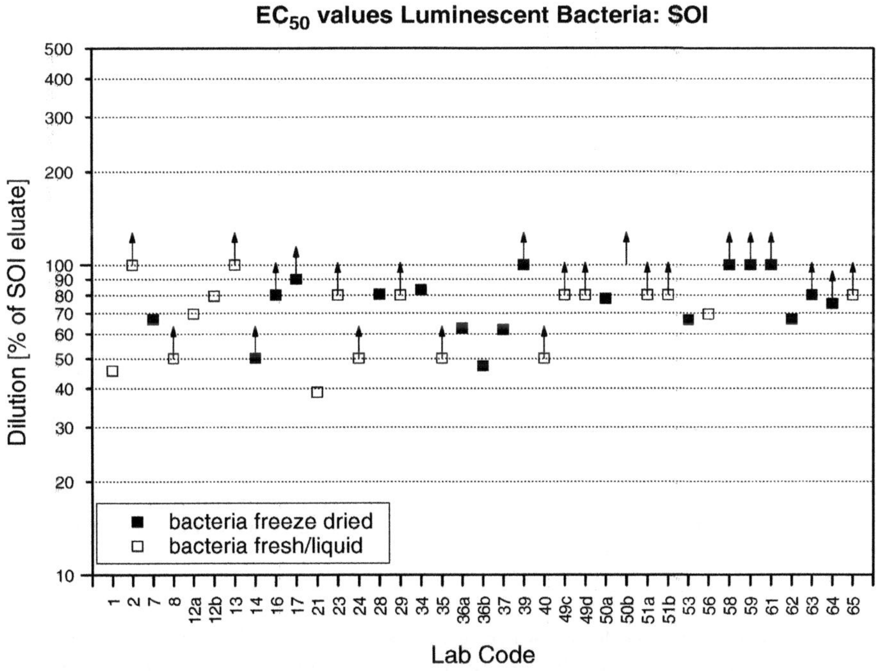

Fig. 9.4 EC_{50} values (% waste eluate) for the waste substrate SOI with *Vibrio fischeri*

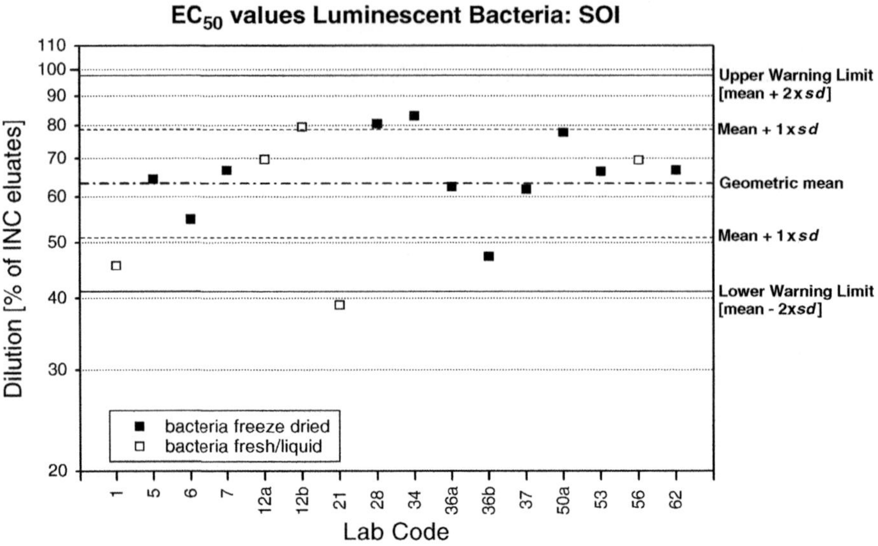

Fig. 9.5 EC_{50} values (% waste eluate) with warning limits for the waste substrate SOI with *Vibrio Fischeri*

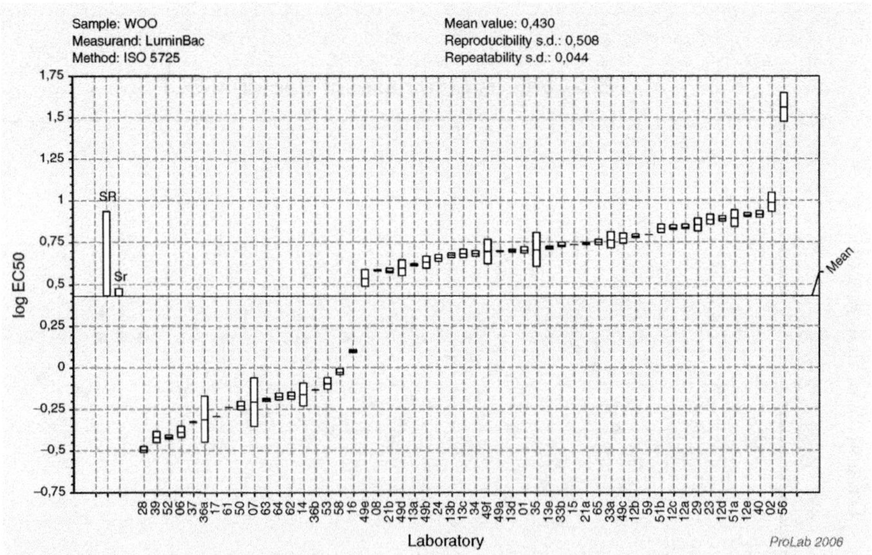

Fig. 9.6 Results (log EC_{50} as % waste eluate) for the waste substrate WOO with *V. fischeri* (presentation according to ISO 5725–1(1994)) - all sources

out of 53 tests was outside the warning limits (Fig. 9.7). The freeze dried bacteria were significantly more sensitive to heavy metals than the other two preparations. For the freeze-dried bacteria the factor between reproducibility and repeatability was 6.5 (Figs. 9.8 and 9.9), for freshly prepared bacteria 2.6 (Fig. 9.9) and for liquid dried bacteria 5.3 (Fig. 9.10). The results of the two latter two groups are shown in combination in Fig 9.12.

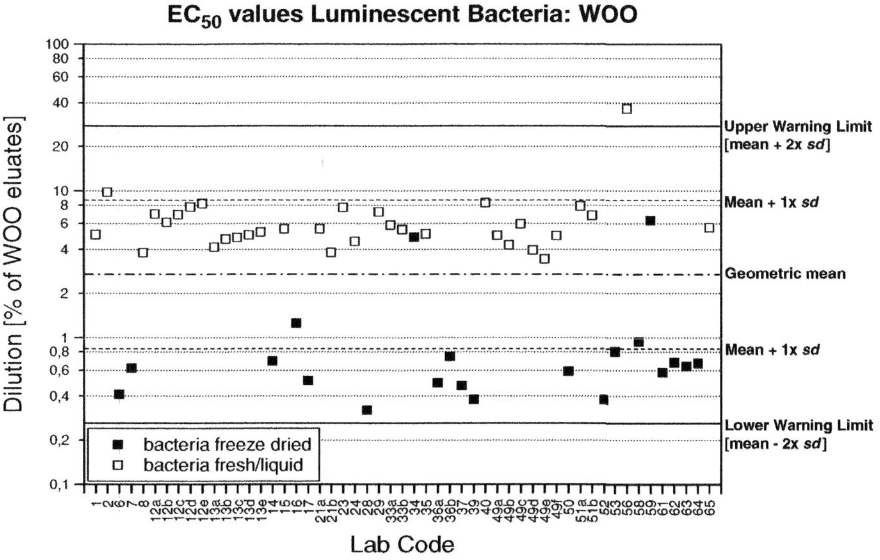

Fig. 9.7 EC_{50} values (% waste eluate) with warning limits for the waste substrate WOO with *Vibrio fischeri* - all sources

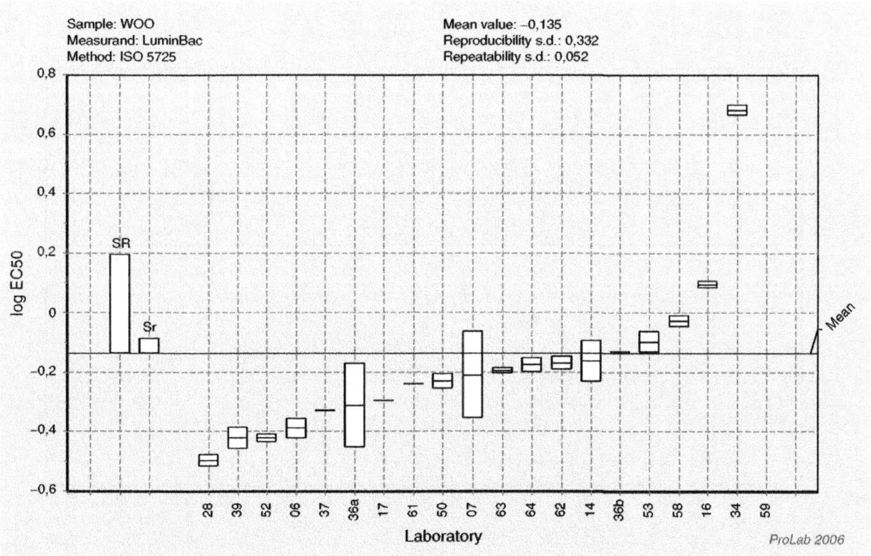

Fig. 9.8 Results (log EC_{50} as % waste eluate) for the waste substrate WOO with *Vibrio fischeri* (presentation according to ISO 5725-1(1994)) - only for freezed dried bacteria

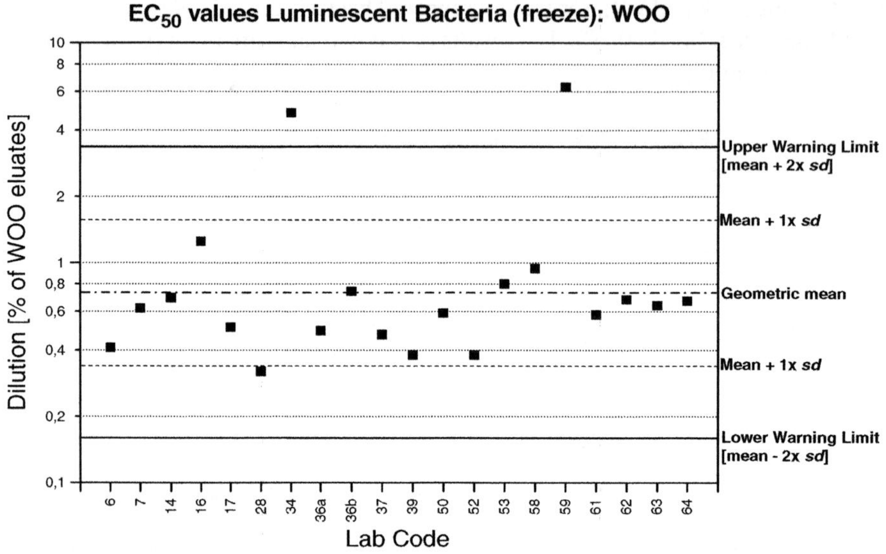

Fig. 9.9 EC_{50} values (% waste eluate) with warning limits for the waste substrate WOO with *V. fischeri* - only freezed dried bacteria

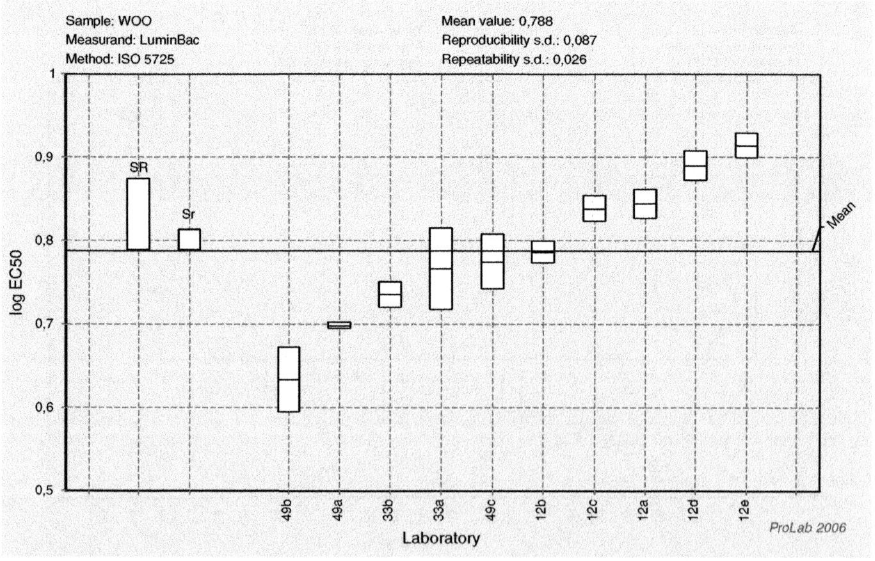

Fig. 9.10 Results (log EC_{50} as % waste eluate) for the waste substrate WOO with *V. fischeri* (presentation according to ISO 5725-1(1994)) - only for freshly prepared bacteria

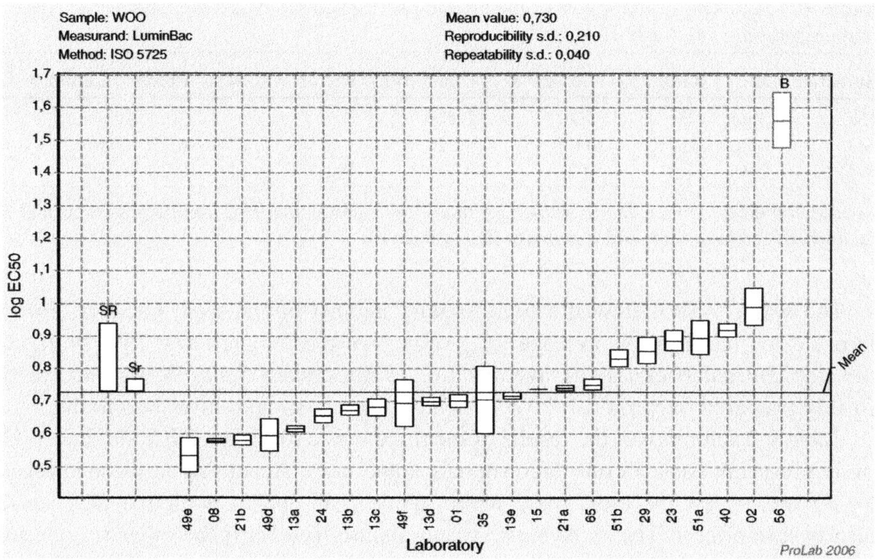

Fig. 9.11 Results (log EC_{50} as % waste eluate) for the waste substrate WOO with *V. fischeri* (presentation according to ISO 5725-1(1994)) - only for liquid dried bacteria. C: statistical outlier

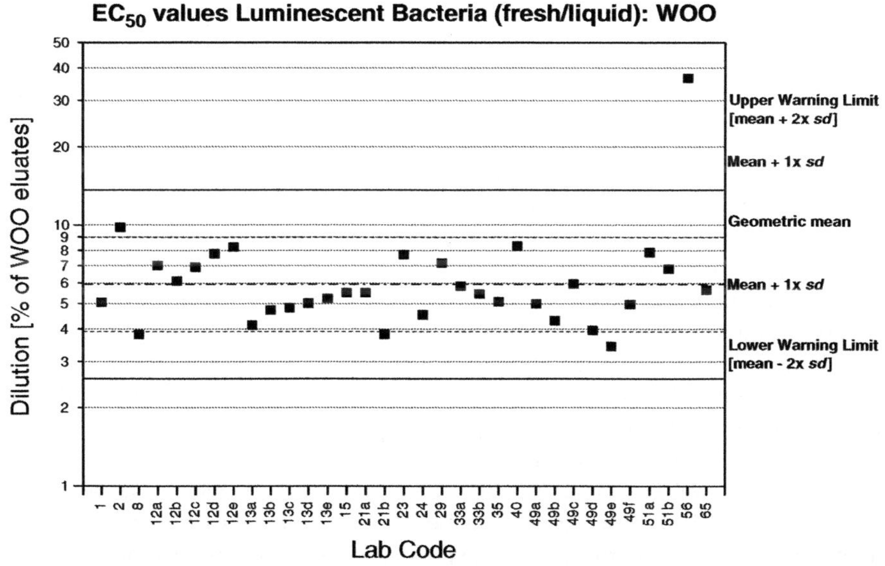

Fig. 9.12 EC_{50} values (% waste eluate) with warning limits for the waste substrate WOO with *Vibrio fischeri* - freshly prepared/liquid dried bacteria

Table 9.2 Summary of the results of the luminescent bacteria tests and the three waste types (data expressed as % waste eluate)

Substrate	EC_{50}	Ratio s_R/s_r	Lower WL	Upper WL	Factor Min-Max	Freeze	Liquid/fresh
INC	35.4	4.7	10.7	110.9	7.8	30.8	n.a.
SOI	65.5.	3.2	41.1	97.7	2.1	65.8	n.a.
WOO	2.56	11.5	0.28	27.9	121.7	0.73	5.6

Final (geometrical) mean EC_{50} values, s_R/s_r ratio, lower and upper 95% confidence limits and the factor between minimum and maximum EC_{50} values

The waste WOO showed a high toxicity against luminescent bacteria. For the freeze dried bacteria, the average EC_{50} values were calculated as 0.73% WOO, for the liquid dried bacteria the average EC_{50} values were 5.4% WOO and for the freshly prepared bacteria 6.1% WOO. Only one statistical outlier was found.

Table 9.2 summarises the results of the three waste materials INC, SOI and WOO on luminescent bacteria (all three sources combined). According to these numbers, WOO was by far the most toxic waste, while INC and in particular SOI caused clearly less effects. The three waste materials are listed just the other way around when looking at the Min-max-Factor, which was extremely high in the WOO tests (121.7) and far lower for the other two wastes INC (7.8) and SOI (2.1).

9.4 Discussion

The methodology of the luminescent bacteria test did not cause any problems during the ring test. Only a few laboratories reported difficulties in keeping the validity criteria. The total ratio of acceptance (90.9%) is considered good. As regularly reported in the literature, freeze-dried bacteria are more sensitive towards chemical stress, especially against heavy metals (Ibanez et al. 2001; Farré et al. 2004; LANUV-NRW 2006). In the waste tests this correlation could be confirmed only for WOO, where freeze-dried bacteria were eight times more sensitive than freshly prepared or liquid dried bacteria (the latter two sources differed only slightly in their EC_{50} values). In contrast to the tests with WOO, no difference in sensitivity was found between the two preparations in the tests with INC and SOI. For the characterization of waste material, it will be necessary to define the kind of bacteria (freshly prepared/liquid dried or freeze-dried) as it is required, for example, in the German waste water regulation: it insists on tests with freshly prepared bacteria/liquid dried bacteria.

The reproducibility and robustness of the test results were high: out of the 140 tests that were methodologically acceptable, only two tests were identified as statistical outliers. However, while in INC and SOI the factor between minimum and maximum EC_{50} values was considered reasonable (2–7), the factor in WOO was relatively high (>100). Considering that the minimum-maximum factor for tests with a reference substance, performed in one laboratory, is at least two and between laboratories, at least four (Environment Canada 1999), the results obtained were in an acceptable range, considering the sometimes challenging substrates such as wastes.

The sensitivity of the bacteria test is moderate compared to the other standard aquatic tests (Chap. 1) and follows the expected order (i.e. highest toxicity in WOO, lowest in SOI and intermediate in INC). While in SOI and WOO the sensitivity between luminescent bacteria and Algae did not differ considerably, the bacteria clearly reacted less to the INC eluates.

When comparing the results from the literature with the results from the ring test, it appears that the order of toxicity (and in the case of INC also the EC_{50} values) are in an acceptable range (Wundram et al. 1996; Farré et al. 2004). The results of the bacteria tests, as well as from the other test systems, clearly show that the detection of hazardous waste needs to be based on biological testing and chemical analysis. From the scientific point of view, more experience regarding the reaction of the single test species against different waste types is needed to get a better understanding and, in the long run, to predict the ecotoxicity, of waste. In this context it is also necessary to study the influence of the waste properties themselves (i.e. not the contaminants) on the test organisms.

9.5 Recommendations

It is recommended that a bacteria test should be part of the final test set (CEN 2005). The test gives rapid results, but it is necessary to give clear definitions for the luminescent bacteria sources to be used because of the different reaction, especially with heavy metals.

Chapter 10
Plant Tests

B. Foerster, C. Firla, and T. Junker

Abstract Plants are the most important terrestrial primary producers, forming the basis of all ecosystems. Therefore, they are an indispensable part of any test set for the ecotoxicological characterization of wastes. In the ring test, following ISO standard guideline 11269-2, the plant growth test was performed with three waste types INC, SOI, and WOO. A total of 107 tests (55 with *Avena sativa* (oat) and 52 with *Brassica rapa* (turnip)) were classified as acceptable out of 124 tests conducted. Methodologically, almost no problems occurred but further guidance on adjusting the moisture of the test substrate mixtures is needed. After statistical evaluation, only one further test result per species was rejected. The min-max factors (2–11) between individual EC_{50} values and the CVs (22–68%) varied widely, showing that the plant growth test has medium robustness. The toxicity differed considerably among the three waste types but relatively little between the two plant species, which is exemplified by the final EC_{50} values for *A. sativa* (INC = 29.4%, SOI = 57.8% (based on just six values), WOO = 10.0%) and *B. rapa* (INC = 23.9%, SOI = 63.0%, WOO = 2.64%). The plant growth test is more sensitive than the other basic terrestrial test (earthworm acute). Therefore, it is recommended to retain the plant growth test with both species in the final test set. In addition, the sensitivity of the test system should be proved by conducting a reference test regularly (e.g. with boric acid).

Keywords *Avena sativa, Brassica rapa*, Growth, Ring test, Seedling emergence, Waste

10.1 Introduction

Plants are the most important terrestrial primary producers, forming the basis of all ecosystems. By means of photosynthesis they transform bioavailable energy in biomass. In addition, they fulfill ecosystem functions such as providing food and

B. Foerster (✉)
ECT Oekotoxikologie GmbH, Boettgerstr. 2-14, 65439, Floersheim, Germany

H. Moser and J. Römbke (eds.), *Ecotoxicological Characterization of Waste*,
DOI: 10.1007/978-0-387-88959-7_10,

habitat for many animal species and protecting the soil against erosion by covering the soil (Gurevitch et al. 2002). Besides nutrients, plants also mobilize and accumulate pollutants, making them available for terrestrial food webs. As plants are (usually) immobile and in close contact with the soil and soil pore water, they can be easily used as indicators of the respective soil and site properties. Finally, many higher plants (especially crop species) can easily be cultured and bred in the laboratory. For these reasons, they are perfectly eligible test organisms for the assessment of the toxicity of chemicals in soil (Keddy et al. 1994).

There are several characteristic ecotoxicological test methods with higher plants (Kalsch et al. 2006a). In standardized tests, crop species are almost exclusively used, often with at least one monocotyledonous and one dicotyledonous species (OECD 2006a,b). Normally, the duration of acute tests is between a few days and three weeks, depending on the respective species (Heiden et al. 2000). The most important measurement endpoints are biomass and shoot length, as well as semiquantitative parameters such as the occurrence of chlorosis or necrosis, while the emergence rate is less suitable due to its low sensitivity. Life-cycle parameters such as reproduction are rarely used because of the greater effort needed to measure them (ISO 2004). In general, testing the growth of two species of higher plants is considered to be sufficient for the protection of these organisms in the terrestrial environment. However, while the testing of individual chemicals or contaminated soils with plants is quite common (e.g. Hund-Rinke et al. 2002), solid waste materials have rarely been used in plant tests so far (Kostka-Rick 2004).

10.2 Method

The test measuring the seedling emergence and growth of terrestrial plants, according to the ISO-Guideline No. 11269-2 (1995) as part of the basic test set, was described in a Standard Operation Procedure (SOP). The test was performed with two substrates: a control soil of known good properties for plant growth as a reference and the test substrate (a mixture of control soil and waste material in different ratios; for details see Chapter 3). Preferably, a natural standard soil (e.g. LUFA Standard Soil 2.3) was used as control, but any soil having the following characteristics may be used: organic carbon content $\leq 1.5\%$, pH between 5.0 and 7.5, and fine fraction (<0.02 mm) comprising less than 20% of the soil dry weight. The emergence and growth of seedlings of one monocotyledonous (*Avena sativa* (oat)) and one dicotyledonous (*Brassica rapa* (turnip)) plant species were observed over a time period of 14–21 days following the emergence of at least 50% of the seedlings in the control. On Day 14 (earliest) or on Day 21 (latest) the shoot length and shoot biomass (dry or wet weight) of each pot were assessed. The main assessment endpoint was the shoot biomass of the plants, evaluated as EC_{50} value (median effect concentration). These values were determined by the probit analysis method, using the statistical program TOXRAT® (2006). Test results were considered valid when the emergence rate in the control vessels was at least 70% (at least seven seedlings should have emerged per control pot). While in the newest version of the ISO guideline 11269-2 (2004) boric acid is recommended as a reference substance, no details concerning the expected effect range for the two species used in the ring test are provided. According to literature (Stephenson et al. 1997), its EC_{50} value (shoot weight) should be in the range of 100 to 400 mg/kg (dry weight).

The overall assessment of the test results from the various participants was performed in several steps (for details see Chap. 4). First, the total number of tests that fulfilled the acceptance criteria was determined. Secondly, all data sets in which no EC_{50} could be determined (i.e. where they had to be given as < or > a tested dilution) were excluded from the statistical evaluation. The remaining data sets were checked for passing the outlier tests (i.e. following the ISO approach) and being outside the warning limits (Environment Canada 2005). The 'final (geometric) EC_{50} values' for each substrate were calculated excluding all outliers irrespective of the evaluation method used.

10.3 Results

A total of 22 participating laboratories provided 124 test data sets, for both species together (Tables 10.1 and 10.2). The results of 17 tests were invalid (14%), as either the validity criteria were not met (e.g. the emergence rate was too low (=10%) or the SOP was not followed (=4%). No biological outliers were presented, meaning that the statistical assessment was based on 107 tests. Two other test results were identified as statistical outliers and the same data (plus one other data set with *A. sativa*) were outside the warning limits, which defined the range of acceptable tests. Nine laboratories presented results from reference tests with boric acid. However, the delivery of such data was not considered an acceptance criterion. 86.3% of tests were accepted.

In the following sections, the results of the plant tests with INC, SOI and WOO are presented, following the same scheme as in the other chapters of Part II. The number

Table 10.1 Summary of the acceptance of plant tests with *Avena sativa* (oat)

Substrate	No. labs	No. tests	Invalid tests	Accepted tests
INC	22	22	3	19
SOI	21	21	3	18
WOO	21	21	3	18
Sum	n.a.	64	9	55

n.a. Not applicable

Table 10.2 Summary of the acceptance of plant tests with *Brassica rapa* (turnip) tests

Substrate	No. labs	No. tests	Invalid tests	Accepted tests
INC	21	21	3	18
SOI	20	20	3	17
WOO	19	19	2	17
Sum	n.a.	60	8	52

n.a. Not applicable

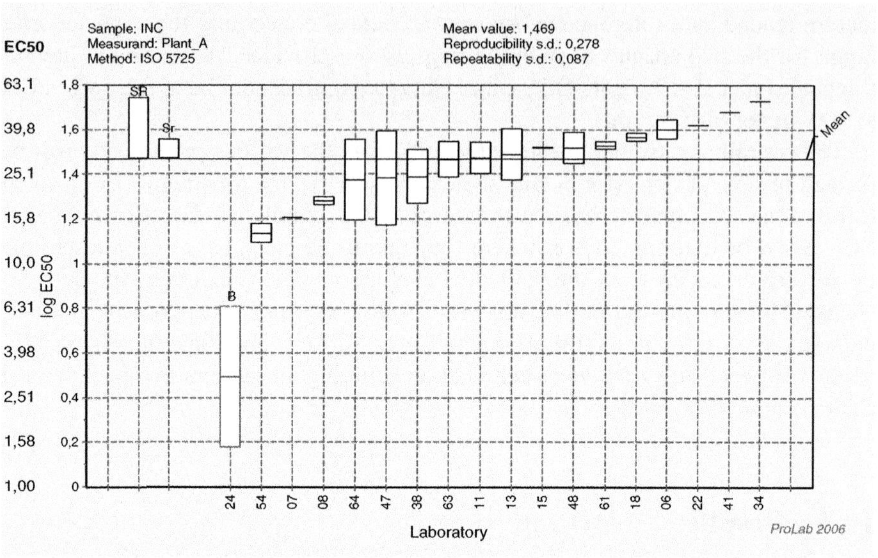

Fig. 10.1 log EC_{50} (% waste in field soil) for the waste substrate INC with *Avena sativa* (oat) (presentation according to ISO 5725-1 (1994)). B: statistical outlier

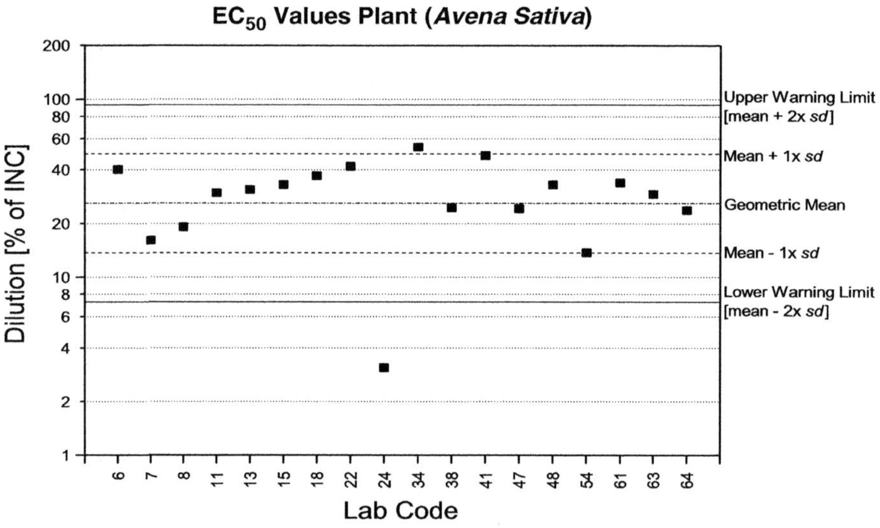

Fig. 10.2 EC_{50} values (% waste in field soil) for the waste substrate INC with *Avena sativa* (oat)

of acceptable tests performed with INC was 19 in the case of *A. sativa* (Figs. 10.1 and 10.2), and 18 in the case of *B. rapa* (Figs. 10.3 and 10.4).

In the tests with INC, only EC_{50} value for *A. sativa* was statistically identified as an outlier (Fig. 10.1). Accordingly, the factor between reproducibility and repeatability was small (3.3). With one exception, all EC_{50} values were within the warning limits (Fig. 10.2). Actually, with one more exception, all results were within a range

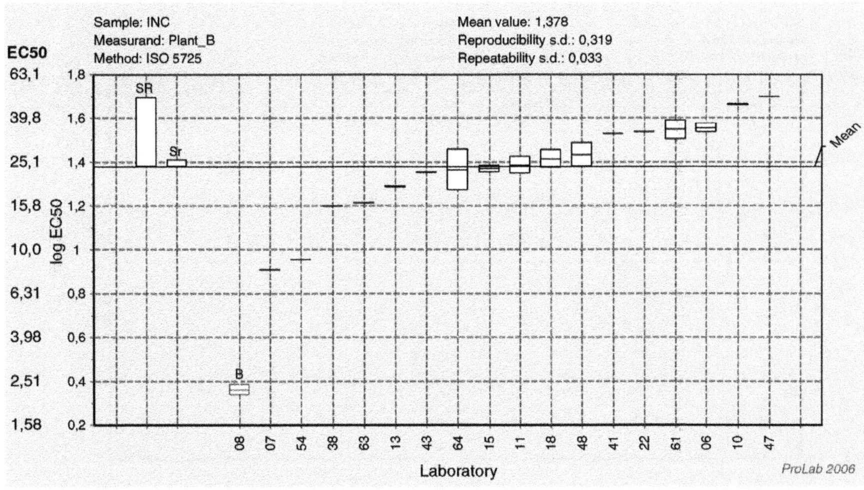

Fig. 10.3 log EC_{50} (% waste in field soil) of plant tests with *Brassica rapa* (turnip) and the waste substrate INC (presentation according to ISO 5725-1 (1994)). B: statistical outlier

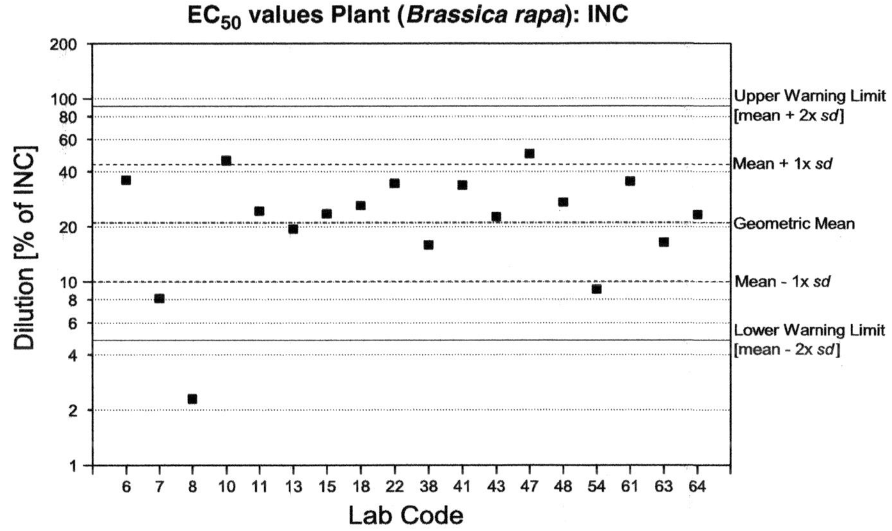

Fig. 10.4 EC_{50} values (% waste in field soil) of plant tests with *Brassica rapa* (turnip) and the waste substrate INC

of the mean ±1SD, i.e. the individual data sets differed by less than a factor of four. No reason can be given for the one outlier (No. 24). The effects of INC on *B. rapa* were similar (Fig. 10.3). The factor between reproducibility and repeatability was clearly higher (9.7). Again, only one test (No. 8) was statistically identified as an outlier and was also outside the warning limits (Fig. 10.4). The light intensity in test No. 8 was relatively low (but still in an acceptable range), which may have had an influence on the test results.

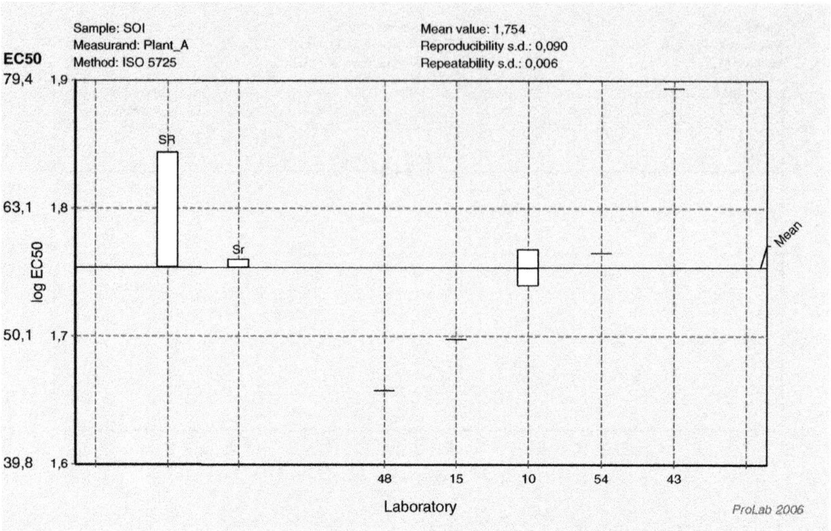

Fig. 10.5 Results (log EC_{50} as % waste in field soil) of plant tests with *Avena sativa* (oat) and the waste substrate SOI (presentation according to ISO 5725-1(1994)

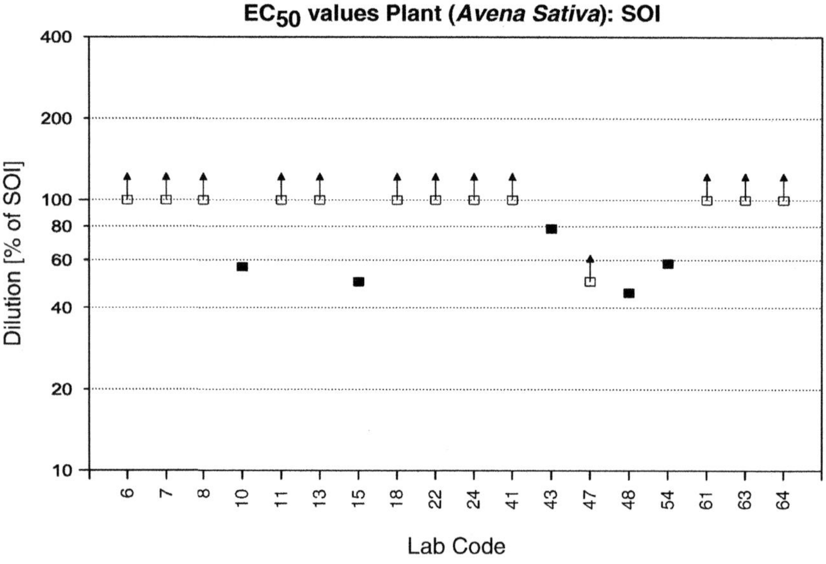

Fig. 10.6 Results (EC_{50} as % waste in field soil) of plant tests with *Avena sativa* (oat) and the waste substrate SOI

In SOI, the effects were clearly lower. For *A. sativa* a factor of 15 was observed between reproducibility and repeatability (Fig. 10.5). However, 13 out of 18 tests resulted in an EC_{50} above 100% (Fig. 10.6), meaning that the calculation of warning limits is not useful. Based on just six values, an EC_{50} of 57.8% was calculated.

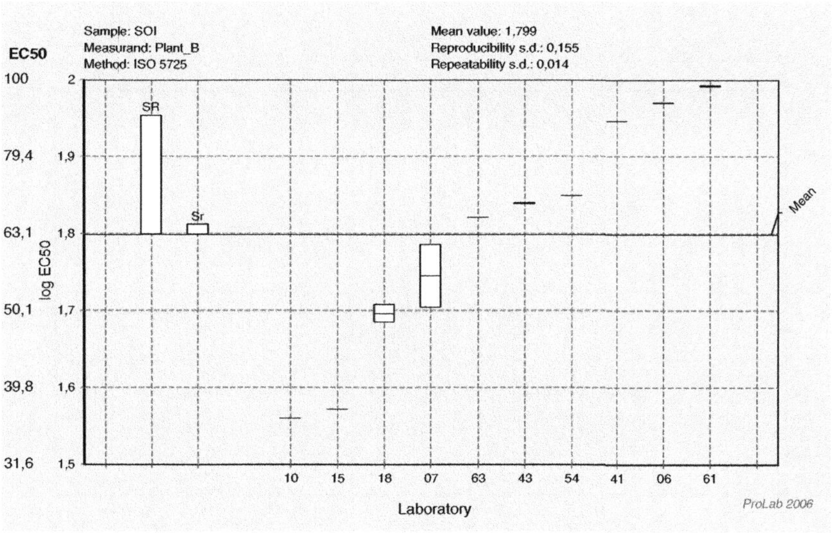

Fig. 10.7 Results (log EC_{50} as % waste in field soil) of plant tests with *Brassica rapa* (turnip) and the waste substrate SOI (presentation according to ISO 5725-1 (1994))

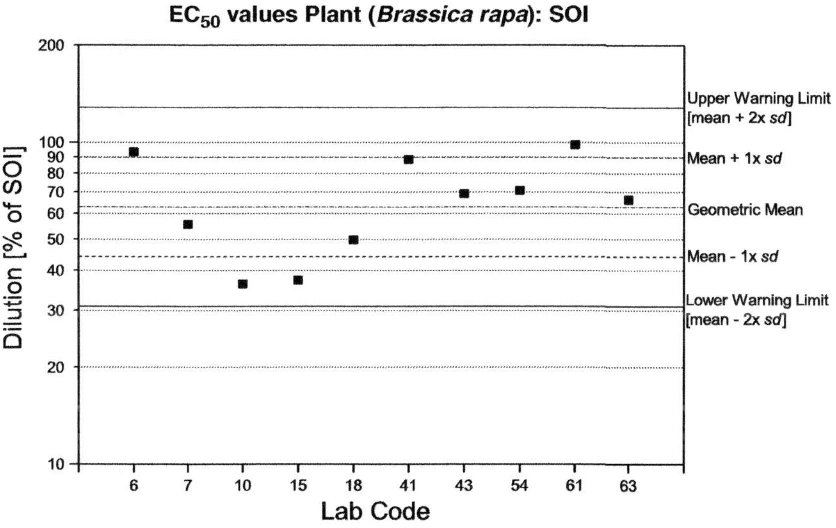

Fig. 10.8 Results (EC_{50} as % waste in field soil) of plant tests with *Brassica rapa* (turnip) and the waste substrate SOI

B. rapa appeared to react more sensitively to SOI. The observed factor of 11 between reproducibility and repeatability was, although still high, lower than in the tests with *A. sativa* (Fig. 10.7). In ten out of seventeen cases, the EC_{50} was lower than 100% (Fig. 10.8). For these ten tests the warning limits were calculated. All test results were inside the limits.

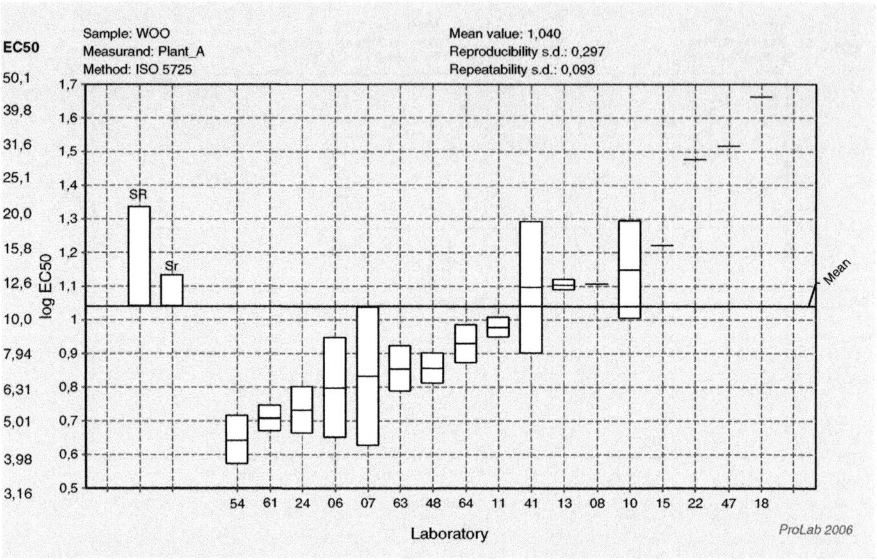

Fig. 10.9 Results (log EC_{50} as % waste in field soil) of plant tests with *Avena sativa* (oat) and the waste substrate WOO (presentation according to ISO 5725-1 (1994))

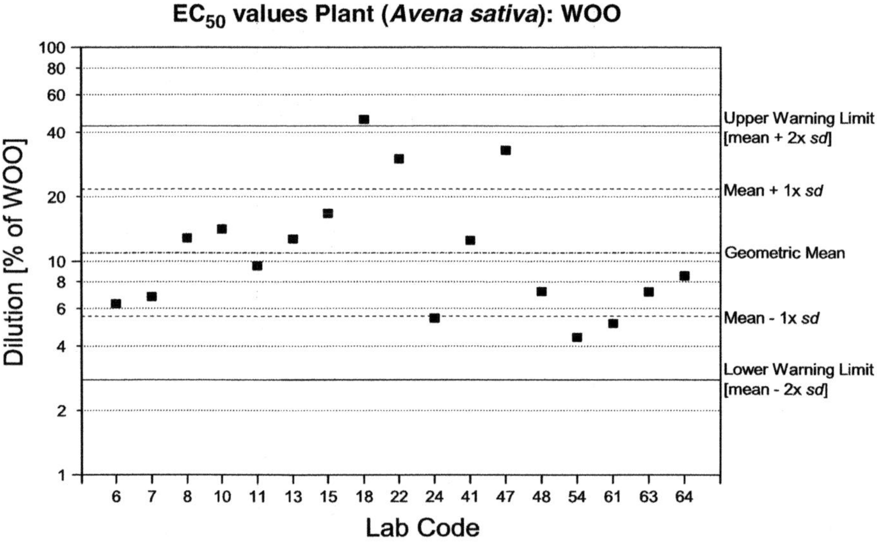

Fig. 10.10 Results (EC_{50} as % waste in field soil) of plant tests with *Avena sativa* (oat) and the waste substrate WOO

In tests with WOO and *A. sativa,* the factor between reproducibility and repeatability was very low (3) (Fig. 10.9), indicating a reasonable robustness of the test results. Only one out of 17 tests was (slightly) outside the warning limits (Fig. 10.10). Five test results were outside the mean ±1SD but lower than the range of the mean ±2SD. In the

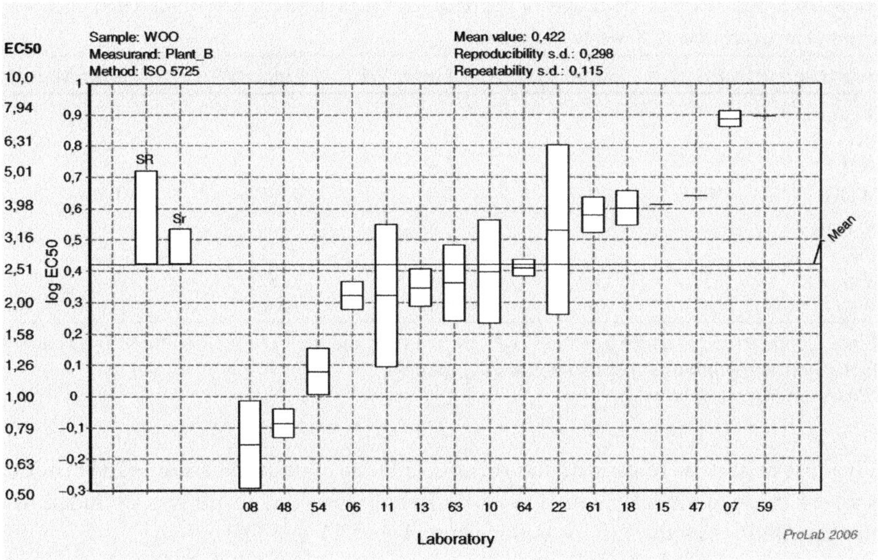

Fig. 10.11 Results (log EC_{50} as % waste in field soil) of plant tests with *Brassica rapa* (turnip) and the waste substrate WOO (presentation according to ISO 5725-1 (1994))

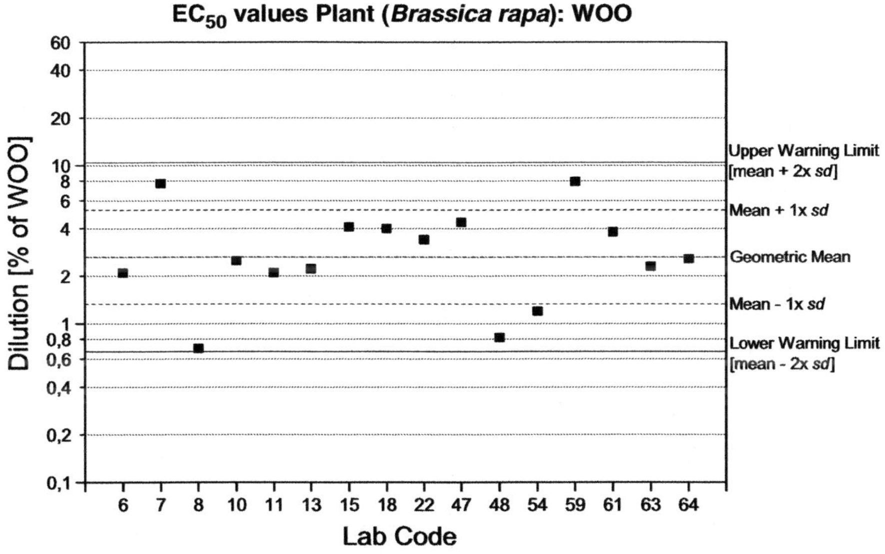

Fig. 10.12 Results (EC_{50} as % waste in field soil) of plant tests with *Brassica rapa* (turnip) and the waste substrate WOO

WOO tests with *B. rapa*, the factor of 2.6 between reproducibility and repeatability was even lower than the one determined for *A. sativa*, confirming the high robustness of the plant test (Fig. 10.11). No test result was outside the warning limits (Fig. 10.12).

Table 10.3 Summary of the results of the plant tests with *A. sativa* and *B. rapa* and three waste types (data expressed as % waste in field soil)

Substrate	EC_{50}	Ratio s_R/s_r	Lower WL	Upper WL	Factor Min-Max
A. sativa					
INC	29.4	3.2	7.2	93.3	17.3
SOI	57.8[a]	15.0	n.a.	n.a.	n.a.
WOO	10.0	3.2	2.8	43.0	10.5
B. rapa					
INC	23.9	9.7	4.8	91.3	21.7
SOI	63.0	11.1	30.9	128.3	2.7
WOO	2.6	2.6	0.7	10.4	11.3

Final (geometrical) mean EC_{50} values, s_R/s_r ratio, lower and upper 95% confidence limits and the factor between minimum and maximum EC_{50} values

[a]Based on just six values

Five test results were outside the mean ±1SD but within the mean ±2SD (not the same as those in the tests with *A. sativa*). In both tests one result was excluded from the calculations as the EC_{50} was determined as <3.1% WOO.

The results of the plant tests are summarized in Table 10.3. WOO is always the most toxic substrate, followed by INC and SOI. INC caused almost the same effect for both species, but the toxicity of the other two substrates was higher in *B. rapa* than in *A. sativa*. The variability of the test results differs strongly between substrates and, partly, species. The s_R/s_r ratio is lower than four in both tests with WOO and in the *A. sativa* test with INC, indicates that the test method can be fairly robust even according to standards usually required for ring tests with chemical parameters. However, in both SOI tests and the *B. rapa* test with INC, ratios between 10 and 15 were determined. While the small number of SOI tests is probably responsible for these high numbers, it is not clear why the results of the remaining test are so variable.

Only three out of 83 test results were outside the warning limits. On the other hand, the min-max factor between individual test results was relatively high (10–20) when considering the tests with INC and WOO (the low factor in the SOI tests with *B. rapa* is biased by the low number of data sets). These values are clearly higher than the range recommended by Environment Canada for interlaboratory comparisons (2–4; Chapman 1995; Environment Canada 1999).

10.4 Discussion

Almost no problems were reported with the practical test performance. In two laboratories, emergence collapsed completely. While in one case no explanation can be given, it seems that in the other case the moisture levels were too high from the very beginning. Because of the properties of the waste substrates, more guidelines are needed to adjust the moisture conditions for the plants. While in soil the correct range of moisture (40–60% of the WHC) during seedling emergence is usually easy to obtain, the correct moisture of field soil - waste mixtures can only be estimated.

However, looking at the high emergence rates and high acceptance rate in general, these estimations seem to have been performed very well.

The total ratio of acceptance (86.3%) is considered good. However, the evaluation of the acceptability of the test results was impeded by the fact that only a few participants performed tests with a reference substance (for individual results see Chap. 6). In previous versions of the ISO-Guideline No. 11269-2 (1995) such a requirement was not obligatory, but according to recent discussions this situation is likely to change. In accordance with a proposal from the literature (Römbke & Ahtiainen 2007) it is recommended that the performance of tests with the reference substance boric acid be required. First results from this ring test support this recommendation, as with one exception (EC_{50} = 78.5 mg/kg) six tests resulted in EC_{50} values in the expected range of 100–400 mg/kg boric acid (Stephenson et al. 1997).

As regularly reported in the literature, *B. rapa* (turnip) is more sensitive towards chemical stress than *A. sativa* (oat) (Wilke et al. 1998; Kalsch et al. 2006b). However, in the waste tests, this relationship could only be confirmed in the case of WOO where *B. rapa* was four times more sensitive than oat, and SOI. With INC, no difference in sensitivity was found between the two species. The relatively high variability as indicated by the high min-max factors may have been caused by a combination of physical effects (the different structure) and chemical stress (the toxic chemicals) of the test substrates.

The sensitivity of the plant test in comparison with the other standard terrestrial test (the earthworm acute test) is high (Chap. 1) and follows the expected order (i.e. highest toxicity in WOO, lowest in SOI and intermediate in INC). The lowest EC_{50} value of all standard tests with solid waste was found in the WOO test with *B. rapa* (2.6%). Compared to the results of the additional tests, the situation becomes more complex: At least the earthworm avoidance and reproduction tests, as well as the *Arthrobacter* test, are equally or even more sensitive than the two plant species.

So far, very few plant tests with wastes have been performed. Most often, ashes from waste to energy plants have been studied. For example, in tests with another *Brassica* species (*B. chinensis*) Wong & Wong (1989) determined EC_{50} values in the range of 10–25% fly ash when mixed into sandy control soil, depending on test duration and endpoint. In tests with twelve German ashes and both species (*A. sativa, B. rapa*) a wide range of EC_{50} values (3–30%) was found, which could not be correlated with the age of these ashes or their content of heavy metals (Römbke & Moser 2007). As in the ring test, both species showed a similar sensitivity and they reacted more strongly than earthworms in the acute test. All three different types of ashes were classified as toxic for three plant species when tested in a modified plant growth test according to OECD guideline 208A (Deventer et al. 2004). However, Quilici et al. (2004) determined a broader effect range (EC_{50}s between 24 and >100%) when testing French bottom ashes, indicating that these ashes differ considerably in their physico-chemical properties and content of contaminants. In several of these studies with solid ash samples (e.g. Ferrari et al. 1999) the plants reacted more sensitively than aquatic organisms in leachates when comparing mixture ratios.

The toxicity of PAH-contaminated soil was studied several times in plant tests; however, because of the influence of the respective soil parameters and the potential impact of other contaminants, the results can hardly be compared. For example, after mixing a PAH-contaminated soil into LUFA or OECD artificial soil, EC_{50} values of 15–25% were determined in a chronic plant test (Kalsch et al. 2006b), which indicates clearly a higher toxicity than in the tests described here. However, as this soil contained high concentrations of chromium also (750 mg/kg), the observed effect was probably caused by the heavy metal and not the PAHs. It seems that plants are usually reacting not very sensitively to PAHs in soil (EC_{50} values determined in laboratory tests after spiking PAHs often >1,000 mg/kg for both species (Römbke et al. 1995; Henner et al. 1999)). This impression is backed up by the results of this ring test in which the SOI substrate was the least toxic.

Results from plant tests with copper-contaminated wood substrate are not known. However, plant tests with copper-contaminated soils or soils spiked with copper salts have been performed. Based on these studies it is likely that the heavy metal copper, while essential for plant physiology, can also be toxic. In agriculture, a content of 100 mg/kg is considered tolerable for most crop species (Hock & Elstner 1984). Higher concentrations of copper occurred in the tests described here, assuming a copper content of WOO of 2,000 mg/kg (see Chap. 12). Further evaluations are difficult as the comparability between soil and waste tests is limited, probably because of the completely different availability of copper in the two substrates.

Comparing the (few) results from the literature with the results from the ring test, it appears that the order of toxicity (and, in the case of INC, also the EC_{50} values) is in an acceptable range. The results of the plant tests, as well as the other test systems, clearly show that the detection of hazardous waste needs to be based on biological testing and chemical analysis. From a scientific point of view, more experience regarding the reaction of the single test species against different waste types is needed to get a better understanding and, in the long run, to predict the ecotoxicity of waste. In this context it is also necessary to study the influence of the waste properties themselves (i.e. not the contaminants) on the test organisms.

10.5 Recommendations

It is recommended that a higher plant test with two species differing considerably in taxonomy and physiology should be part of the final test set (EN 2005) as these organisms form not only the basis of terrestrial food webs but are also often quite sensitive due to their close interaction with the pore water as well as the substrate itself. The sensitivity should be checked regularly by performing reference tests with boric acid. In addition, more guidelines regarding the moisture regime of the mixtures are necessary.

Chapter 11
Earthworm Tests

T. Moser, C. Firla, A. Haller, and A. Scheffczyk

Abstract Earthworms are considered an important part of the soil community, so they represent soil organisms in the test set for the ecotoxicological characterization of wastes. In keeping with the ISO standard guideline 11268-1, the acute earthworm test was performed with three waste types: INC, SOI, and WOO. A total of 44 tests out of the 52 tests conducted were classified as acceptable. Methodologically, almost no problems occurred, but further guidelines on determining the moisture of the test substrate mixtures are needed. After statistical evaluation, only one other test result was rejected. The min-max factors (2–5) between individual LC_{50} values and the CVs (20–31%) were low, showing that the acute earthworm test is a robust method. Toxicity differed considerably between the three waste types, as seen in the final LC_{50} values: INC = 46.3%, SOI ≥ 50%, WOO = 21.0%. These results are in agreement with the few data known from the literature. The sensitivity of the earthworm acute test was low compared to other plant and invertebrate tests with solid wastes. Therefore, it is recommended that a more sensitive alternative be found: either an earthworm test with a chronic endpoint such as reproduction, or another invertebrate species (e.g. a collembolan) should replace this test. In addition, an alternative for the currently used reference substance chloroacetemide (boric acid) should be selected.

Keywords *Eisenia fetida/andrei*, Acute test, Ring test, Waste, Ash, Wood, PAH

11.1 Introduction

Earthworms are important members of the soil community because of their ability to change or modify their habitat through various activities. Thus, some species are considered "ecosystem engineers" (Lavelle et al. 1997). Their activities can improve soil structure, stabilize soil aggregates, increase water infiltration and water-holding capacity (Edwards & Shipitalo 1998), form a humic layer close to

T. Moser (✉)
ECT Oekotoxikologie GmbH, Boettgerstrasse 2-14, 65439, Floersheim, Germany

H. Moser and J. Römbke (eds.), *Ecotoxicological Characterization of Waste*,
DOI: 10.1007/978-0-387-88959-7_11,

the soil surface (Doube & Brown 1998), and increase yield in orchards or grasslands (Blakemore 1997). In temperate regions, most of the earthworms responsible for these activities belong to the family Lumbricidae (class Clitellata, phylum Annelida). Although several species of earthworms have been used in ecotoxicological testing, only two closely related species were included in standard ISO and OECD test protocols: *Eisenia fetida and Eisenia andrei* (Spurgeon et al. 2003). The effects of chemicals on them are relatively well known, because tests with these animals have been required for the registration of plant protection products for more than two decades (Edwards & Bohlen 1996). Still, in most tests, mainly acute effects are measured in artificial soil (ISO 1993; OECD 1984). On the basis of the experiments conducted so far, the earthworm acute test has been selected as a representative method to assess the ecotoxicological hazard potential of wastes to terrestrial invertebrates (EN 2005).

11.2 Method

The performance of the 14-day test with earthworms (*E. fetida* or *E. andrei*) according to the ISO-Guideline No. 11268-1 (1993) was described in a SOP. The test was performed with artificial soil (mixture of 70% quartz sand, 20% kaolinite clay, 10% peat, calcium carbonate to adjust the pH to 6.0 ± 0.5 and deionised water to achieve a soil moisture of 40–60% of the WHC_{max}) as control and mixtures of three waste types (INC, SOI, WOO) and artificial soil. Tests were performed under the following conditions: temperature 20 ± 2°C, light cycle 16/8 h (400–800 lux) and without feeding. Mortality, biomass and morphological or behavioral changes were determined after 7 and 14 days. The main assessment endpoint was the mortality of the worms after 14 days, evaluated as LC_{50} value (lethal median concentration). These values were determined by the probit analysis method, using the statistical program TOXRAT® (2003). Test results were considered valid when the mortality in the control was ≤10% at the end of the test. The decrease in the biomass of the test animals in the control should be less than 20% at the end of the test. In addition, the results of a reference test with the chemical chloroacetamide should be in the range of 20–80 mg/kg (ISO 1993). One participant tested the fungicide carbendazim, which is required as reference substance in the earthworm reproduction test (ISO 1998). As the result was in the range asked for in that guideline, the test results were accepted.

The overall assessment of the test results from the various participants was performed in several steps (for details see Chap. 4). First, the total number of tests that fulfilled the acceptance criteria were determined. Secondly, all data sets in which no LC_{50} could be determined (i.e. where they had to be given as < or > a tested dilution) were excluded from the statistical evaluation. The remaining data sets were checked for passing the outlier tests (following the ISO approach) and whether they were outside the Warning Limits (Environment Canada 2005). The 'final mean (geometric) LC_{50} values' for each substrate were calculated excluding all outliers irrespective of the evaluation method used.

11.3 Results

In all, 18 participating laboratories provided 52 test data sets. Only four tests (8%) were not valid, as the SOP was not followed. Two laboratories did not present results from reference tests, resulting in another four tests (8%) not being accepted. The results of the individual reference tests are provided in Chap. 6. There was no other reason to exclude data sets. The final statistical assessment (ISO 1994; EC 2005) was based on 44 tests, and the overall acceptance was 84.6% (Table 11.1).

In the following sections, the results of the earthworm tests with INC, SOI and WOO are presented, using the same scheme as in the other chapters of Part II. With INC, in five tests out of 13 no mortality was observed and no LC_{50} could be calculated. In these cases the LC_{50} must be given as >50% INC, the highest percentage tested; i.e. only the eight tests from which the LC_{50} value could be calculated are presented in graphs (Figs. 11.1 and 11.2).

Table 11.1 Summary of the acceptance of earthworm tests

Substrate	No. labs	No. tests	Invalid	No reference	Accepted
INC	18	18	2	2	14
SOI	16	17	1	1	15
WOO	16	17	1	1	15
Sum	n.a.	52	4	4	44

n.a. Not applicable

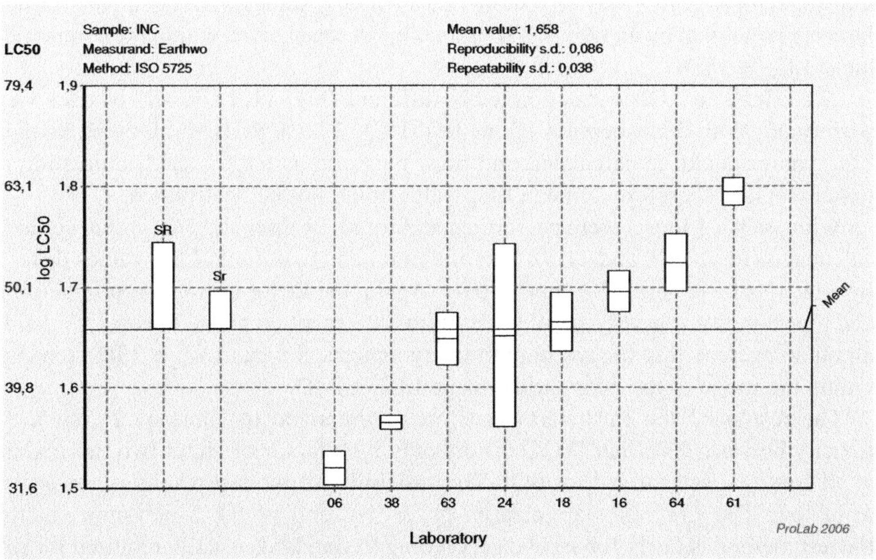

Fig. 11.1 Results (log LC_{50} as % waste in artificial soil) of the earthworm tests with the waste substrate INC (presentation according to ISO 1994)

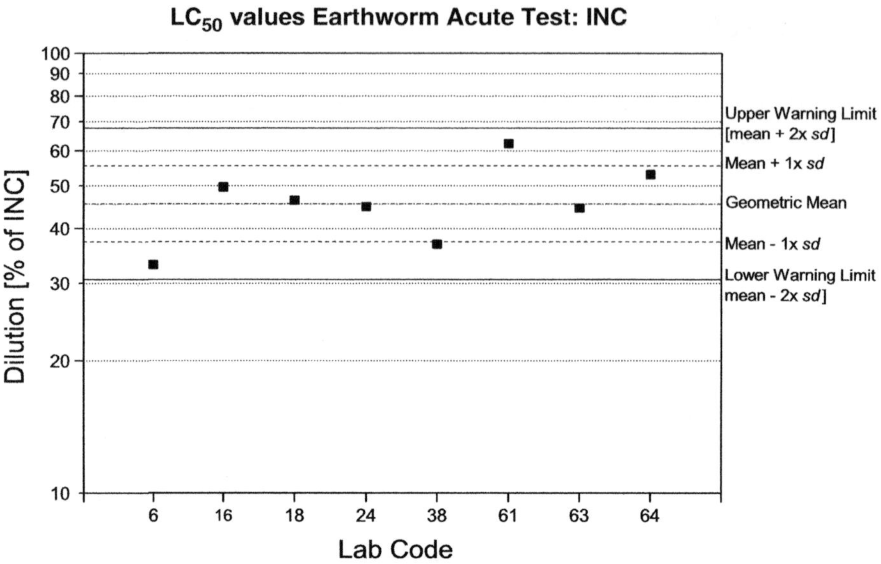

Fig. 11.2 Results (LC_{50} as % waste in artificial soil) of the earthworm tests with the waste substrate INC

The minimum and maximum LC_{50} values determined for INC differed only by a factor of two (min: 35%; max: 65%). Accordingly, the factor between reproducibility and repeatability was very small (2.3). All LC_{50} values were within the warning limits (geometrical mean LC_{50} ± 2SD), six (almost seven) even within the geometrical mean LC_{50} ± 1SD.

The effects in SOI were completely different (Fig. 11.3): In all, 14 tests were performed. With the exception of one test (LC_{50}: 71.7% SOI), in all other cases no LC_{50} values could be calculated and must be given as >100% SOI, indicating no toxicity at all. Therefore, a mean LC_{50} value could not be determined.

With WOO, 14 tests were performed and for all of them the LC_{50} value could be calculated (Figs. 11.4 and 11.5). The minimum and maximum LC_{50} values differed by a factor of five (min: 7%; max: 35%). Again, the factor between reproducibility and repeatability was very small (2.7). With the exception of the lowest LC_{50} value, all others were within the warning limits (geometrical mean LC_{50} ± 2SD), ten even within the range of the geometrical mean LC_{50} ± 2SD.

The results of the earthworm tests are summarized in Table 11.2. While the toxicity between INC and WOO differs only by a factor of about two, no toxicity at all could be determined for SOI. The variability of the test results is considered to be low: The s_R/s_r ratios are clearly <4 for INC and WOO, which indicates that the test method is fairly robust even according to standards usually required for ring tests with chemical parameters. Only one out of 29 test results was outside the warning limits and the min-max factor between individual test results was also very

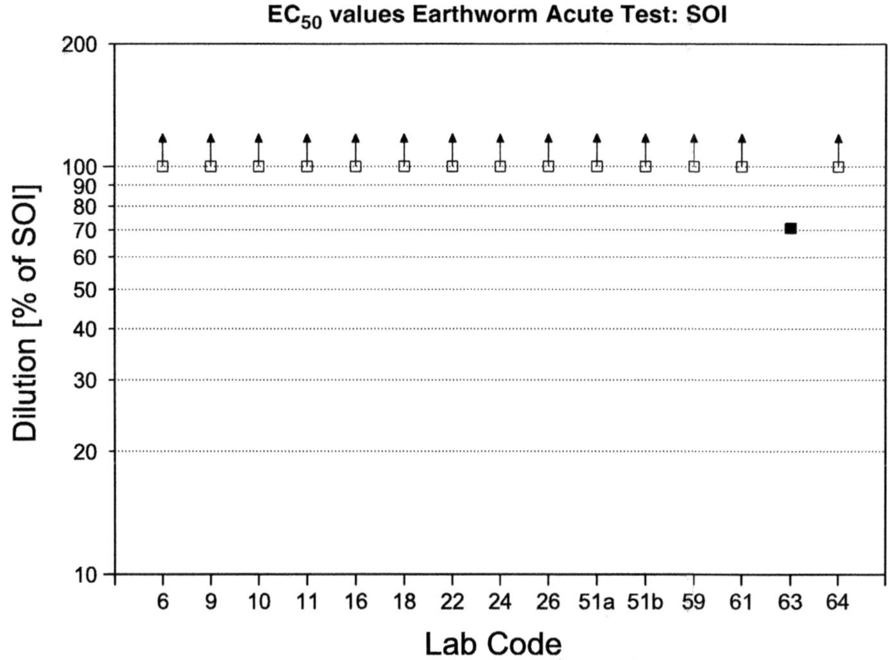

Fig. 11.3 Results (LC_{50} as % waste in artificial soil) of the earthworm tests with the waste substrate SOI. Neither reproducibility/repeatability nor warning limits assignable

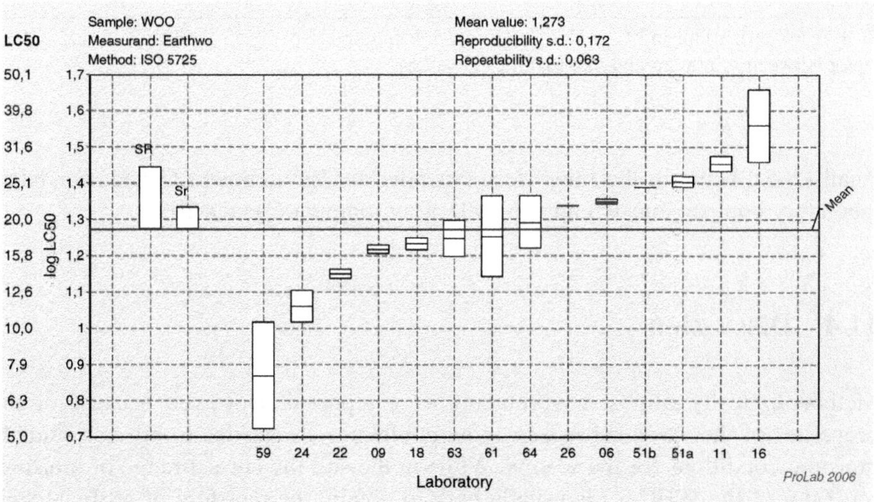

Fig. 11.4 Results (log LC_{50} as % waste in artificial soil) of the earthworm tests with the waste substrate WOO

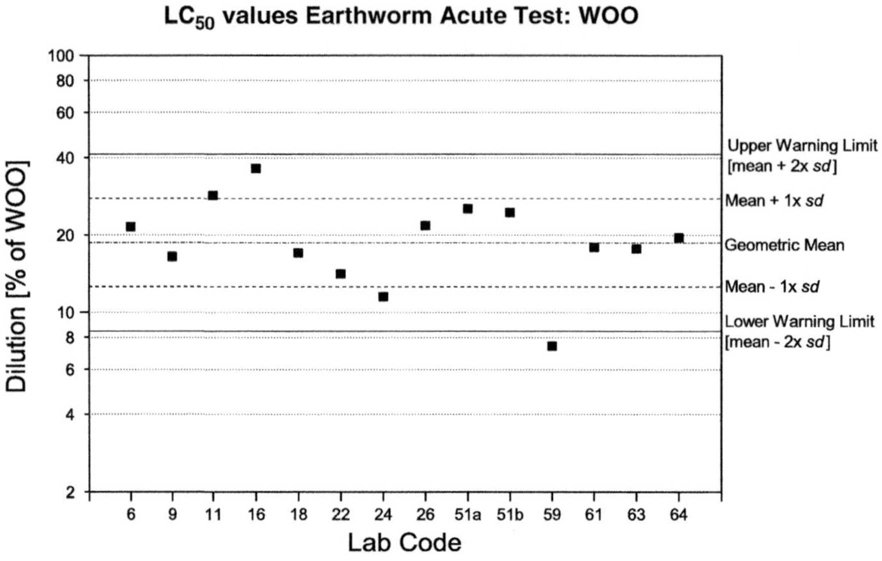

Fig. 11.5 Results (LC_{50} as % waste in artificial soil) of the earthworm tests with the waste substrate WOO

Table 11.2 Summary of the results of the earthworm acute tests and the three waste types (data expressed as % waste in artificial soil)

Substrate	LC_{50}	Ratio s_R/s_r	Lower WL	Upper WL	Factor Min-Max
INC	45.5	2.3	30.6	67.7	1, 9
SOI	n.a.	n.a.	n.a.	n.a.	n.a.
WOO	20.1	2.7	8.5	41.2	3, 2

Final (geometrical) mean LC_{50} values, sR/sr ratio, lower and upper 95% confidence limits and the factor between minimum and maximum LC_{50} values

small (2–5), being in the range recommended by Environment Canada for interlaboratory comparisons (Chapman 1995; Environment Canada 1999).

11.4 Discussion

Methodologically, almost no problems were reported. However, because of the properties of the waste substrates, more guidelines are needed to provide suitable moisture conditions for the worms. While in the soil the correct range of moisture (40–60% of the WHC_{max}) is usually easy to obtain, the moisture of artificial soil/waste mixtures cannot always meet the requirements because of the physical properties of the different waste types (i.e. other waste types with contrasting properties have to be tested in order to determine which wastes are suitable for bulk soil testing). Therefore, the moisture was determined and adjusted "manually": the moisture was considered correct if after gently squeezing the mixtures, small drops of water appeared (but did not drip off) between the fingers. Another modification to be considered is an exchange of the currently used reference substance chloroacetamide (which is possibly carcinogenic) by a substance less toxic for humans, e.g. boric acid (Römbke & Ahtiainen 2007).

In general, the variability of the test results was low, as indicated by the sr/sr ratio being lower than four and min–max factors between two and five. In fact these results are "better" than those determined in the ring test with a single chemical, performed when the test method was developed (Edwards 1984).

The sensitivity of the earthworm acute test is low compared to other tests with solid wastes. This is most obvious in SOI, where there was almost no reaction of the worms. Compared to the plant tests, the LC_{50} values for INC and WOO were about twice as high as the respective plant EC_{50} values. The lack of sensitivity becomes even more obvious compared to the results of non-standard tests (in particular the earthworm reproduction (ISO 1998) and avoidance tests (ISO 2007) and to a lesser extent, the collembolan and enchytraeid reproduction tests (ISO 1999; ISO 2004)).

So far, the experiences with acute earthworm tests in waste testing have been limited. Recently, the toxicity of twelve incineration ashes was studied in Germany (Römbke & Moser 2007). The toxicity (LC_{50} values) of these ashes ranged between 11.5 and 43.6% ash, but no clear correlation could be found between their processing or aging status and their content of heavy metals. These results are in very good agreement with those obtained in the ring test. Comparable incineration ashes from France with very low heavy metal contents showed a lower toxicity in the earthworm acute test (LC_{50} values 40–76% waste), but no details of the test conditions are known (Quilici et al. 2004).

The toxicity of PAH-contaminated soil was tested more often in earthworm tests; however, because of the influence of the respective soil parameters, the results are difficult to compare. As a general tendency, these soils usually had only low effects on earthworms. For example, after mixing a PAH-contaminated soil (sum of EPA-PAH: about 3.000 mg/kg) with artificial soil an LC_{50} of 32.9% was determined (Potter et al. 1999). This result seems to be in agreement with the results of the ring test where SOI contained about 840 mg/kg PAH.

Results from tests with copper-contaminated wood substrate are not known so far. On the other hand, the heavy metal copper is highly toxic for earthworms. Reproduction and, to a lesser extent, mortality, are the most sensitive endpoints (Rundgren & Van Gestel 1998). For example, in the field, significant effects on earthworm populations were found starting at copper concentrations somewhere between 30 and 100 mg/kg soil (Belotti 1998). Assuming a copper concentration of 2,000 mg/kg in WOO, it becomes clear that the observed effects occurred in an order of magnitude comparable to the one determined in soil studies.

Comparing the (few) results from the literature with the results from the ring test, it appears that the respective toxicity of the three waste substrates is in agreement with them.

11.5 Recommendations

It is recommended that a worm (i.e. oligochaete) test should be part of the final test set as these are in many soils the most important invertebrates, which are also often quite sensitive due to their close interaction with the pore water as well as the substrate itself (feeding). However, because of the low sensitivity of the acute endpoint mortality, alternative methods with more sensitive endpoints have to be checked (e.g. the chronic earthworm reproduction test (ISO 1998) or, probably mainly for screening purposes, the earthworm avoidance test (ISO 2006)). Even the chronic enchytraeid reproduction test may be a better option (ISO 2004). Independent of the oligochaete test that is finally selected, more guidelines are necessary regarding the moisture regime and the pH of the mixtures. To avoid problems due to moisture and pH, other soil-dwelling invertebrates like collembolans (ISO 1999) or predatory mites (Bakker et al. 2003) which are not as sensitive as earthworms to these parameters could also be an option. Furthermore, the influence of the waste properties themselves (not the contaminants) on the test organisms has to be studied as it has recently been done for different soil types (Jaensch et al. 2006). In addition, an alternative for the currently used reference substance chloroacetamide (boric acid) should be selected.

Chapter 12
Lemna Growth Inhibition Test

H. Moser and M. Pattard

Abstract The *Lemna* growth inhibition test was part of the additional aquatic test battery within the ring test. The number of reported data sets was sufficient for statistical assessment. The methodology and the results show that the *Lemna* (Duckweed) test, using a higher plant as test organism, can be successfully applied to assess waste eluates. The results for the different substrates are within the same range as the other biological test systems, with only weak effects in SOI eluates, medium ecotoxicity for INC and the detection of a clear ecotoxic response for WOO eluates. The phytotoxicity of waste eluates cannot be covered only by the Algae test and the use of unicellular organisms. By using the higher plant *Lemna*, the test results enable a better understanding of the impact of waste material on macrophytes in the ecosystem. More information on the environmental risk assessment of waste to be reused in the environment makes the *Lemna* test valuable for an enlarged biotest battery. For samples with specific limitations to application of biotests, such as colour, turbidity, high Ammonia content or a non-neutral pH value, the *Lemna* test can be recommended even for a limited test set, such as waste classification.

Keywords Lemna minor, Growth, Ring test, Eluate, Aquatic test, Macrophyte

12.1 Introduction

Aquatic plant toxicity tests are frequently used in environmental risk assessments to determine the potential impact of contaminants on primary producers. The plant test organism *Lemna minor* (duckweed) is used as a model organism for higher water plants. Duckweeds are monocotyledonous, free-floating angiosperms, with an enormous growth potential and the ability to react to nutrients and pollutants dissolved in water or to accumulate compounds of the water fraction. As primary

H. Moser (✉)
Environment Agency, Woerlitzer Platz 1, Dessau, 06844 Germany

H. Moser and J. Römbke (eds.), *Ecotoxicological Characterization of Waste*,
DOI: 10.1007/978-0-387-88959-7_12, © Springer Science+Business Media, LLC 2009

producers and part of the aquatic environment in many climate zones, they play an important role in ecotoxic test batteries (Landolt & Kandeler 1987).

The ecotoxicological assessment of waste or waste eluates is a regulatory requirement in Europe (EEC, Directive 91/689/EEC). The use of a highly organized test organism like duckweed is recommended to get a more direct and integrated estimation of ecotoxicity. The duckweed growth inhibition test shows a biological response to a complex mixture of compounds and integrates different factors like pH and solubility, and antagonistic or synergistic interactions of eluate compounds, and covers the bioavailability of substances during a long-time exposure of 7 days. This macrophyte test system can be applied in environmental samples or eluates of waste or soil, unaffected by the turbidity or colour of the sample, within a wide range of pH from 6–8 (Fomin et al. 2000).

12.2 Method

The *Lemna* test is based on the influence of pollutants on the growth and development of duckweed plants. Different physiological endpoints can be used, such as a decrease in growth, measured as frond (=leaf) number or area, a decrease in fresh or dry weight, or the alteration of the pigmentary content of chlorophyll or anthocyane (Lewis 1995). In the ISO 20079 guideline, frond number is the obligatory endpoint of the test, with one additional parameter such as frond area, dry weight or chlorophyll content. The macrophyte *L. minor* is cultured as stock culture, often under axenic and mixotrophic conditions. Before starting the test, the plants are grown in a pre-culture under conditions similar to the exposure conditions, e.g. level of light or nutrient supply (Christen & Theuer 1996).

As laid down in the guideline ISO 20079 (2005) a dilution row of the waste eluate and the nutrient solution is prepared, while the volume of the exposure vessels can vary from 100 to 400 mL. At the beginning of the exposure, young, green and healthy plants are inoculated and grown for 7 days under standardized conditions for light and temperature. The control consists of a nutrient solution and at least three replicates have to be tested. After 7 days, the growth rate of the plants is determined by counting the frond number or measuring the frond area with image analysis systems. The test is considered valid if the growth rate for the parameter frond number in the controls is at least 0.275/*d*, which means a sevenfold increase of the number of fronds within 7 days.

Regarding the guideline and experiences with the *Lemna* test system 3,5-Dichlorophenol is tested as reference substance, meaning that the EC_{50} value should be in the range of 2.2 and 3.8 mg/L. In the ring test, the main assessment endpoint was frond number, but some participants of the ring test also provided data on frond area.

The *Lemna* test was part of the additional aquatic test battery, but because the required number of data sets for a statistical assessment was achieved, the EC_{50} values based on the individual lab raw data were determined by probit analysis.

The overall assessment of the test results from the various participants was performed in several steps (for details see Chap. 4). First, the total number of tests that fulfilled the acceptance criteria was determined (Table 12.1). Secondly, all data sets in which no EC_{50} could be determined (i.e. where they had to be given as $<$ or $>$ a tested dilution) were excluded from the statistical evaluation. The 'final mean (geometric) EC_{50} values' for each substrate were calculated excluding all outliers, irrespective of the evaluation method used.

12.3 Results

The *Lemna* growth inhibition test was one of the additional aquatic test systems applied voluntarily by the participants. The number of data sets for the additional tests was lower than that for the basic test set. Only for the *Lemna* test the required number of results was achieved with 51 data sets, to enable a meaningful statistical assessment. The acceptance and validity of the results have been limited to data sets containing raw data and the test conditions had to be within the limits required by the guideline ISO 20079. In case an EC_{50} value could not be determined, this result was excluded from further statistical assessment. The occurrence of valid results of reference tests was not used as an exclusion criterion (Table 12.1).

The duckweed *L. minor* was used in all the laboratories and most of them used vessels of 120 mL for exposure. None of the labs used macroplates or other miniaturized exposure systems in the test. Except one, all the participants conducted the eluate test once and the majority of the laboratories reported raw data and the results of their reference test with 3,5-Dichlorophenol. The guideline ISO 20079 for the *Lemna* test specifies the range of the reference test results for 3,5-Dichlorophenol from 1.8 to 3.6 mg/L. As all reported reference results lay within the given range, no data set had to be excluded. From 15 accepted data sets of the WOO substrate, 11 data sets were verified by reference test results.

The results of the *Lemna* test for the eluates of the substrate INC did not allow a clear classification in the ring test. In six tests, the participants reported EC_{50} values of 90% or above. Three laboratories reported EC_{50} values of 50%, because of a limitation of the substrate concentration in the test. These results were supported by the results of three laboratories, which measured EC_{50} values of 62.0%, 68.8% and 95% in an adapted dilution row. The results shown in Fig. 12.1 are based

Table 12.1 Number of acceptable and valid test results of the *Lemna* test

			Acceptance		EC_{50} determined?	
Substrate	No. Labs	No. Tests	No	Yes	No	Yes
INC	15	16	6	10	7	3
SOI	15	16	4	12	12	0
WOO	15	19	6	13	0	13
Sum	15	51	16	35	19	16

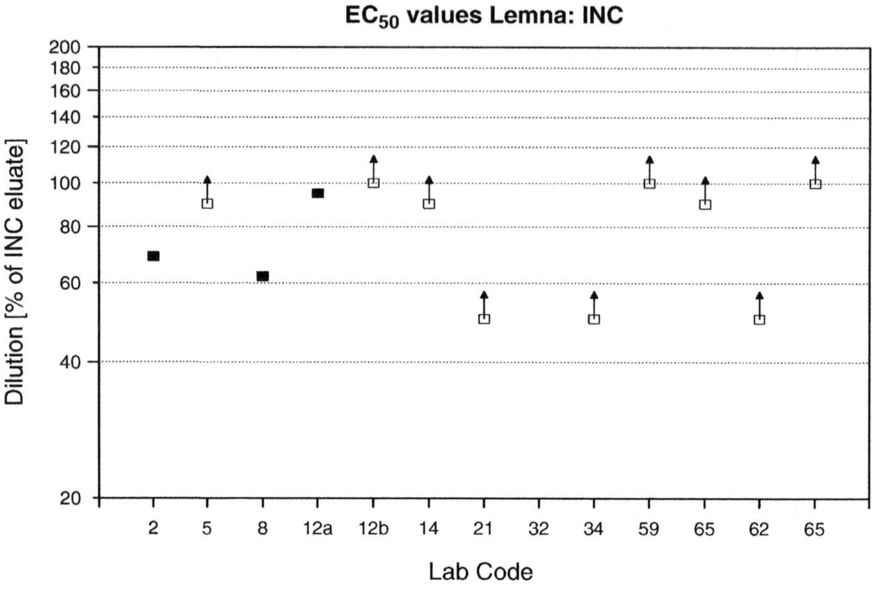

Fig. 12.1 Results (EC_{50} as % waste eluate) of the *Lemna* tests with the waste substrate INC (Environment Canada 2005)

on the measurement of the frond number. Using the endpoint front area, these results were confirmed for most of the data sets, with a tendency toward a slightly lower EC_{50} value.

The phytotoxicity of the waste eluates of SOI was the lowest of all three waste substrates. Ten out of twelve laboratories reported EC_{50} values of 90% or above. Two participants tested a sample content of 50% as the highest concentration and reported EC_{50} values above 50%. Therefore the calculation of statistical parameters such as the confidence limits (Environment Canada 2005) was not applicable. Figure 12.2 shows the results for the *Lemna* test in SOI for the physiological endpoint frond number. The results for the frond area of the plants were similar in all laboratories.

In all the systems of the ring test, the substrate WOO showed the strongest ecotoxic effects. The high content of water soluble heavy metals enabled the transfer of the contaminants into the water fraction during the leaching process and resulted in clear test responses in the aquatic test systems, e.g. the *Lemna* test. The twelve EC_{50} values for the endpoint frond number are within the range of 0.42 to 2.98%, with a mean value of 1.70%. By using the Warning Limit Approach (see Chap. 4) it was shown that all laboratories except one are within the warning limits. Eleven data sets are within the range of the mean ±1SD (see Fig. 12.3). The factor between the minimum and maximum of all valid data is determined as five, which can be considered acceptable - or even about four, excluding the value which is outside the lower warning limit.

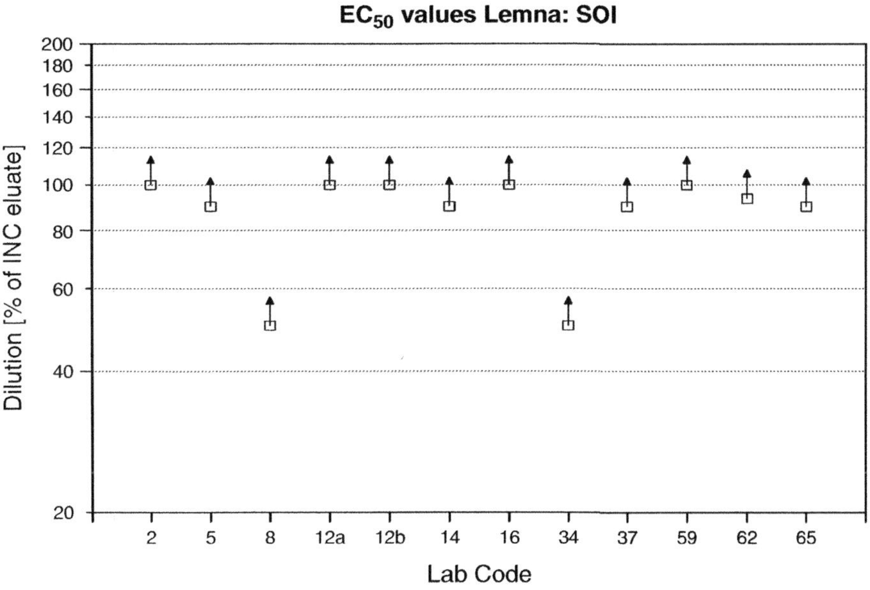

Fig. 12.2 Results (EC_{50} as % waste eluate) of the *Lemna* tests with the waste substrate SOI (Environment Canada 2005)

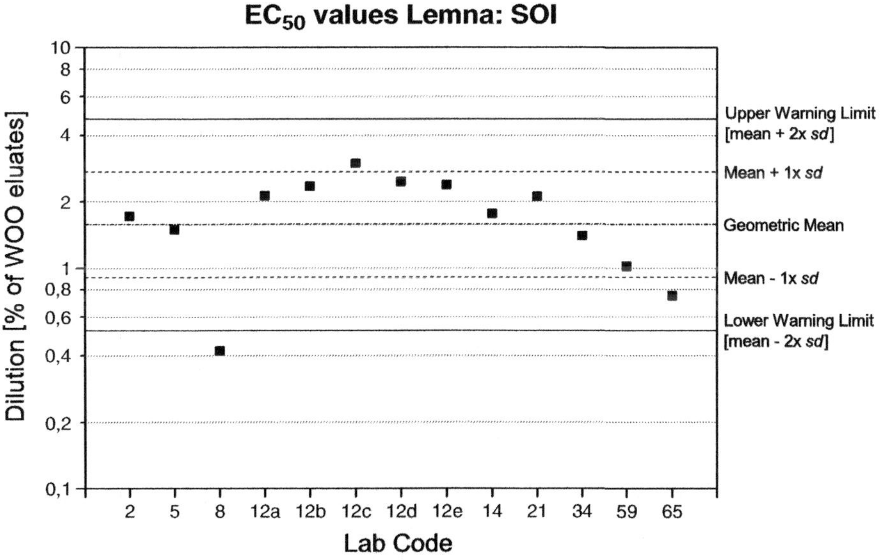

Fig. 12.3 Results (EC_{50} as % waste eluate) of the *Lemna* tests with the waste substrate WOO (Environment Canada 2005)

Table 12.2 Comparison of EC_{50} values determined for the endpoints number (*left side*) and the area (*right side*) of fronds after exposure to WOO eluates

EC_{50} (% dilut.) number		95% CL		EC_{50} (% dilut.) Area	
</>		Lower	Upper	</>	
–	1.72	1.08	2.91	–	1.19
–	1.49	0.94	2.40	–	0.87
–	0.42	0.30	0.58	–	0.62
–	2.12	1.47	3.03	–	0.79
–	2.34	1.73	3.18	–	0.75
–	2.98	2.36	3.80	–	0.98
–	2.46	2.06	2.95	–	0.90
–	2.38	1.83	3.12	–	0.84
–	1.76	1.40	2.20	–	n.d.
–	1.20	n.d.	n.d.	–	1.10
–	5.26	4.46	6.06	–	n.d.
–	2.11	1.49	2.99	–	n.d.
–	1.40	n.d.	n.d.	–	n.d.
–	n.d.	n.d.	n.d.	–	1.43
–	1.02	n.d.	n.d.	–	0.95
–	0.75	0.44	1.17	–	0.52
–	1.96	–	–	–	0.91

Contrary to the results for SOI and INC, the assessment of frond area as physiological endpoint shows a higher sensitivity of the test plants against the WOO eluate (Table 12.2). The EC_{50} value for the impact of WOO on the frond number was 1.96% (13 data sets) for the frond area 0.91% (12 data sets) (Fig. 12.4; it should be noted that here 12 and not 13 valid data sets are available). The main pollutant of the WOO eluate consists of the heavy metal copper, which could influence the growth physiology of the plants by leading to a reduced size of the fronds, but without influence on the formation of daughter fronds (Huebert & Shay 1993).

12.4 Discussion

The results of the ring test show that the *Lemna* test can be applied for the ecotoxicological characterization of waste eluates. The results for the different substrates are within the same range as in the other biological test systems, with only weak effects in SOI eluates, medium ecotoxicity for INC and the detection of a clear ecotoxic response for WOO eluates. However, looking closely at the EC_{50} values, it becomes clear that the *Lemna* test was less sensitive than the other aquatic tests for the INC eluate while there was no difference in SOI. In the tests with WOO eluates, the sensitivity of the *Lemna* test was at the higher end of the spectrum but indicated the high toxicity of this substrate (EC_{50} = 1.70%). The eluates of SOI and INC did not cover the ecotoxicity of the waste material satisfactorily, because the main pollutants of the wastes were not transferred into the water fraction during the leaching

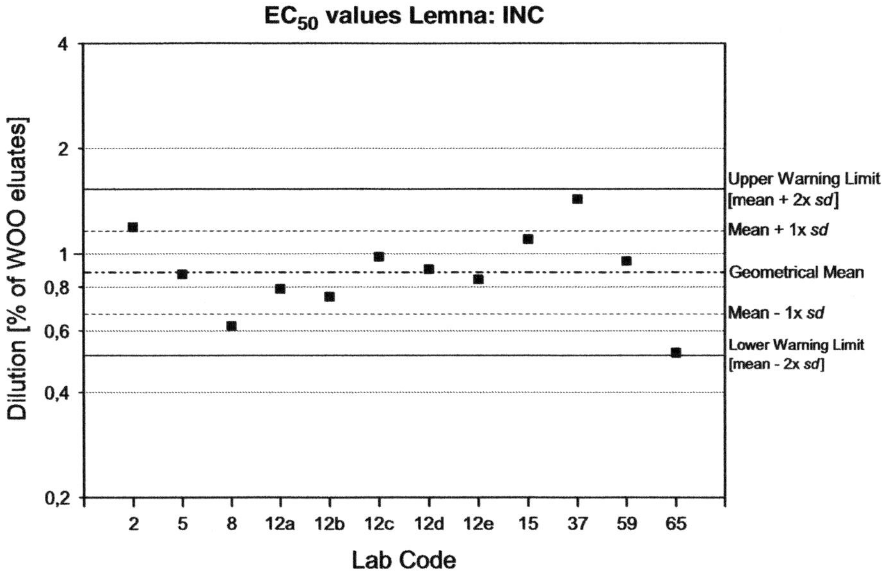

Fig. 12.4 Results (EC_{50} as % waste eluate) of the *Lemna* tests (aendpoint area) with the waste substrate WOO (Environment Canada 2005)

process. In the *Lemna* test, as for all ecotoxicity tests for eluates, only the soluble and non-volatile fraction of the waste can be assessed.

Regarding the methodology, no limitations or necessary modifications were reported by the participating laboratories and the results of the reference tests showed that the performance of the *Lemna* test met the given validity criteria. The results and reports of the various participants indicate that in laboratories where the test is well established and the assessment of frond number and frond area is possible, the validity of the test and the result of the reference tests are acceptable without limitations. Due to the comparatively low number of participants the statistical assessment can only be preliminary. In any case more experience with eluates from other waste types is needed. Attention should be paid to the fact that in cases where the volume of the eluate is limited, the exposure of *Lemna* can be difficult.

Most of the experience in testing environmental samples with *Lemna* was gained in the assessment of waste water (Wang 1991) and the monitoring of leachates from landfill sites (Clement & Merlin 1995, Becker-van Slooten et al. 1999). In many investigations, the plant-related toxicity of eluates was monitored only with algal test organisms. The assessment of phytotoxicity with a unicellular organism and the extrapolation of the results to macrophytes are unavoidably limited, showing that testing with duckweed is preferable. The *Lemna* test is recommended (Fomin et al. 2000, Cleuvers & Ratte 2002) especially for the assessment of turbid or coloured samples or for eluates with pH values outside the neutral range and for waste eluates with Ammonia content.

12.5 Recommendations

The *Lemna* test is not the most sensitive aquatic test method, but it could provide information on the plant-related toxicity in eluates (colored, turbid or those outside a neutral pH-value), where an application of the Algae test is not possible.

Chapter 13
Toxkit Tests

K. Wadhia and G. Persoone

Abstract Research on the controlled production and storage of dormant (or immobilised) stages of selected test species resulted in the development of the "Toxkit microbiotests", which bypass the need for continuous culturing/maintenance of live stocks of test species. These culture/maintenance free assays are used today worldwide for a variety of applications in aquatic and terrestrial ecotoxicology. In the framework of the international ring test for the characterisation of waste, several laboratories from eight European countries performed assays with the Algaltoxkit and the Daphtoxkit (both of which abide by the ISO norms for algal tests and acute Daphnia tests); some of the participants also submitted results for the Thamnotoxkit, the chronic Rotoxkit and the Phytotoxkit. In all, 72 Toxkit test data were submitted, 63 of which (87%) can be considered acceptable. In view of the limited number of data for the Thamnotoxkit and the Phytotoxkit, the statistical evaluation of the EC_{50}s was limited to the Algaltoxkit, Daphtoxkit and chronic Rotoxkit results, which revealed that for these assays all EC_{50}s were within the "warning limits" of 2SD from the mean. Variation coefficients, however, varied substantially from one test to another and for different types of waste, ranged from 16% to as high as 92%. A comparison of these latest figures with those of national and international ring tests performed over the previous years with pure chemicals using the Algaltoxkit and the Daphtoxkit indicated that the CVs in the former ring tests reported here were much lower than those obtained with the solid waste eluates. This may be attributed to the homogeneity of the waste samples and/or the preparation of the eluates. In view of the advantages pertaining to the Toxkit microbiotests compared with the conventional assays, and the associated robustness exemplified in the ring test discussed here, it is recommended that the microbiotests should be considered valid alternatives to standard bioassays for the evaluation of the toxicity of solid wastes.

Keywords Waste, Eluates, Copper, PAH, Laboratory tests

G. Persoone (✉)
MicroBioTests Inc, Kleimoer 15, 9030, Mariakerke-Gent, Belgium

H. Moser and J. Römbke (eds.), *Ecotoxicological Characterization of Waste*,
DOI: 10.1007/978-0-387-88959-7_13,

13.1 Introduction

The need for year-round culturing and maintenance of live stocks of test species in good health and in sufficient numbers is a serious burden. This hinders routine testing and restricts bioassays using toxicity assessment to a limited number of highly specialised laboratories worldwide. In the early eighties, research was therefore initiated in the Laboratory for Biological Research in Aquatic Pollution (now renamed Laboratory for Environmental Toxicology) at the Ghent University in Belgium, to overcome this problem. The investigations resulted in the development of acute and short-chronic "Toxkit microbiotests" which make use of dormant or immobilised stages of selected test species which can be stored for a considerable time, and which can be "de-immobilised" or hatched at the time of performing the assays.

A set of acute and short-chronic "culture/maintenance free" microbiotests with species belonging to several phylogenetic groups (micro-algae, protozoans, rotifers and crustaceans) was gradually developed, and made commercially available by a spin-off company (MicroBioTests Inc.). The Toxkits have now been employed worldwide for diverse applications in both aquatic and terrestrial ecotoxicology (Persoone et al. 2000). The laboratories participating in the international ring test for the ecotoxicological characterisation of waste were at liberty to also apply "additional test methods". In addition to the five selected bioassays, a number of laboratories from eight European countries opted to perform assays using different Toxkit microbiotests (Algaltoxkit, Daphtoxkit, Thamnotoxkit, chronic Rotoxkit and Phytotoxkit).

13.2 Methods

The tests using Toxkit microbiotests were conducted using materials (test containers, test media, dormant or immobilised test biota) provided in the respective kits. "Standard Operational Procedure" manuals designed specifically for the bioassay are included in the kits (see references Algaltoxkit F, Daphtoxkit F, Thamnotoxkit F, chronic Rotoxkit F and Phytotoxkit) and convey detailed instructions for test performance. The Algaltoxkit and Daphtoxkit test procedures follow the ISO and OECD Guidelines totally for the 72 h algal growth inhibition test and the acute *Daphnia magna* test, respectively. The only difference is that for the Algaltoxkit assay the micro algae are obtained from the "algal beads" and for the juveniles for the Daphtoxkit test from the dormant eggs (ephippia). The test species used in the Algaltoxkit is *Pseudokirchneriella subcapitata.*

The Thamnotoxkit microbiotest is a 24 h mortality test, performed on larvae hatched from cysts of the anostracan crustacean *Thamnocephalus platyurus*. The chronic Rotoxkit is a 48 h reproduction inhibition assay with the rotifer *Brachionus calyciflorus*, also hatched from cysts. The assay determines chronic toxicity to *B. calyciflorus* in 48 h and closely follows the French standard (AFNOR

NF T90-377 2000); it is in the final stage of evaluation for acceptance by ISO (ISO/CD 20666 2005). The Phytotoxkit microbiotest is a 3-day seed germination and early plant growth microbiotest performed in flat transparent test containers, with measurement of root and shoot growth by image analysis. While the Toxkit plant test with its SOP employs one monocotyledon (*Sorghum saccharatum*) and two dicotyledon species (*Lepidium sativum* and *Sinapis alba),* the assays for the ring test were required to be performed with the *Avena sativa* and *Brassica rapa.* The duration of the pot trials in the ring test was 4 days, with root length measurement alone required. The test results were evaluated by calculating the geometrical means of the EC_{50}s, with standard deviations and coefficients of variance based on the warning limit approach (Environment Canada 2005).

13.3 Results

Table 13.1 shows the number of test results from the 12 laboratories that submitted data for five different Toxkit microbiotests. It also shows that of the 72 test data sets submitted, 63 results (87%) could be considered acceptable. Furthermore, from the table it may be deduced that the large proportion of Toxkit data are results of algal and Daphnia tests (24 h acute test). The number of laboratories that provided results for chronic Rotoxkit assays was four: only two laboratories submitted data for the crustacean Thamnotoxkit. In the case of the Phytotoxkit, one laboratory performed the test using the ring test stipulated species and another laboratory tested the Toxkit species.

This chapter focuses on the evaluation of the "interlaboratory precision" of the Toxkit microbiotests. A separate chapter (Chap. 17) is devoted specifically to the comparison of the results of the Toxkit tests with those of the standard assays. As only two results are available for the Thamnotoxkit and only one for the Phytotoxkit, the statistical evaluations presented below deal only with Algaltoxkit, Daphtoxkit and chronic Rotoxkit data.

In Fig. 13.1 the five acceptable 72 h EC_{50} values of the Algaltoxkit tests performed on the eluate of the INC waste are plotted for the five laboratories (indicated by their code number on the *X*-axis) with horizontal lines for the geometric mean and the ±1 and ±2SD from the mean. The mean 72h EC_{50} for the five data was

Table 13.1 Summary of the number of Toxkit microbiotests and their acceptability

Substrate	Algaltoxkit	Daphtoxkit	Thamnotoxkit	Chronic Rotoxkit	Phytotoxkit
INC	7 (5)	11 (8)	1 (1)	4 (4)	1 (1)
SOI	7 (7)	11 (11)	2 (1)	3 (3)	1 (1)
WOO	7 (6)	11 (11)	2 (1)	3 (3)	1 (1)
Sum	21 (17)	33 (30)	5 (3)	10 (10)	3 (3)

( ) Number of acceptable tests

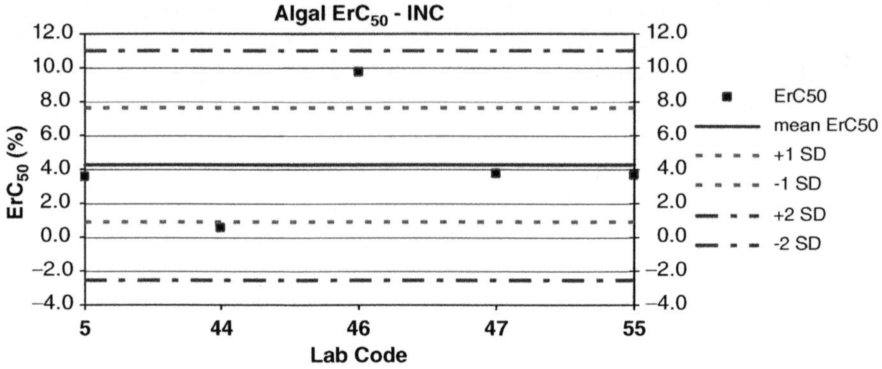

Fig. 13.1 ErC_{50} values (72 h) for the Algaltoxkit tests performed on the eluate of the INC waste

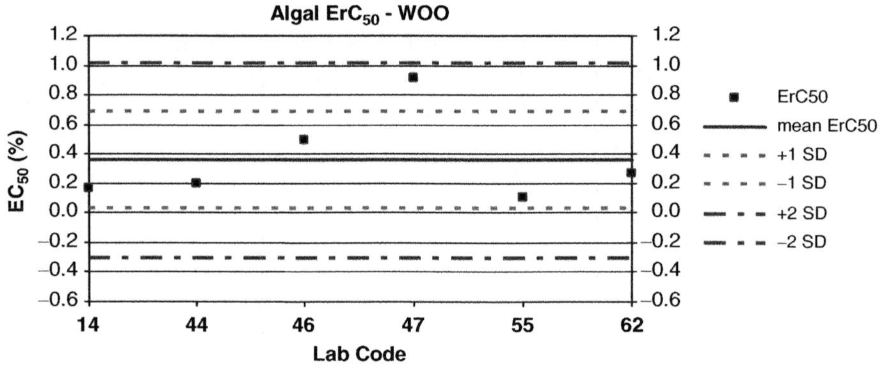

Fig. 13.2 ErC_{50} values (72 h) for the Algaltoxkit tests performed on the eluate of the WOO waste

4.24%, with a SD of 3.37 and a variation coefficient of 79.41%. Figure 13.1 also shows that all the data points are located within the "warning limits" of 2SD and that three EC_{50}s fall within 1SD from the mean value.

The results for the six WOO waste eluate Algaltoxkit results are given in Fig. 13.2. The mean EC_{50} was 0.36% with a SD of 0.33 and a variation coefficient of 92.80%. All six results are within the "warning limits" and five of them are within 1SD from the mean.

In the case of SOI waste no EC_{50} values could be calculated since the percentage effect was below 50% at the highest test concentration.

Figure 13.3 shows the results for the 24 h Daphtoxkit tests for the INC eluate.

The mean for the EC_{50} eight data was 2.41% with a SD of 0.99 and a variation coefficient of 41.09%. All Daphtoxkit results fell within the warning limits, and six

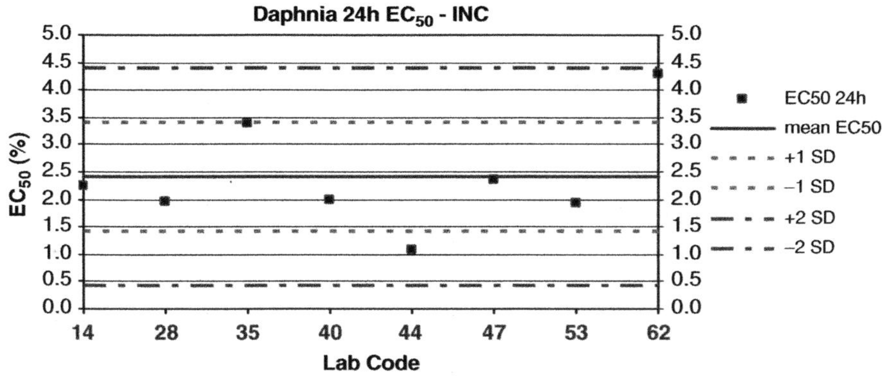

Fig. 13.3 EC_{50} values (24 h) for the Daphtoxkit tests performed on the eluate of the INC waste

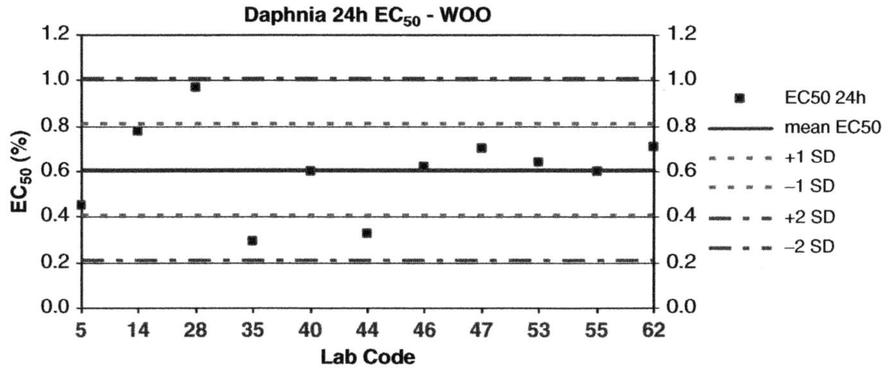

Fig. 13.4 EC_{50} values (24 h) for the Daphtoxkit tests performed on the eluate of the WOO waste

were less than 1SD from the mean. The testing performed with the WOO waste eluate (Fig. 13.4) resulted in a mean EC_{50} of 0.61%, with a SD of 0.20 and a variation coefficient of 32.23%. All the EC_{50}s were within ±2SD from the mean value and 9 were within 1SD of the mean.

As in the case of the Algaltoxkit test with the SOI waste eluate, EC_{50} determination was not applicable for Daphtoxkit results as the toxicity at the highest test concentration was below 50%. The 48 h EC_{50}s for the four chronic Rotoxkit microbiotests on the INC eluate given in Fig. 13.5 corresponded to a mean value of 5.35%; the associated SD was 0.88 and the variation coefficient applicable was 16.49%. It is evident from the figure that all four results are within the warning limits and that three of them are situated within 1SD from the mean.

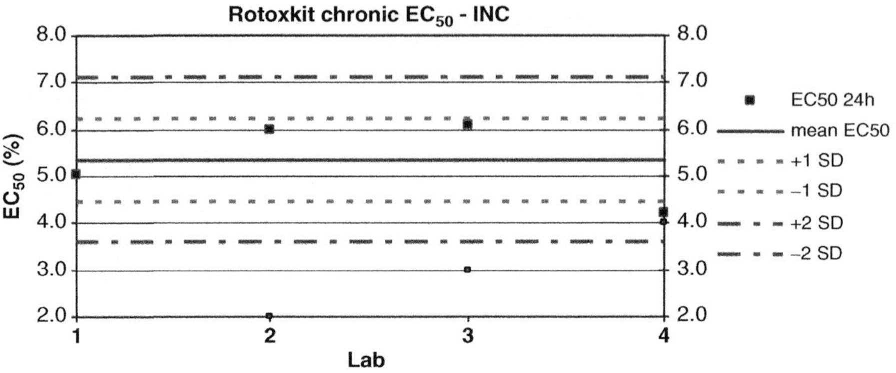

Fig. 13.5 EC_{50} values (48 h) for the Rotoxkit tests performed on the eluate of the INC waste

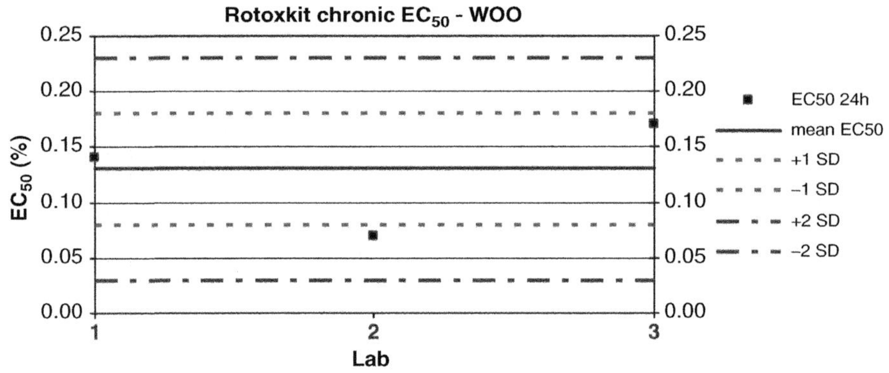

Fig. 13.6 EC_{50} values (48 h) for the Rotoxkit tests performed on the eluate of the WOO waste

Similar to the Algaltoxkit and Daphtoxkit results, no Rotoxkit EC_{50} value could be calculated for the SOI waste, due to the low toxicity of the eluate. The mean value for the three chronic Rotoxkit results for the WOO eluate shown in Fig. 13.6 was 0.13%. The corresponding SD was 0.05 and the variation coefficient was 40.51%. All three EC_{50}s were within the warning limits, and two were within 1SD from the mean.

The EC_{50} values of the only (acceptable) Thamnotoxkit for the INC, SOI and WOO eluates were 13.4%, >100% and 0.31% respectively. The 96 h Phytotoxkit EC_{50} values for the root length with the INC waste were 50% for *B. rapa* and 24.23% for *A. sativa*. In the case of SOI, EC_{50} values were >100%, and 4.38% for *B. rapa* and 32.9% for *A. sativa* were the EC_{50} results obtained for the WOO waste.

13.4 Discussion

Assessment from the methodological point of view indicated that no specific problems were reported by the laboratories that performed tests with the Toxkits. As reported in Sect. 13.3, the overall level of acceptability (87%) of the microbiotests was satisfactory. With respect to the specific Toxkit microbiotests, the proportion of tests fulfilling the acceptability criteria was 85% for the Algaltoxkit and 90% for the Daphtoxkit. All the assays (3) performed with the chronic Rotoxkit were acceptable. The number of tests (2) performed with the other Toxkit assays was too low to draw a meaningful conclusion.

The min–max factor (difference between lowest and highest EC_{50} values) for the different microbiotests was found to exhibit considerable variation, depending on the specific Toxkit: it was 18 for the Algaltoxkit INC EC_{50}s compared with eight for the Algaltoxkit WOO data. In contrast, the Daphtoxkit INC and WOO EC_{50}s differences between the highest and the lowest values were 4 and 3 respectively. In the case of the chronic Rotoxkit data, the factor was <2 for the INC waste and 2.4 for the WOO waste. With the exception of the Algaltoxkit, all the other min–max ranges are thus within the range recommended by Environment Canada for inter-laboratory comparisons (Chapman 1995; Environment Canada 1999).

As for the variability of the test results, all EC_{50} data for the Algaltoxkit, Daphoxkit and chronic Rotoxkit were within ±2SD and the majority of the EC_{50} values were within 1SD from the mean. The variation coefficients for the Algaltoxkit and Daphtoxkit data were, however, relatively high: 79–92% for the algal tests and 32–41% for the Daphnia assays. The chronic Rotoxkit assay CVs were in the 16–40% range. Most of these CV values were thus higher than the 30% coefficient of variance for ring tests indicated by Environment Canada (EC 2005) as the validity criterion for individual test runs. However, as emphasised in other chapters in this volume, it is noteworthy that the 30% CV value has its origins in tests performed with "pure chemicals" and should therefore not strictly apply to "complex" substances such as wastes (or waste eluates).

The validity of the Toxkit microbiotests in terms of acceptable interlaboratory variability and "robustness" has indeed been sufficiently confirmed in national and international ring tests performed in recent years with both the Algaltoxkit and the Daphtoxkit microbiotests. In 2006, for example, an International Intercalibration Exercise on the Algaltoxkit was organised by the Laboratory for Environmental Toxicology and Aquatic Ecology of Ghent University in Belgium, along with 28 laboratories from 14 countries, which generated Algaltoxkit data for the reference chemical potassium dichromate (Persoone and Janssen 2006). The outcome of this ring test eventually revealed a variation coefficient of 31%, marginally above the 30% limit set by Environment Canada for ring tests.

In 2003, a ring test on two "simulated" waste waters was organised in Slovenia by the National Institute of Chemistry on the acute *Daphnia* assay, in which nine laboratories performed Daphtoxkit assays (Venturan and Ros 2003). The CVs for both waste water samples were 11 and 14% (substantially lower than the maximum variability indicated by Environment Canada (EC 2005)). In 2003, a similar ring test on the acute *Daphnia* assay was organised in Italy by the Italian Agency for Environmental Protection, using the reference chemical potassium dichromate. 55 laboratories submitted results for Daphtoxkit assays, generating a CV of 20% (Baudo et al. 2004). This Italian interlaboratory exercise was repeated in 2005 using potassium dichromate and potassium chloride as test compounds. Forty Daphtoxkit data were conveyed for potassium dichromate with a CV of 24%. The CV for potassium chloride for the 38 Daphtoxkit results reported was 16% (Baudo et al. in press). The (relatively) high variation coefficients found in several cases with Toxkit tests performed in the ring test reported here on wastes is discussed in more detail in Chap. 23. The chapter also provides a comparative perspective on CVs of the standard tests and Toxkits for both (*Pseudokirchenriella subcapitata* and *D. magna)* test species. The matter of the sensitivity of the Toxkits to the three types of wastes is also dealt with in more detail in that chapter.

13.5 Recommendations

Considering that the Toxkit microbiotests are not dependent on the culturing/maintenance of live stocks of test species, they have substantial (practical) advantages in comparison with "standard" bioassays. This is essentially the reason for these alternative tests to the standard assays to be employed worldwide for a variety of applications. Three (Algaltoxkit, Daphtoxkit and chronic Rotoxkit) of the five Toxkit microbiotests evaluated in the ring test on wastes follow the test procedures of the standard tests with the corresponding test species. The high coefficients of variation found in a number of cases may be attributed to the nature of the material tested in the ring test; and sample homogeneity may also have had a bearing on the outcome. The high coefficients, falsely reflecting a poor degree of standardisation of the microbiotests, cannot be attributed directly to the test performance per se.

It is thus recommended that the Daphtoxkit and the Algaltoxkit microbiotests be taken into consideration and allowed as "practical, robust and low cost" alternatives to the standard algal and *Daphnia* tests for the evaluation of the toxicity of solid wastes. It would also be prudent to consider the utilisation of other innovative developments such as the Microbial Assay for Risk Assessment (MARA) (Wadhia et al 2007; Blaise et al 2008) to facilitate the toxicity evaluation and classification of wastes.

Chapter 14
Pseudomonas putida Growth Inhibition Test

C. Hafner

Abstract The soil bacterium *Pseudomonas putida* is ubiquitous in soils and surface waters. It has been applied in the growth inhibition test according to ISO guideline 10712 to evaluate the toxicity of the three waste eluates INC, SOI and WOO. Four of the 64 participants in the ring test performed this test, three of them applied the miniaturized version using 96-well microplates (MP) with a volume of 200 µlL test suspension per well. One participant applied the standard version using 100 µlL flasks with a test suspension volume of 50 µlL. The EC_{50} for the reference substance 3,5-Dichlorophenole was determined with 22.5 - 23.6 mg/l L in the valid tests. For INC the EC_{50} was determined with 6.6 - 16.4% dilution (mean: 11.7% dilution). No effect was observed with SOI by any lab. For WOO the EC_{50} was determined with 0.12 - 0.17% dilution (mean: 0.14% dilution). There was no significant difference in the results between the standard version and the miniaturized version using MPs. In comparison with other aquatic test systems the *Pseudomonas* growth inhibition test showed a medium sensitivity: in tests with *Daphnia magna* lower EC_{50} values were determined. On the other hand, Algae and *Lemna* seemed to react more or less similarly.

Keywords Bacteria, Waste, Ring test, PAH, Copper

14.1 Introduction

The *Pseudomonas putida* growth inhibition test was performed according to ISO guideline 10712 (ISO 1995). The test is suitable for determining the inhibitory effects of surface water, waste water or water soluble substances on the soil

---

C. Hafner
Hydrotox - Labor fuer Oekotoxikologie und Gewaesserschutz GmbH,
Boetzinger Strasse 29, Freiburg i.Br., 79111 Germany

H. Moser and J. Römbke (eds.), *Ecotoxicological Characterization of Waste*,
DOI: 10.1007/978-0-387-88959-7_14,

bacterium *Pseudomonas putida*. It is also suitable for evaluating the ecotoxicological effects of soil, especially in abandoned, hazardous sites, compost and waste. In these cases aqueous eluates are applied. Limitations are posed by dark-coloured or turbid samples because the bacterial growth is analysed photometrically.

Pseudomonas putida is a gram-negative rod-shaped saprophytic soil bacterium with polar flagellum that can move freely. With a length of 2 - 4 μm it is ubiquitous in soils and surface waters. Its optimum temperature is between 25 and 30°C. It demonstrates very diverse metabolism, including the ability to degrade organic solvents such as toluene or phenole (Marques & Ramos 1993; Bayly & Wigmore 1973). The diverse metabolism of *P. putida* may be exploited for bioremediation; for example, it is used as a soil inoculant to remedy naphthalene-contaminated soils (Gomes et al. 2005) and to biodegrade oil.

14.2 Method

The basis of the test is that the bacteria grow in the test solution (waste eluates), to which 20% concentrated growth medium has been added. After an exposure time of 16 hours bacterial growth is determined by measuring the optical density (OD) in a photometer at a wavelength of 436 nm. The inhibition of bacterial growth is calculated by comparing the OD of the test solutions with control samples prepared with diluted growth medium. The details are described in ISO 10712 (ISO 1995). For an optimum standardization and to eliminate the device-specific divergency of measurement, the photometer used has to be calibrated using Formazine Nephelometric Units (FNU). The procedure follows ISO 7027 standard (ISO 1999). A well-defined dilution series between 50 and 200 FNUs is prepared and measured in the photometer used for the test at a wavelength of 436 nm. The FNU and OD are correlated, whereby only extinctions between 0.1 and 0.4 are allowed and the curve should cross the origin. In the test a defined bacterial starting concentration of 50 FNU is inoculated. The OD measured at the end of the test (at 436 nm) is converted into FNU again.

The bacteria to be used can be purchased from common stock collections such as the DSMZ - Deutsche Sammlung von Mikroorganismen und Zellkulturen GmbH, Inhoffenstrasse 7 B, 38124 Braunschweig, Germany (DSM 50026). The lyophylized pellet is dissolved in growth medium and cultivated for at least 6 hours. It is recommended to pre-cultivate the bacterial suspension for at least 24 hours before starting the starter culture because otherwise the growth rate of 60 fold, which is required as a validity criterion in the ISO guideline, will be difficult to achieve. This applies to suspensions from self-made working cultures as well. The following flow-chart presents the optimized test procedure.

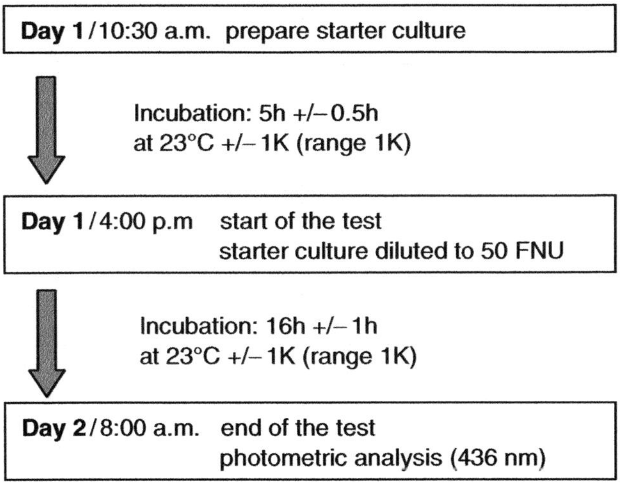

Typically the exposure of the test suspension is performed in a volume of 25 µlL in 100 µlL wide-necked Erlenmeyer flasks incubated in a temperature controlled water bath. Each concentration is tested in triplicate and the negative control is performed in four replicates. At the end of the test a sample is taken from each test and control vessel and analysed photometrically.

The miniaturized version is usually performed in 96-well microplates (MP) with a volume of 200 µlL test suspension per well. Each concentration is tested in triplicate which means that 4 samples with 6 concentrations per plate can be investigated on one plate. The plates are incubated in temperature controlled cabinets and shaken orbitally. The optical density is determined at the end of the incubation period in a microplate reader (Gellert et al. 1996; Schmitz et al. 1998; Schmitz et al. 1999; Eisentraeger et al. 2007). In the ring test 3 of the 4 laboratories applied the test in its miniaturized microplate version.

14.3 Results

Four laboratories of the 60 participants carried out the *Pseudomonas* test. One lab did not achieve the required cell multiplication of 60-fold within 16 hours. Therefore these results could not be accepted for the ring test. Another lab did not provide data concerning reference substances, but these results are presented in the following tables (Tables 14.1–14.3.

One lab performed the test in the standard version with a test volume of 25 µlL, three labs performed the test in its miniaturized version using MP. The tests not accepted were all performed using the MP version.

Table 14.1 Summary of the acceptance of *Pseudomonas* tests

Substrate	No. labs	No. tests	Invalid tests	Accepted tests
INC	4	4	1	3
SOI	4	4	1	3
WOO	4	4	1	3
Sum	4	12	3	9

Table 14.2 Results of reference substance testing

Lab Code	Reference substance	Test version	EC_{50} mg/lL	CI [(95%)]	Calculation method	Statistical software
1	3,5-DCP	MP	22.5	n.a.	Probit analysis	n.a.
16	3,5-DCP	MP	7.4	n.a.	Logistic regression	n.a.
23	3,5-DCP	STD	23.6	22.6 - 24.6	Probit analysis	ToxRat

3,5-DCP: 3,5-Dichlorophenole;, *MP*: Microplate version;, *STD*: Standard version;, *n.a.* Data not available

Table 14.3 Results of waste eluates testing

Lab code	Test version	INC EC_{50} [(% dilut)]	SOI EC_{50} [(% dilutdilute)]	WOO EC_{50} [(% dilut)]	Statistical software
1	MP	6.6	>50	0.17	Probit analysis
		0.4 - 119	n.d.	0.1 - 0.3	n.a.
16	MP	*46.1*	*>80*	*0.19*	Logistic
		30.4 - 63.5	n.d.	0.13 - 0.21	Regression, n.a.
23	STD	12.2/(>80)	>80	0.12	Probit analysis
		n.d.	n.d.	0.12 - 0.12	ToxRat
37	MP	16.4	>80	0.14	Logit
		10.6 - 30.1	n.d.	0.13 - 0.14	ToxCalc
Mean	–	11.7	>72.5	0.14	–

Results of lab 16 invalid due to low multiplication of cells (data given in italics), results of lab 37 not accepted due to lack of reference substance testing

The ISO guideline claims an EC_{50} for the reference substance 3,5-Dichlorophenole within a range of 10 - 30 mg/lL. In a ring test with 21 participants performed in 1989 during the implementation of the ISO guideline an EC_{50} of 21.4 mg/lL was found. The coefficient of variation was found to be 23%. As presented in Table 14.2 one lab failed the EC_{50} for the reference substance. It also failed the validity criteria for sufficient multiplication, which explains the invalid results for the reference substance's EC_{50}. The two other labs, with 22.5 mg/lL and 23.6 mg/lL, respectively, were well within the range, regardless of the test version, MP or standard version used. The detailed results of EC_{50} determination of the reference substance 3,5-dichlorophenole are given in Table 14.2.

Results of waste eluate testing are given in Table 14.3. With INC, three labs found quite similar EC_{50} between 6.6% and 16.4% eluate proportion. One lab did not fulfil the validity criteria and the reported test result was much less sensitive. one lab repeated the test with another eluate because the pH decreased considerably in some dilutions in the first test. With this second eluate no effects could be measured (EC_{50} > 80% eluate proportion, results not presented in detail) but the pH did not change in this experiment.

With the eluate of SOI all 4 four labs found no effect with the highest concentration testable. With the eluate of WOO, all 4 four labs found quite high toxicity and the results of EC_{50} were between 0.12% and 0.19% eluate proportion. The detailed results of lab 23 are presented in Table 14.4 and Fig. 14.1. The eluate in this test was about 140-hours old and did not conform to the SOP provided, which does not allow storing eluates for longer than 72 hours. But because of the strong toxic effect, three tests had to be run.

Table 14.4 Detailed results for WOO achieved by lab 23

		NC	D256	D512	D1024	D2048	D4096	D8192
	Repl.	0%	0.39%	0.19%	0.097%	0.049%	0.024%	0.012%
FNU 0h		7						
FNU 16 h	a	424	7	26	326	426	475	437
	b	423	5	21	344	438	446	432
	c	419	6	25	361	427	444	441
	d	406						
Mean FNU		418	6	24	344	430	455	437
Mean inhibition [(%])		–	99	94	18	0	0	0
pH 0 h		7.0	7.0	7.0	7.0	7.0	7.0	7.0
pH 16 h		6.9	6.8	6.6	6.7	6.7	6.7	6.8

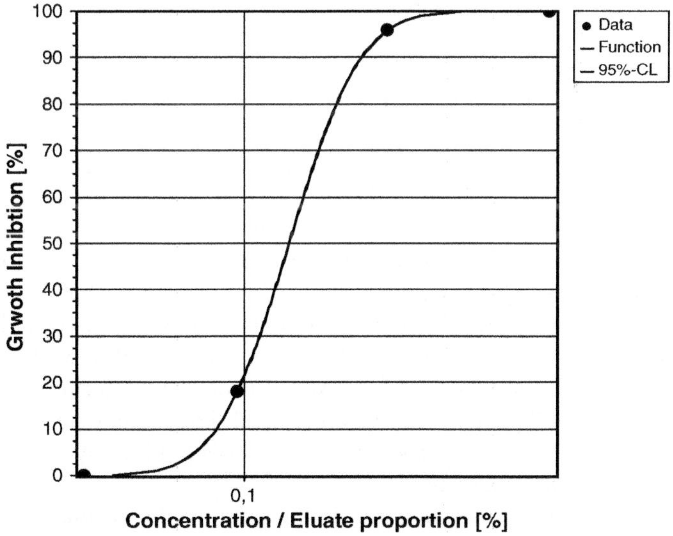

Fig. 14.1 Dose-response curve for WOO achieved by lab 23

14.4 Discussion

In the ring test, two different versions of the *Pseudomonas* growth inhibition test were applied. One lab performed the standard version, incubating a volume of 25 µlL in 100 µlL flasks. Three labs applied the miniaturized test version using MP. Most important, both test versions produced comparable results (relating to the valid tests). This is true for all three eluates and shows the high standardization and good reproducibility of the method even though difficult samples like waste eluates were tested.

The test seems to be especially suitable for the testing of wastes, and soil samples, particularly with regard to abandoned hazardous sites and compost, because the bacteria applied are of specific ecological relevance. The test is quite easy to perform and the demands on laboratory equipment are low as far as the standard version is concerned. The MP version allows a high sample throughput and reduces the costs considerably.

In comparison with other aquatic test systems, the *Pseudomonas* growth inhibition test showed medium sensitivity; in tests with *Daphnia magna,* lower EC_{50} values were determined. On the other hand, Algae as well as *Lemna* reacted more or less similarly.

The *Pseudomonas* growth inhibition test was first described more than 30 years ago (Brinkmann & Kuehn 1977). Further development of the test for waste water evaluation ceased during the early 1980s because other tests seemed to be more promising (Knie 1992). For bleaching effluents the test is described to be less sensitive than the luminescent bacteria inhibition test or the Algae growth inhibition test (Ahtiainen et al. 1994). Gellert et al. (1996) found that the MP version was less sensitive for organic compounds and waste water compared to the standard version performed in glass bottles. It was assumed that this can be attributed to the adsorptive effects of lipophilic compounds to the plastic material of the MP. This finding could not be confirmed in the ring test results. For testing the potential toxicity of pharmaceuticals in sewage treatment plants, the *Pseudomonas* test is evaluated as less suitable because *Pseudomonas* is no model organism for activated sludge (Kuemmerer & Alexy 2006).

Anyway the test is described in the OSPAR background document concerning Whole Effluent Assessment (WEA) (OSPAR 2000). Recently several labs have applied the *Pseudomonas* test to evaluate soil and compost samples (Prokop et al. 2003; Juvonen et al. 2000), pharmaceuticals (Al-Ahmad et al. 1999; Kuemmerer et al. 2000; Zounková et al. 2007; Suetterlin et al. 2007) and other chemical substances in the environment (Xu et al. 2005). In addition, the test is recommended especially for the examination of abandoned hazardous sites (LFUG 1995), while it is pointed out that problems might arise by the growing of unwanted germs from unsterile samples, mimicing cell growth. On the other hand, infections with bacteriophages could mimic growth inhibition. The decrease of susceptibility by genetic variation has also to be taken into account.

The application of the *Pseudomonas* test system for the on-line surveillance of surface waters has been described before (Fritz-Langen et al. 1994). Here cell growth has been measured by oxygen consumption of the bacteria. Slabbert & Grabow (2006) found that the *P. putida* growth inhibition test is more sensitive than the oxygen uptake assay. In recent years there have been attempts to develop ready-to-use test kits with lyophylized bacteria (Mallau et al. 2001). Spiller (2006) describes an automated test system using a 1,536 well MP with miniaturized conductivity to measure cells for high throughput testing. Other novel developments are genetically engineered bioluminescent *Pseudomonas* strains for testing heavy metal toxicity (Ren & Frymier 2003; Ren & Frymier 2005). But so far to the author's knowledge no commercially available ready-to-use test kit exists.

It can be concluded that the *Pseudomonas putida* growth inhibition test, especially the miniaturized version, is a suitable means to assess environmental hazards and test wastes. In the ring test the results were promising, but the low number of samples and the impairments have to be taken into account. Improving the result analysis of the standard method by calculating the growth rate could help to increase the acceptance of this assay (Schmitz et al. 1998).

Chapter 15
Ceriodaphnia dubia Chronic Toxicity Tests

C. Bazin, P. Pandard, A.-M. Charissou, and Y. Barthel

Abstract Crustaceans are considered an important part of the soft water zooplankton community. Therefore they represent water organisms in the test set for the ecotoxicological characterization of wastes. Following ISO/CD guideline 20665, the chronic *Ceriodaphnia dubia* test was performed with three waste types: INC, SOI, and WOO. All the ten tests performed were classified as acceptable. Methodologically, almost no problems occurred. The min-max factors (2.08–2.54) between individual EC_{50} values and the CVs (31–33%) were low for waste eluates, showing that the chronic *Ceriodaphnia dubia* test is a robust method. These results are in agreement with the few data on substances known from the literature. The obtained data show that the test is able to discriminate different toxicity levels: INC = 4.8%, SOI > 90%, WOO = 0.08%. The sensitivity of the *Ceriodaphnia* chronic test is equivalent to the rotifer chronic test and slightly higher than the Algae test for the wood eluate. Therefore it is recommended for the toxicity assessment of waste.

Keywords *Ceriodaphnia dubia*, Chronic test, Reproduction, Ring test, Eluate, Waste

15.1 Introduction

Cladocera *Ceriodaphnia dubia* is considered representative of the zooplankton species and widely used in North America (USA & Canada) and France for measuring the chronic toxicity of waste waters (Costan et al. 1993; Vindimian et al. 1999). The shortness of this chronic toxicity test (7 days) and the use of relatively low volumes compared with usual chronic tests are major assets for obtaining results on samples that may be subject to changes during the storage period. This test is currently under standardization in ISO/TC 147. Recently, the *C. dubia* chronic test was selected as a sensitive and representative method to assess the ecotoxicological hazard potential of

---

C. Bazin (✉)
INSAVALOR - POLDEN, BP 521321, 69601, Villeurbanne cedex, France

H. Moser and J. Römbke (eds.), *Ecotoxicological Characterization of Waste*,
DOI: 10.1007/978-0-387-88959-7_15,

wastes to aquatic invertebrates. In addition, Pandard et al. (2006) retained *C. dubia* in their minimum test set, on the basis of the results obtained on 40 wastes belonging to 27 waste types.

15.2 Method

The *C.* dubia chronic toxicity test was performed according to ISO/CD 20665 (ISO 2006) guidelines. As this test was not included in the minimal set, no SOP was established. Therefore, this test is briefly described here.

According to EN 14735, young *C. dubia*, aged less than 24 h at the start of the test, are exposed individually, under semi-static conditions, to dilutions of the waste eluates obtained. The *C. dubia* used for the test should be taken from a brood comprising at least eight new-born animals. The reproduction and mortality of the adult females are monitored at least five times during exposure. The test ends typically on day 7 when 60% of the adult females in the control solutions have produced young during their third brood. The exposure time can be prolonged to 8 days if this criterion is not satisfied. Mineral water (Evian®) was used as control and for diluting water. The tests were performed under controlled conditions: temperature 25 ± 2°C, diffuse light (<300 lux, photoperiod 16 h light/8 h dark), and no pH adjustment of the test solutions.

To allow growth and reproduction, the organisms have to be fed during the test. The food is composed of commercial fish foods, *Chlorella vulgaris* Algae and *Pseudokirchneriella subcapitata* Algae. The following constituents should be added to each test solution before the transfer of organisms (a)12×10^6 cells per litre of *C. vulgaris* and (b) 6×10^6 cells per litre of *P. subcapitata*; 500 µL per litre of the fish food suspension. The health of the culture of *C. dubia* and the sensitivity of the test organisms were checked using sodium pentachlorophenolate (NaPCP). EC_{50} were in the range of 0.253–0.400 mg/L. These values are consistent with the results obtained during the European ring test performed in 1999.

The test is considered valid if the following conditions are met in the control solutions (a) the mean mortality rate of the adult females at the end of the test does not exceed 20%; (b) the proportion of adult males does not exceed 10%; (c) 60%, or more, of the adult females produce three broods by the end of the eighth day (or earlier); (d) the average number of offspring born per alive adult female at the end of the test is greater than or equal to 15.

15.3 Results

Four laboratories performed the *C. dubia* reproduction test. The participating laboratories provided ten data sets for the three eluates (Table 15.1). The small number of results did not allow an overall statistical assessment as described in Chap 4.

Table 15.1 Summary of the acceptance of *Ceriodaphnia dubia* chronic toxicity test

Substrate	No. labs	No. tests	Invalid	No reference	Accepted
INC	4	4	0	0	4
SOI	3	3	0	0	3
WOO	3	3	0	0	3
Sum	n.a.	10	0	0	10

n.a. Not applicable

Table 15.2 Summary of the results of the *Ceriodaphnia dubia* chronic toxicity test and the three waste types

Substrate	EC_{50}	Min value	Max value	CV (%)	Factor Min-Max
INC	4.78	2.4	6.1	31.3	2.54
SOI	n.a.	n.a	n.a	n.a	n.a.
WOO	0.08	0.048	0.100	33.1	2.08

Data expressed as % waste eluate: mean EC_{50} values, min and max values, CV and the factor between minimum and maximum EC_{50} values

The results of the *C. dubia* chronic toxicity tests with INC, SOI and WOO are presented (see Table 15.2).The mean EC_{50} value of the four results for INC is 4.78%. The maximum and minimum EC_{50} values differed by a factor of 2.54 (min: 2.4%; max: 6.1%). The inter-laboratories C.V. for INC is 31.3%.

No effect on reproduction was observed in the tests with SOI. Consequently, no EC_{50} could be calculated. For this waste, the EC_{50} is given as >90% SOI, the highest percentage of eluate tested. These results were consistent with those obtained with the other aquatic toxicity tests.

The mean *E* value of the three results with WOO is 0.08%. The maximum and minimum EC_{50} values differed by a factor of 2.08 (min: 0.048%; max: 0.100%). The inter-laboratories C.V. for WOO is 33.1%.

The results of the *C. dubia* toxicity test are summarized in Table 15.2.

15.4 Discussion

From a methodological point of view, no significant problems were reported during the ring test. For the contaminated wood eluate, the overall results showed that the *C. dubia* was the most sensitive species. On the other hand, for INC, the *C. dubia* chronic toxicity test was slightly less sensitive than other aquatic tests (algal growth inhibition test, *daphnia* mobility test, and *Brachionus calyciflorus* reproduction test) except the inhibition of light emission of *Vibrio fischeri*.

The number of data sets is limited. Nevertheless, these results (min-max factors; CVs (reproducibility)) are consistent with those obtained with chemicals in Europe and Canada (NaPCP, $CuSO_4$, NaCl, $ZnSO_4$) (ISO 2006). However, for waste eluates, the variability of the results presented in this chapter integrates not only the variability of the ecotoxicity test for chemicals, but also the variability of the entire procedure: sampling of the test portion, eluate preparation (leaching procedure, separation of the solid and liquid phases) and ecotoxicity test.

So, it can be concluded that the *Ceriodaphnia* chronic test is a robust and convenient method to assess waste toxicity.

15.5 Recommendations

The *C. dubia* test is a rapid, chronic toxicity test, which does not need large volumes compared to the usual chronic tests on crustaceans. The obtained results show good reproducibility for a chronic bioassay. These conclusions, based on a limited number of data sets, allow the inclusion of this assay for toxicity evaluation of waste.

Chapter 16
Genotoxicity Tests

J. Römbke and H. Neumann-Hensel

Abstract Genotoxic substances in waste or waste eluates may be hazardous for aquatic and terrestrial organisms, but may also bear a human toxic risk. Therefore, the water extractable genotoxic potential has to be assessed by using standardized tests. The most widely used genotoxicity test is the umu test (so-called because of its dependence on *umuC* gene induction) (ISO 13829). The genotoxicity of waste materials has rarely been studied and it was not part of the basic test set in the EU waste ring test. The tests reported here were performed voluntarily by six labs. to give an idea of the water-extractable genotoxicity of three different types of waste substrates (INC, SOI and WOO). In addition, information was collected regarding a new ring test for instance concerning the number of participants. No indication of water-extractable genotoxic potential could be observed in concentrations (usually) up to 66.7% eluate. In fact this result was expected, as the (known) main contaminants in the waste samples are mostly heavy metals and PAHs which are not known to show strong genotoxicity. However, it is recommended that such tests be included in the basic test battery at least for the time being, considering the limited amount of data available.

Keywords Umu test, Ring test, Waste, PAH, Copper, Assessment

16.1 Introduction

Genotoxic substances in contaminated soils may be hazardous for soil organisms. As they are exposed to such contaminants, most probably via the soil pore water, the water extractable genotoxic potential has to be assessed by testing water extracts. Therefore, several such test methods have been developed and, partly, standardized in the last few years (Eisentraeger et al. 2005). The most widely used test of genotoxicity is known as the umu test (because of its dependence on *umuC* gene induction) (ISO 13829, 2000). The *Salmonella*/microsome test (Ames test)

J. Römbke (✉) and H.N.-Hense
ECT Oekotoxikologie GmbH, Boettgerstr. 2-14, 65439, Floersheim, Germany

H. Moser and J. Römbke (eds.), *Ecotoxicological Characterization of Waste*, 165
DOI: 10.1007/978-0-387-88959-7_16,

according to DIN 38415 T4 (1998) should be carried out additionally, but only if the umu test is negative and there are strong hints from chemical analysis or site history that mutagenic compounds are present. The approach presented here is a screening method to identify substances that can cause gene mutations and cannot be used to identify clastogenic substances.

So far, the genotoxicity of waste materials has rarely been studied, but for the assessment of substrates like contaminated soils, the performance of such tests is widely accepted (Rila and Eisentraeger 2003; Eisentraeger et al. 2004). For example, the umu-test, and if needed, the Ames-test, is required for the ecotoxicological characterization of soils and soil materials according to ISO guideline 17616 (2006). The determination of the genotoxicity was not part of the basic test battery in the waste ring test, but it is listed in the Annex of EN 14735. The test results reported here were voluntarily performed by six labs. Due to the small number of available results this exercise can be seen as a screening effort. Its main purpose is twofold:

- To give an idea about the water-extractable genotoxicity of three different types of waste substrates (INC, SOI or WOO; for details see Chap. 3).
- To provide information which may be helpful in designing another ring test. On the basis of such detailed information, it can be decided whether the determination of water-extractable genotoxicity should routinely be required for the ecotoxicological characterization of wastes. In any case the performance of the umu-Test should always be accompanied by residue analysis of those contaminants which will most probably be found in the respective sample. Otherwise it will be difficult to evaluate the results.

16.2 Method

As no Standard Operation Procedures (SOPs) were prepared for the additional test systems and the documentation of these tests was less detailed than other tests, the following description focuses on the guideline (ISO 2000) and a description published by one of the participants (Eisentraeger et al. 2005).

The umu-test is performed with *Salmonella choleraesius subsp. chol.* (strain TA1535/pSK1002) (formerly: *Salmonella typhimurium*) in 96-well microplates according to ISO 13829 (2000). Stock cultures are stored at −80°C in 10% dimethylsulfoxide (DMSO). An aliquot of the stock culture is incubated at 37°C for about 2 h in a nutrient solution in a Duran® flask. Bacteria are harvested during logarithmic growth period at a cell density of 10^8/mL and incubated with the sample in a 96-well microplate for 2 h at 37°C on a shaker. After this, an aliquot is diluted tenfold with fresh substrate on a new microplate and incubated for 2 h at 37°C to enrich the bacterial mass. The absorbance at OD_{600} is measured photometrically with a microplate photometer. From the data obtained, the growth factor of the bacteria is calculated. After cell lysis with β-mercaptoethanol the enzyme reaction is initiated by the addition of 2-nitrophenyl-β-D-galactopyranoside solution. The reaction is terminated after

30 min incubation at 28°C by adding 1 M Na_2CO_3. The absorbance at OD_{420} is measured and β-galactosidase units are calculated. The test is performed in parallel without (–S9) and with (+S9) metabolic activation. By adding an extract of rat liver (S9 Mix) it is possible to simulate enzymatic processes of eukaryotic cells, i.e. the activation of promutagenes contained in the samples after their splitting by the enzymatic extract.

The extent of induction of the SOS-repair system, indicated by *umuC-lacZ* gene, which encodes for β-galactosidase, is calculated from the growth factor (GW) and the β-galactosidase units (induction rate = IR). According to the guideline ISO 13829, a sample is supposed to be genotoxic if this induction rate exceeds 1.5. To avoid false positive interpretation of the results caused by cytotoxic effects, the growth factor must exceed 0.5.

As mentioned above, the evaluation of the test results is based on a comparison between a measured induction rate and a trigger value (here: 1.5). Due to this very specific design no EC_{50} values could be calculated, and so no statistical evaluation of the results, as for all other tests, will be presented.

16.3 Results

In all, seven participating labs provided 23 test data sets (Table 16.1). Two of these tests, both from the same lab, were not accessible (the data documentation was not complete); so the final assessment was based on 21 data sets coming from six labs. All these 21 tests fulfilled the validity criteria concerning the performance of the test. In addition, the positive control, i.e. the reference tests with 4-nitroquinoline-N-oxide (4-NQO) and/or aminoanthracene (2-AA), also provided results confirming the validity of the data.

In the following sections, the results of the genotoxicity tests with INC, SOI and WOO are presented, following the same scheme as in the other chapters of Part III. However, no figures will be presented. In contrast, the original data, i.e. the dilution step at which an IR > 1.5 is observed, are presented in Table 16.2. In two cases (Lab code 16) the test +S9 was not performed due to technical difficulties. However, there is no indication that an effect could have occurred.

In almost all tests performed up to an eluate dilution of 33.3% (three tests) or 66.7% (18 tests) the trigger value for the IR (=1.5) was not breached. Only in one test with WOO waste and without metabolic activation, an IR of 1.8 was determined at

Table 16.1 Summary of the acceptance of genotoxicity tests

Substrate	No. Labs	No. tests	Invalid tests	Accepted tests
INC	5	7	0	7
SOI	6	7	1	6
WOO	6	9	1	8
Sum	n.a.	23	2	21

n.a. Not applicable

Table 16.2 2 Compilation of the results of the 21 genotoxicity tests with three wastes; always the dilution step at which – with or without activation – an IR > 1.5 was observed

Lab Code substrate	IR > 1.5 ? (– S9) (% dilution)		IR > 1.5 ? (+ S9) (% dilution)		Remark
INC					
1	>	66.7	>	66.7	
12a	>	66.7	>	66.7	
12b	>	66.7	>	66.7	
16	>	66.7		n.d.	+S9 not tested
23	>	33.3	>	33.3	
33a	>	66.7	>	66.7	
33b	>	66.7	>	66.7	
SOI					
1	>	66.7	>	66.7	
12a	>	66.7	>	66.7	
12b	>	66.7	>	66.7	
16	>	66.7		n.d.	+S9 not tested
23	>	33.3	>	33.3	
33	>	66.7	>	66.7	
WOO					
1		66.7	>	66.7	IR = 1.8 in the test –S9
12a	>	66.7	>	66.7	
12b	>	66.7	>	66.7	
12c	>	66.7	>	66.7	
16	>	66.7		n.d.	+S9 not tested
23	>	33.3	>	33.3	
33a	>	66.7	>	66.7	Cytotoxicity at 66.7%
33b	>	66.7	>	66.7	Cytotoxicity at 66.7%

a concentration of 66.7%. Despite this one outlier it can be stated that according to these tests, there is no genotoxic potential identifiable in the waste samples tested. Except the one case already mentioned, no difference between the three waste substrates was observed. However, one laboratory (lab code 33) reported that in two tests with the substrate WOO cytotoxicity occurred at the highest concentration tested (66.7%).

16.4 Discussion

According to the preliminary results presented here, no indication of water-extractable genotoxic potential could be observed when testing the three waste types INC, SOI and WOO in concentrations (usually) up to 66.7% eluate in the umu-genotox test. Therefore, no further assessment, for example concerning the comparison between the three waste substrates, is necessary or useful. In fact this result was expected, since the (known) main contaminants of the waste samples are mostly heavy metals. These substances are not known to show strong genotoxicity. In the literature data also no genotoxicity was found, either in ash samples, the properties of which resemble those of the INC substrate (Römbke & Moser 2007) or in soil samples contaminated with PAHs (Brinkmann & Eisentraeger 2008).

16.5 Recommendations

Despite the fact that no genotoxicity could be found in the umu-tests performed as part of the additional test set in the European waste ring test, it is recommended that such tests be included in the basic test set, at least for the time being. The recommendation is based on the following considerations:

- The available amount of data seem to indicate that wastes do not have a genotoxic potential. However, with just a few tests from three waste substrates not containing substances known to be genotoxic, it is clearly too early to skip this kind of testing.
- Further research with more waste substrates, covering the whole range of possible contaminations, is necessary to provide a robust answer concerning the question whether the umu-test is necessary for the routine characterization of the ecotoxicological hazard of wastes.

Chapter 17
Earthworm Reproduction Tests

F. Riepert, J. Römbke, and T. Moser

Abstract Earthworms are considered an important part of the soil community. The acute earthworm test is part of the basic test set for wastes, but the earthworm reproduction test is considered ecologically more relevant; however, its duration is much longer (56 instead of 14 days). Therefore, only 17 chronic tests were performed with the three waste types INC, SOI, and WOO (12 of them were classified as acceptable). Methodologically, almost no problems occurred but further guidelines on determining the moisture of the test substrate mixtures is needed. The toxicity of the three waste types differed considerably as demonstrated by the final EC_{50} values: INC = 16.1%, SOI => 50%, WOO = 4.1%. It should be kept in mind that these results are based on only four tests per substrate. Compared to other plant and invertebrate tests with solid wastes, the sensitivity of the earthworm reproduction test is similar to the Collembolan reproduction test. Therefore, it is worth taking into consideration, whether this chronic test could be an alternative to the rather insensitive earthworm acute test. In addition, an alternative should be selected for the currently used reference substance Carbendazim. Reference substances (e.g. boric acid) that might replace pesticides are under discussion in the corresponding working groups of the standardization bodies.

Keywords *Eisenia fetida/Eisenia andrei*, reproduction test, Ringtest, Waste

17.1 Introduction

Earthworms, whose important ecological role has already been described in Chap. 11, are used as representatives of the soil fauna in many fields of application, including the characterization of intrinsic properties of chemicals (Edwards & Bohlen 1996).

F. Riepert (✉), J. Römbke, and T. Moser
Julius Kuehn-Institut, Bundesforschungsinstitut fuer Kulturpflanzen
Koenigin-Luise-Str. 19, 14195, Berlin, Germany

H. Moser and J. Römbke (eds.), *Ecotoxicological Characterization of Waste*,
DOI: 10.1007/978-0-387-88959-7_17, © Springer Science+Business Media, LLC 2009

In general, soil fauna tests have seldom been applied for characterizing wastes (Wilke et al. 2007), even though they are routinely used in chemical testing and soil quality assessment (Spurgeon et al. 2003; EC 2004). The laboratory test on acute toxicity as published by OECD and ISO (OECD 1984; ISO 1993) has been proposed for being part of a test battery. Mortality is the main endpoint in the acute test but biomass change also has to be recorded and is used as an indicator of sub-lethal toxicity. Due to obvious shortcomings with respect to sub-lethal effects, a test on reproduction toxicity was developed, resulting in a BBA-Guideline addressing the requirements of the evaluation of side-effects of pesticides (BBA 1994; Riepert & Kula 1996). In addition, a modified version addressing the requirements of soil characterization was standardized shortly thereafter (ISO 1998).

In general, the earthworm reproduction test proved to be much more sensitive than the earthworm acute test when studying pesticides or metals (Schaefer 2003). Therefore, in current regulations, e.g. covering the risk assessment of veterinary pharmaceuticals, the earthworm acute test is skipped and the reproduction test is required (VCH 2005). However, in order to assess the ecotoxicological hazard potential of wastes to terrestrial invertebrates, the earthworm acute test has been selected (EN 2005). Consequently, it was studied as part of the basic test set in the European ring test (see Chap. 11). However, in order to clarify whether the earthworm reproduction test may be an alternative, several participants performed this chronic test as part of the additional test set.

17.2 Method

Ring test participants performed the 56-day earthworm reproduction test according to ISO-Standard No. 11268-2 (ISO 1998) but did not have a SOP. Artificial soil was used as test substrate (mixture of 70% quartz sand, 20% kaolinite clay, 10% peat, calcium carbonate to adjust the pH to 6.0 ± 0.5 and deionised water to achieve a soil moisture of 40–60% of the WHC_{max}). This medium, introduced first in the Earthworms Acute Toxicity Test (OECD 1984) and applied in most of the standardized terrestrial tests, is recommended by EN 14735 for use as dilution medium for testing solid waste materials. The environmental conditions of the test correspond to the acute test with the exception that worms are fed in the chronic test (temperature 20 ± 2°C, light cycle 16/8 h (400–800 lux)). At the start of the test, ten adult worms from a synchronized culture are introduced into each test vessel. The duration of the test period is eight weeks. After four weeks the adult worms are removed from the test vessels and mortality as well as biomass development are recorded. Afterwards, the test vessels are kept for another four weeks under the same test conditions to enable the offspring to hatch from the cocoons placed by the adults during the first four weeks of the test. The main assessment endpoint is the reproduction expressed as number of juvenile worms per test vessel. The results are presented as % waste in artificial soil (dry weight (dw)). The results are used to calculate, if possible, an EC_{50} by using the probit analysis method and the statistical program ToxRat® (2003). In addition, $NOEC_{Biomass}$ (adults) values are determined after 4 weeks. Results are considered valid if the percent mortality of the adults in the control(s) is ≤10%, the mean number of juveniles is at least 30 and the coefficient of variance of reproduction does not exceed 30%. A reference test with the fungicide Carbendazim should be reported. Effects on reproduction should be observed at concentrations between 1 and 5 mg Carbendazim/kg soil dw.

17.3 Results

A total of six participating laboratories provided 17 test data sets (Table 17.1). Two laboratories did not present results from reference tests (for details of the results of the reference tests see Chap. 6). As the data base is restricted, these data sets were excluded for the final assessment.

In the following sections, the results of the earthworm reproduction tests with INC, SOI and WOO are presented. However, in several tests, no definite EC_{50} could be determined, usually because the test substrate (mainly SOI) did not have an effect. To indicate the responses of the sublethal endpoints available in the earthworm reproduction test, EC_{50} values for reproduction are plotted as % waste in artificial soil for the waste substrates INC, SOI and WOO.

With INC, four of the six tests were acceptable (Fig. 17.1), showing EC_{50} values between 8 and 27% (i.e. the results differed by a factor of about 3). In the four accepted tests with the substrate SOI (six tests performed) no effect was found at 50 (once) or 100% (twice) waste (Fig. 17.2). Only in one test an EC_{50} could be determined (56%).

In the four acceptable tests with WOO (five tests performed) a mixed picture emerged (Fig. 17.3): in two tests, the EC_{50} could not be determined and had to be regarded as >6.3% and 8.3% respectively (the lowest dilution tested). In the two other tests EC_{50} values of 3.4% and 4.9% were calculated.

A summary of the results is given in Table 17.2. Despite the low number of tests, it is obvious that WOO was by a factor of about four more toxic than INC, while almost no effect could be determined in the tests with SOI. The min-max factors are very low (1.4–3.2), but these numbers are not very meaningful since they are based on only four and two test results, respectively. However, the effects on reproduction are backed by the impact of these wastes on the other chronic endpoint of this test, the development of adult biomass ($NOEC_B$: INC=63%, SOI=63.0%, WOO=8.8%).

Table 17.1 Summary of the acceptance of earthworm reproduction tests

Substrate	No. labs	No. tests	Valid	Reference	Accepted
INC	6 (5)	6	2	2	4
SOI	6 (5)	6	2	2	4
WOO	5	5	1	2	4
Sum	n.a.	17	5	6	12

n.a. Not applicable

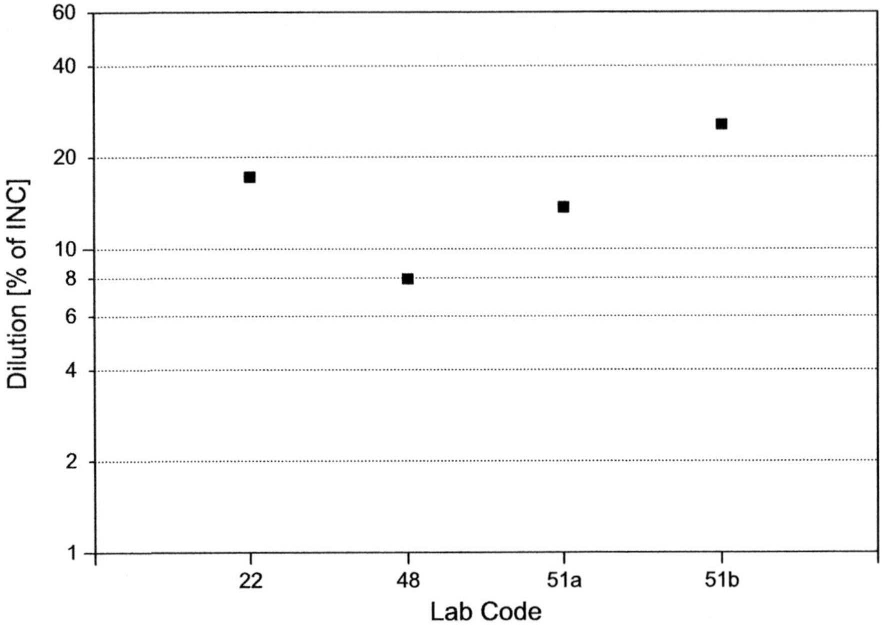

Fig. 17.1 Results (EC_{50} as % waste in artificial soil) of the earthworm reproduction tests with the waste substrate INC

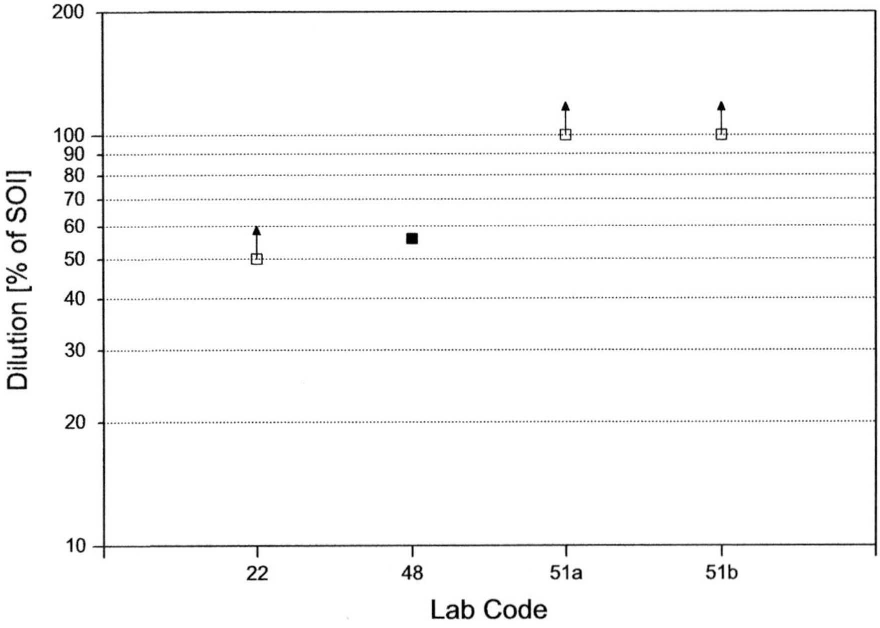

Fig. 17.2 Results (EC_{50} as % waste in artificial soil) of the earthworm reproductiontests with the waste substrate SOI

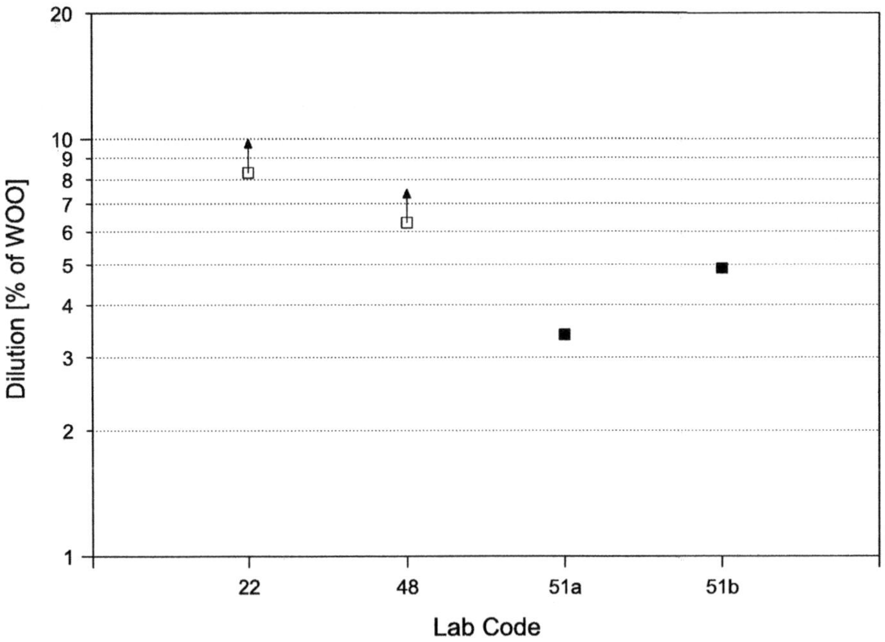

Fig. 17.3 Results (EC_{50} as % waste in artificial soil) of the earthworm reproduction tests with the waste substrate WOO

Table 17.2 Summary of the Earthworm reproduction test results with the three test wastes (data expressed as % of weight of waste in artificial soil)

Substrate	Mean EC_{50}	SD	CV	Factor Min-Max
INC	16.1	n.a.	n.a.	3.2
SOI	n.a.	n.a.	n.a.	n.a.
WOO	4.1	n.a.	n.a.	1.4

Due to the low number of test results no standard deviation or CV has been calculated. Arithmetic mean EC_{50} values and the min-max factor are given.

17.4 Discussion

According to the reports of the participating laboratories, some methodological shortcomings occurred, for example, problems with maintaining environmental conditions, use of non-synchronized test animals and lack of results of reference tests. The laboratory using test animals from a non-synchronized population reported rather low reproduction rates for the controls in the tests with the wastes INC and SOI and no reproduction at all for the test with the waste WOO. Consequently, these results were not accepted. A more general challenge, not solved in a satisfying manner here, is the fact that the dilution steps applied did not cover the required response range to determine an EC_{50} value including confidence limits.

Accepting the data with all their shortcomings and comparing them with the data of the earthworm acute test, it appears that the test substrate SOI did not have an effect on mortality but effects on reproduction could occur at about 56% waste. More important (and more peculiar) are the effects of the wastes INC and WOO on reproduction, resulting in EC_{50} values about four times lower than those determined in the acute test. Another striking observation is the response of the other sublethal parameter tested (biomass of adults: $NOEC_{biomass}$), revealing an almost similar sensitivity. The high importance of sublethal endpoints is also highlighted by the results of the earthworm avoidance test performed as part of this ring test (see Chap. 20): a high sensitivity comparable to those of the earthworm and collembolan reproduction tests was found for all three waste substrates.

Compared to the Collembolan reproduction test, no clear difference could be found (see Chap. 18): While INC did affect earthworms more than springtails, the situation was the opposite with SOI: clear effects on collembolans at about 46% waste but just hints of an impact on earthworm reproduction at about 56% waste. WOO caused almost identical effects on both organisms. Wilke et al. (2007) give a number of practical advantages as reasons why they prefer the 28-day collembolan reproduction test for the assessment of wastes: simple rearing, a relatively short test period (4 instead of 8 weeks in the earthworm reproduction test), a much more constant high reproduction rate in the control(s), and last but not least, the availability of a computerized image analysis system providing GLP-compliant documentation. However, long-term experience with this test system has revealed a high variability between replicates, leading very often to a high CV. Another potential alternative could be the Enchytraeid Reproduction Test, which, however, showed a lower sensitivity in this ring test (in particular with WOO; see Chap. 18).

17.5 Recommendations

The earthworm reproduction test has proved in many cases with spiked pesticides that it is working well as a higher tier test system. Meanwhile comparable experience is growing when using this test for the assessment of the quality of contaminated soils. According to these experiences, the sensitivity of this chronic test is as high as, or even higher than, behavioural responses (ISO 2007). Therefore, it is recommended that the usefulness of the earthworm reproduction test to be investigated as an alternative to the currently used earthworm acute test. Its sensitivity and practicability should also be investigated in comparison to the earthworm avoidance, the enchytraeid reproduction and the Collembolan reproduction tests. Based on the results of the ring test, the earthworm reproduction test will not be part of a minimum test set for waste classification purposes, but it may become an ecotoxicological and soil-related instrument in the risk assessment for an environmentally sound reuse of waste.

Chapter 18
Enchytraeid Reproduction Tests

M.J.B. Amorim, R. Kuperman, and J. Römbke

Abstract The Enchytraeid Reproduction Test was added to the test set for the ecotoxicological characterization of wastes as a method for assessing the hazard potential of wastes to terrestrial invertebrates. The method used was an adaptation of the ISO 16387 Guideline. This test was performed with three waste materials identified as INC, SOI, and WOO, diluted with OECD artificial soil to prepare the exposure treatments. Eleven of the twelve tests conducted were classified as acceptable for evaluation of the results. The ecotoxicological data established by the participating groups through these tests were consistent. The minimum-maximum factors for individual EC_{50} values were small (2.8 and 3.3), confirming that the Enchytraeid Reproduction Test is a robust method. The Enchytraeid Reproduction Tests showed that the toxicity order for waste materials was (from greatest to least) WOO > INC > SOI, based on the respective mean EC_{50} values for juvenile production of 13.2, 28.7, and >100% (not toxic). While the EC50 values were about 37% lower than the earthworm acute test, the sensitivity of the two tests to the three materials was statistically similar; the enchytraeid reproduction test can therefore be considered for inclusion in the future test set. Guidelines for determining the appropriate hydration levels for mixtures of test materials with soil will be needed to further improve data quality in future tests.

Keywords *Enchytraeus crypticus*, Reproduction toxicity test, Ring test, Waste material, Ash, Wood

18.1 Introduction

Enchytraeids (*Oligochaeta*) are common and important members of many soil ecosystems, and contribute to the vital processes of this environmental compartment. They contribute both directly and indirectly to the degradation of organic matter by

---

M.J.B. Amorim (✉), R. Kuperman, and J. Römbke
CESAM and Department of Biology, University of Aveiro3810-193 Aveiro, Portugal

H. Moser and J. Römbke (eds.), *Ecotoxicological Characterization of Waste*,
DOI: 10.1007/978-0-387-88959-7_18,

grazing on litter material and interactions with soil microorganisms, thus affecting key soil processes that may be important to the regulation, flow, and internal cycling of carbon and nutrients in the ecosystems. The activity of enchytraeids in soil also contributes to improving the soil pore structure (Didden 1993), and can thus affect soil hydrology and aggregation. Enchytraeids are annelid worms belonging to the family Lumbricidae, Class Clitellata, Phylum Annelida, and are the closest relatives to earthworms (Erséus and Kaellersjoe 2004). Guidelines for the Enchytraeid Reproduction Tests (ISO 2005; OECD 2004) have been developed and optimized for the test species *Enchytraeus albidus*. These guidelines also permit the use of alternative species of enchytraeids, including *Enchytraeus crypticus* and *Enchytraeus luxuriosus,* when ecotoxicological testing requires experimental conditions beyond the environmental optimum (e.g., soils with low pH and organic matter content) of *E. albidus*. The Enchytraeid Reproduction Test (ERT) has been used extensively since the release of the Draft OECD Guideline in 1999 and contributed to the ecotoxicological investigations of many chemicals (Amorim et al. 2005; Dodard et al. 2005; Kuperman et al. 2006a,b; Römbke 2003). The Enchytraeid Reproduction Test was used as on additional method for assessing the ecotoxicological hazard potential of wastes to terrestrial invertebrates (CEN14735 2005).

18.2 Method

The Enchytraeid Reproduction Test was used to assess the effects of waste materials on the reproduction of the enchytraeid worm *Enchytraeus crypticus*. The test was an adaptation of the ISO 16387 Guideline (ISO 2005), which was modified for use with waste material mixtures with OECD artificial soil (OECD 1984). The modifications included (a) the different hydration levels required for waste material mixtures with OECD artificial soil, that had different water holding capacity (WHC) compared with OECD artificial soil, and (b) shorter test duration for *E. crypticus* (28 days vs. 42 days) due to shorter generation time of this species, compared with *E. albidus* for which ISO 16387 test conditions were originally optimized. The tests were performed with OECD artificial soil (mixture of 70% quartz sand, 20% kaolinite clay, 10% peat, calcium carbonate to adjust the pH to 6.0 ± 0.5 and deionised water to achieve a soil moisture of 40–60% of the WHC_{max}) used as control treatment and individual mixtures of the three waste materials with artificial soil. A natural soil, Sassafras sandy loam (SSL: 55% sand, 28% silt, 17% clay, 2.3% organic matter, 9.3 cmol/kg CEC, and pH 4.9), was used by one participating laboratory for testing the toxicity of a reference chemical, boric acid, used as the positive control in that laboratory to validate the condition of the test species, and reliability and precision of results. Tests were performed in an environment-controlled incubator under a 16 h-light:8 h-dark photoperiod cycle with a mean light intensity ranging from 400 to 800 lux, and mean temperature of 20 ± 2°C. Oat flakes were added to test substrates as food for the enchytraeids. The reproduction of juveniles after 28 days was the main assessment endpoint. Juvenile production data were analyzed using linear and nonlinear regression models to determine the EC_{50} values (effect median concentration) and the 95% Confidence Limits (CL) associated with the point estimates. These values were determined by the Probit Analysis Method, using the statistical program ToxRat® (2003) or nonlinear regression models using SYSTAT 11.0 (SYSTAT® Software, Inc., Point Richmond, CA, USA) or Statistica programs.

Validity criteria were included in the tests as part of the Quality Control (QC) procedures. They included the following performance parameters recommended by ISO 16387 for the controls: the adult mortality does not exceed 20% after 14 days, the average number of juveniles is greater than 25 per test vessel at the end of the test assuming that ten adult worms per test vessel were used, and the coefficient of variation (CV) for the mean number of juveniles does not exceed 50% at the end of the test. In addition, the EC_{50} values were determined for the reference chemical Carbendazim in OECD soil should be within a proposed range of 20–50 mg/kg soil dry weight (dw). The EC_{50} values were determined for reference chemical boric acid in SSL soil and plotted on the Boric Acid Warning Chart should be within both the warning limits (WL) of 43–66 mg/kg and 95% CL of 37–71 mg/kg established for *E. crypticus* culture in previous tests with boric acid and SSL soil.

Based on the test results provided by the participating laboratories the EC_{50} values and corresponding 95% CL, minimum-maximum factors, the geometric and arithmetic means, standard deviations (SD), and CV for the EC_{50} estimates were determined. Mean EC_{50} and SD values were used to calculate the upper and lower WL and prepare the warning charts (Environment Canada 2005) for the final assessment of this test system.

18.3 Results

Four laboratories contributed to the ring test of waste materials using the Enchytraeid Reproduction Test (Table 18.1). They provided 12 data sets that were evaluated according to previously established test acceptance criteria. Results of one test with SOI material were rejected because there was no concentration-response relationship. Although one laboratory failed to provide test results for the reference toxicant, its data were accepted because they were consistent with the results reported by the remaining three laboratories.

A concentration-response relationship was established in all four tests with INC material. The EC_{50} values estimated in these tests ranged from 12 to 41% INC in

Table 18.1 Data source parameters for the Enchytraeid Reproduction Test used in toxicity testing of waste materials INC, SOI, and WOO

Waste material	Labs participated	Tests conducted	Tests rejected	Tests accepted
INC	4	4	0	4
SOI	4	4	1	3
WOO	4	4	0	4
Total	4	12	1	11

Table 18.2 Summary of the results of the Enchytraeid Reproduction Tests with the three waste materials; data expressed as percent waste in artificial soil

Waste material	Geometric mean EC_{50}	Lower warning limit	Upper warning limit	SD	CV	Min-Max factor
INC ($n = 4$)	28.7	5.1	58.4	13.3	42	3.3
SOI ($n = 3$)	ND (>100)	ND	ND	ND	ND	ND
WOO ($n = 4$)	13.2	−0.3	29.5	7.5	51	2.8

ND Not determined; no adverse effect up to and including 100% SOI material in three out of four tests

artificial soil. The minimum and maximum EC_{50} values determined for INC differed by a factor of 3.3. All EC_{50} values were within the WL calculated as mean $EC_{50} \pm$ SD. In addition, the LC_{50} values of 38 and 61% INC were established for adult *E. crypticus* survival in the experiments performed by two laboratories, respectively.

The results of the Enchytraeid Reproduction Tests with SOI material contrasted with either INC or WOO materials. All accepted tests showed no adverse effects on juvenile production by *E. crypticus* up to and including 100% of SOI material.

A concentration-response relationship was established in all four tests with WOO material. The EC_{50} values estimated in these tests ranged from 8 to 22% WOO in artificial soil. The minimum and maximum EC_{50} values for WOO differed by a factor of 2.8. All EC_{50} values were within the WL.

The results of the interlaboratory comparison of the Enchytraeid Reproduction Test data for INC, SOI, and WOO waste materials are summarised in Table 18.2. The respective arithmetic means are slightly higher at 31.8% (INC) and 14.6% (WOO). These results show that the toxicity data for each of the waste materials were consistent among the participating laboratories, based on the EC_{50} values and the respective WL (lower and upper), SD, CV, and minimum-maximum factors for INC or WOO effects on juvenile production by *E. crypticus*. Similarly consistent results (no toxicity) were determined for SOI material in the three accepted tests.

18.4 Discussion

The Enchytraeid Reproduction Test was sufficiently robust to establish the concentration-response relationships for INC or WOO wastes and production of juveniles by the enchytraeid *E. crypticus,* based on toxicity data established by the four participating laboratories. It also proved its utility in ascertaining that a waste material can have no adverse effects on the enchytraeid worms as was demonstrated for SOI by data from three out of the four participating laboratories. The acceptability of the test results was 92%, although this number includes a dataset from one laboratory that failed to provide test results for a reference toxicant.

All EC_{50} values established by the Enchytraeid Reproduction Tests with INC or WOO waste materials were within the WL. Calculated 95% CL values ranged from 2.6 to 160% for INC and from 6.6 to 23.6% for WOO material. The minimum-maximum factors were small (2.8 and 3.3) and were in the range recommended by Environment

Canada for interlaboratory comparisons (2–4; Chapman 1995; EC 1999). Similarly, the SD values for the EC_{50} estimates were relatively low and were 13.3 and 7.5 for INC and WOO, respectively. The respective CV values of 42 and 51% established for the EC_{50} estimates of INC and WOO effects on *E. crypticus* exceeded the 30% level recommended for ring test data by Environment Canada (EC 2005). However, this level of variance was recommended for well-standardized tests usually conducted with a single-chemical-spiked soil, and can be difficult to attain when testing mixtures of materials like wastes.

Available data show that the development of new test systems or modification of the standardized exposure conditions can initially result in a greater variance in the performance metrics of the test species, including the CVs (the enchytraeid ring test results; Römbke & Moser 2002). The level of variance is expected to decrease with increasing experience in test performance and with further standardization of a newly designed or modified test system. Accordingly, the CV values determined for the EC_{50} estimates can be considered reasonable or acceptable, considering that they were established for a relatively small sample size ($n = 4$ participants) and for the EC_{50} estimates calculated by the participants using different models (probit analysis of linearized data vs. nonlinear regression analysis of untransformed data), which often affect the endpoint estimates.

The sensitivity of the Enchytraeid Reproduction Test to the effects of INC or WOO waste materials was statistically similar (based on the 95% CL) to that of the 14-day earthworm toxicity test (ISO 1993) used in the basic test set. Neither enchytraeids nor earthworms were adversely affected by exposure to SOI material, while in one of the phytotoxicity tests (*Brassica*) and in the Collembolan reproduction test, adverse effects of SOI were observed. The sensitivity of enchytraeids to INC material was similar to that of collembolans, but enchytraeids were approximately three times less sensitive to WOO compared to the arthropods. However, the enchytraeids were clearly less sensitive to both INC and WOO compared to the earthworms, according to the results of the chronic earthworm tests.

Information on the effects of waste materials on enchytraeids is limited. Sverdrup et al. (2007) used the enchytraeid test to investigate the effects of Benzo-a-pyrene on *E. crypticus*. They found no effects on survival or reproduction at the greatest concentration of 1,000 mg/kg tested in that study. This finding is consistent with our results for SOI effects on *E. crypticus,* where SOI contained about 840 mg/kg PAH (in fact a mixture of 14 PAHs including Benzo-a-pyrene (Chap. 3).

Although data on the toxicity of copper-contaminated wood to enchytraeids are not available, exposure to copper has been demonstrated to be harmful for enchytraeids. In a study with Cu-spiked OECD standard soil, Amorim et al. (2005c) established an EC_{50} of 65 mg/kg for reproduction of *E. luxuriosus*. Based on copper content in WOO of 2000 mg/kg (Chap. 3), an EC_{50} of 13% WOO determined in the ring test is equivalent to copper concentration of 273 mg/kg. The difference between the two EC_{50} estimates is likely a result of differences in copper bioavailability in the two types of exposure; copper was strongly adsorbed to the wood particles used in the ring test, but was highly bioavailable in the soil test with water-soluble copper chloride salt.

The main limitation of the ring test was the small number of participants conducting the Enchytraeid Reproduction Test. Nonetheless, data developed in these tests were sufficient to allow the required evaluation. However, as has been mentioned in Part II for the earthworm test, more guidelines are needed for determining the appropriate moisture levels for individual mixtures of the test material with OECD artificial soil. The mixture hydration rates depend on the proportion of the components, which have different WHC. The final WHC of each mixture may have to be determined before treatment preparation, to provide the optimal moisture conditions for the test species.

18.5 Recommendations

As stated in CEN guideline 14735 (2005), a worm (i.e., Oligochaete) test should be included in the final test set. Such worms are present worldwide and often are among the most common soil invertebrates. When used as surrogates, their sensitivity to chemical contaminants in soil is generally representative of the exposure effects on other soft-bodied taxa in the soil invertebrate community (Römbke 2003; Kuperman et al. 2006c). The EC_{50} values established in the Enchytraeid Reproduction Tests with INC and WOO materials were approximately 37% lower (showing greater sensitivity) compared with those from the acute earthworm tests. At the same time, the enchytraeids were less sensitive than the earthworms when the results of the chronic tests were compared. However, the Enchytraeid Reproduction Test requires considerably smaller quantities of the test materials, which is an important practical consideration in selecting the future test set. In any case, additional tests, including the earthworm reproduction test (ISO 1998) and other chronic tests with collembolans or mites, should be evaluated for possible inclusion in a test set for wastes.

Regardless of the final selections for future test sets, the effects of soil properties on the bioavailability and resulting toxicity of waste materials requires special consideration. The current ring test evaluated the effects of waste materials on enchytraeids in mixtures with OECD artificial soil that was adjusted to near-neutral pH. The exposure conditions in such substrates can differ greatly from those in field soils. Therefore, the authors recommend that natural soil types, which have properties that support relatively high bioavailability of chemicals of concern (e.g., low organic matter and clay contents, and low pH for testing the effects of metal-containing waste materials), should be used instead of, or in addition to, OECD artificial soil for preparation of treatment concentrations for toxicity testing. Such an approach will ensure a greater relevance of ecotoxicological data generated by the future test set for assessing the risks associated with the potential release of toxic compounds from waste materials into the environment.

Chapter 19
Collembolan Tests

A. Scheffczyk, T. Moser, and T. Natal-da-Luz

Abstract Collembolans constitute one of the most numerous groups of soil communities. So they were used as representatives for soil organisms, namely arthropods, in the ring test to define a test battery for the ecotoxicological characterization of waste. More specifically, the Collembola reproduction test with the species *Folsomia candida*, following ISO standard guideline 11267, was performed with the three waste types INC, SOI, and WOO. The results of 18 out of 21 tests were classified as acceptable. Regarding the methodology, no problems were found, except for moisture determination of the substrate. For this, further guidelines on determining the moisture of the test substrate mixtures, namely wastes, is needed. According to the min-max factors between individual EC_{50} values (4.5–8.9), the Collembola reproduction test was classified as a robust method. The toxicity observed in each of the three wastes was considerably different as it can be seen by the mean of EC_{50} values: 26.0% for INC, 47.9% for SOI and 5.0% for WOO. The sensitivity of the Collembola reproduction tests towards soil waste mixtures is high when compared to other plant and invertebrate tests. In particular, it is worth mentioning that the Collembola test was one of only five bioassays used in the ring test, which regularly showed an effect in the tests with SOI waste. This fact highlights the need to include the *F. candida* reproduction test in the battery for the assessment of wastes.

Keywords Folsomia candida, Reproduction test, Ring test, Waste

19.1 Introduction

The Collembola, commonly known as springtails, are one of the most numerous and widely distributed organisms of the soil fauna. Population densities commonly reach $10^5/m^2$ in soil and leaf litter layers. Despite their small size (adults: 0.5–5 mm long) and low contribution to the total biomass and respiration of the soil communities,

---

A. Scheffczyk, T. Moser, and T. Natal-da-Luz
ECT Oekotoxikologie GmbH, Boettgerstr 2-14, 65439 Floersheim, Germany

H. Moser and J. Römbke (eds.), *Ecotoxicological Characterization of Waste*,
DOI: 10.1007/978-0-387-88959-7_19,

they have a very important role as regulators in soil decomposition processes through feeding on microbes and microfauna. They may also contribute to the organic matter breakdown in acidic soils, where earthworms and diplopods are absent (Wiles and Krogh 1998). Several species of springtails have been tested in ecotoxicological studies, such as *Hypogastrura assimilis* (Folker-Hansen et al. 1996), *Isotoma viridis* and *Orchesella cincta*. Several methods have been developed and published using *Isotoma tigrina* (Kiss and Bakonyi 1992) and *Folsomia fimetaria* (Løkke 1995; Wiles & Krogh 1998). For an overview on the use of Collembolan species in soil toxicity tests see Achazi et al. (2000).

However, *Folsomia candida* has been used in most ecotoxicological studies performed with collembolans, (Achazi et al. 2000; Phillips et al. 2002). In addition to the fact that this species is the one proposed by the ISO guideline for reproduction tests (ISO 1999), its high sensitivity, short life-cycle (Hopkin 1997), possibility to be easily kept in laboratory cultures, and wide distribution (Wiles and Krogh 1998) have contributed to its common use as a test species. When developing an ecotoxicological test with this species, several methodological aspects of its performance were studied in a ring test, but these results are not comparable to the one described here (Riepert & Kula 1996). *F. candida* Willem 1902 (Collembola: Isotomidae) is a blind, unpigmented euedaphic collembolan that reproduces parthenogenetically. Usually, the species is classified as microsaprophagous, but it can also feed on nematodes (Hopkin 1997). It is widely distributed throughout Europe and although it is not common in most natural soils, it often occurs at humus-rich sites in very high numbers (Wiles and Krogh 1998).

19.2 Method

Reproduction tests with the springtails *F. candida* were performed according to ISO-Guideline No. 11267 (1999). In short, the test was conducted using artificial soil (OECD 1984) as control and mixtures of the wastes and the artificial soil as test substrates. At a temperature of 20 ± 2°C and a 16:8 h light: dark photoperiod cycle (400–800 lux), the Collembola reproduction tests run for 28 days. The organisms were fed with 2–3 mg of dry yeast at the beginning and at day 14 of the test. The number of surviving adult and juveniles was determined for each test container at the end of the tests. The NOEC and EC_{50} values were defined on the basis of the number of offspring determined in each treatment. These values were determined by the probit analysis method or by applying parametric tests (e.g. Williams or Dunnetts test), respectively, using the statistical program TOXRAT® (2003). Tests were considered valid when in the control vessels the adult mortality was ≤20%, the mean number of juveniles was ≥ 100 and the coefficient of variation was ≤30% at the end of the test period. In addition, the results of a reference test, e.g. with the pesticide formulation Betosip (150 mg Phenmedipham/L), should show an effect in juvenile reproduction (e.g. between 100 and 200 mg/kg = 15–30 mg active ingredient/kg (ISO 1999).

Based on the test results provided by the participating laboratories, the EC_{50} values and corresponding 95% CL, minimum-maximum factors, the arithmetic means, standard deviations (SD), and CV for the EC_{50} estimates were determined. Mean EC_{50} and SD values were used to calculate the upper and lower WL and prepare the warning charts (Environment Canada 2005) for the final assessment of this test system.

19.3 Results

In all, seven participating laboratories provided 21 test data sets. Only one laboratory did not fulfill the validity criteria. Eighteen tests from six laboratories were valid (Table 19.1). Four laboratories presented results from reference tests; while three of them used Phenmedipham as reference chemical (ISO 11267), one laboratory used the insecticide Dimethoate which is often included in ecotoxicological studies with Collembolans (e.g. Løkke & Van Gestel 1998).

The results obtained in the Collembola reproduction tests with the wastes INC, SOI and WOO are presented, following the same scheme used in the other tests reported in Part III.

Six tests were performed with INC waste. NOEC values varied between <3.3% and 25.0%. The EC_{50} values differed by a factor of 7.2, min: 9.1%; max: 65.8% (Table 19.2, Fig. 19.1). In most cases the NOEC and EC_{50} values differed by a factor of 2–3, except for the results obtained by laboratories No. 18 and 58. In the first case, only a small difference between NOEC and EC_{50} values was found, but in the second

Table 19.1 Summary of the valid data of Collembola reproduction tests

Substrate	No. Labs	No. Tests	Invalid	Reference	Accepted
INC	7	7	1	2	6
SOI	7	7	1	2	6
WOO	7	7	1	2	6
Sum	n.a.	21	3	6	18

n.a. Not applicable

Table 19.2 NOEC and EC_{50} values with 95% confidence limits (95% CL) and respective upper (UL) and lower levels (LL) for reproduction of *F. candida* with INC waste treatments obtained by each participating laboratory (Lab Code)

Lab Code	NOEC	EC_{50}	95% CL: UL	95% CL: LL
11	8.0	16.6	11.8	25.9
15	<3.3	10.1	8.6	11.7
16	6.3	20.2	15.4	25.0
18	25.0	34.3	17.5	51.0
22	<4.2	9.1	7.2	11.9
58	5.0	65.8	63.6	67.9

All values are given as % of dry weight of waste in artificial soil

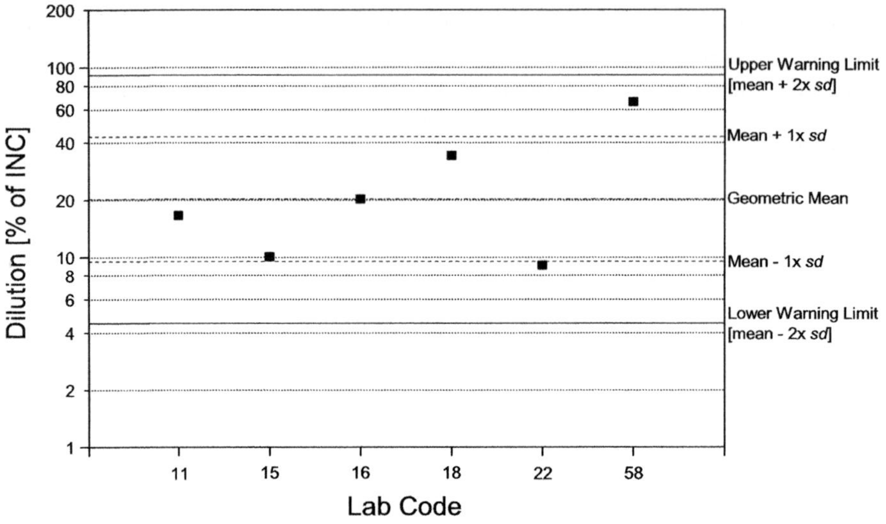

Fig. 19.1 Results obtained by each participating laboratory (log EC_{50} as % of dry weight of waste in artificial soil) of the Collembola reproduction tests with the INC waste treatments

Table 19.3 NOEC and EC_{50} values with 95% confidence limits (95% CL) and respective upper (UL) and lower levels (LL) for reproduction of *F. candida* with SOI waste treatments obtained by each participating laboratory (Lab Code)

Lab Code	NOEC	EC_{50}	95% CL: UL	95% CL: LL
11	50.0	75.1	69.9	80.9
15	<12.5	29.0	n.d.	n.d.
16	<25.0	73.4	19.6	>100
18	<50.0	72.4	50.4	94.3
22	<16.7	16.6	n.d.	n.d.
58	25.0	20.7	18.1	23.5

All values are given as % of dry weight of waste in artificial soil. *n.d.* Not determined

case, a large difference between these values was detected. With the exception of this last example, all the test results were close to each other, providing EC_{50} values between 10 and 35%.

Six tests were performed with SOI waste. In this case, the NOEC values varied between <12.5% and 50.0%, while the EC_{50} values differed by a factor of 4.5, min: 16.6%; max: 75.1% (Table 19.3, Fig. 19.2). From these four test data sets, it was not possible to determine the NOEC value, because of a flat dose-response curve. Two groups of test results can be identified; three of them indicating a low toxicity level (EC_{50} values about 70–75%) and the other three showing a higher toxicity (EC_{50} values around 15–30%). So far, no reason for this difference has been identified.

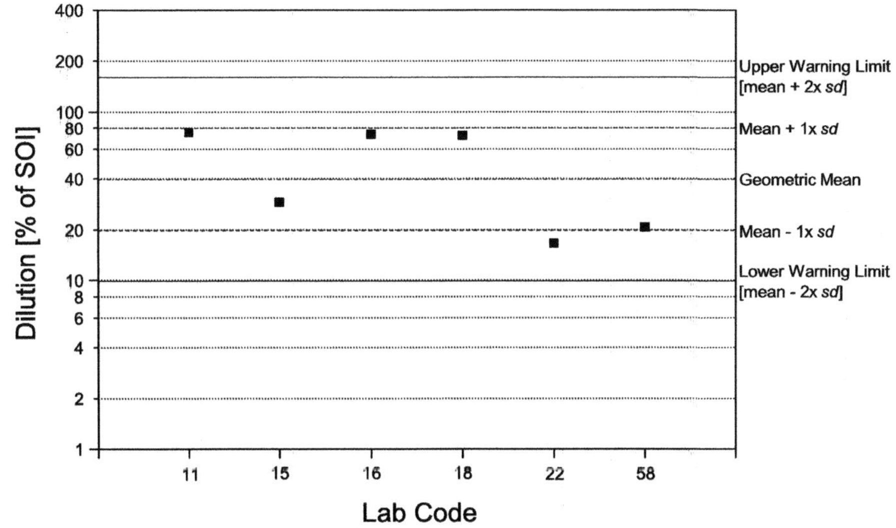

Fig. 19.2 Results obtained by each participating laboratory (log EC_{50} as % of dry weight of waste in artificial soil) of the Collembola reproduction tests with the SOI waste treatments

Table 19.4 NOEC and EC_{50} values with 95% confidence limits (95% CL) and respective upper (UL) and lower levels (LL) for reproduction of *F. candida* with WOO waste treatments obtained by each participating laboratory

Lab Code	NOEC	EC_{50}	95% CL: UL	95% CL: LL
11	<6.8	5.2	4.0	6.0
15	<3.1	4.7	n.d.	n.d.
16	4.5	10.7	10.1	11.2
18	0.8	1.2	0.4	2.1
22	<3.1	3.3	2.4	4.0
58	1.0	4.9	4.0	9.1

All values are given as % of dry weight of waste in artificial soil. *n.d.* Not determined

With WOO waste, six tests were performed. The NOEC values varied between 0.8% and < 6.8% and the EC_{50} values differed by a factor of 8.9, min: 1.2%; max: 10.7% (Table 19.4, Fig. 19.3). In three of the tests performed, the NOEC value could not be determined. In the other cases the NOEC and EC_{50} values differed by a factor between 1.5 and 5. The EC_{50} values of four test data sets were quite similar (3.3–5.2%).

The results of the Collembolan tests can be summarised as follows (Table 19.5). The table shows clearly the differences in toxicity of the three waste substrates: WOO is by far the most toxic one, but INC and SOI also caused severe effects.

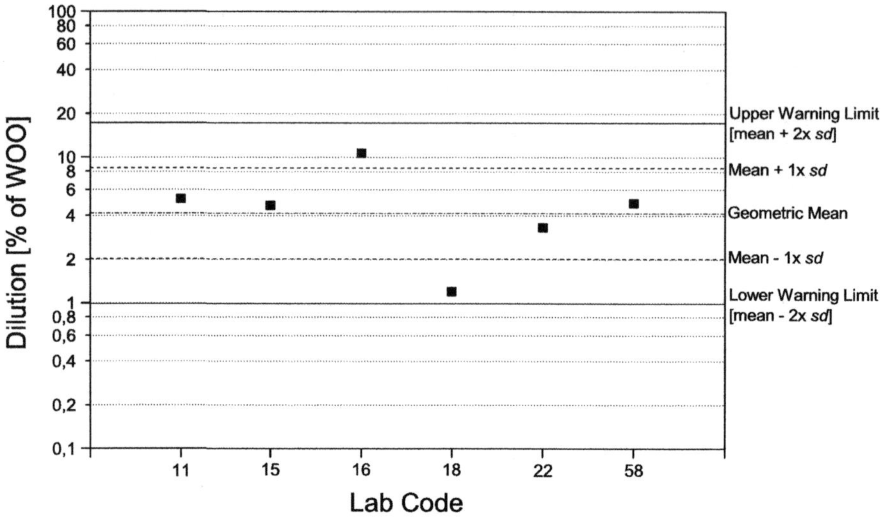

Fig. 19.3 Results obtained by each participating laboratory (log LC_{50} as % of dry weight of waste in artificial soil) of the Collembola reproduction tests with the WOO waste treatments

Table 19.5 Summary of the Collembola reproduction test results with the three test wastes (data expressed as % of weight of waste in artificial soil)

Substrate	Mean EC_{50}	SD	CV	Factor Min–Max
INC	26.0	21.5	82.7	7.2
SOI	47.9	28.5	59.6	4.5
WOO	5.0	3.2	63.2	8.9

Arithmetic EC_{50} mean values and respective standard deviation (SD), coefficient of variance (CV) and the difference between the higher and the lower EC_{50} values (factor Min-Max) are given

19.4 Discussion

Regarding methodology, no problems were reported in the main steps of the test. However, due to the properties of the waste substrates, more guidelines are needed to define a better method to evaluate the suitable moisture of soil-mixtures for collembolans. The range of moisture that is commonly used in an artificial or natural reference soil (40–60% of the WHC) seems to be unsuitable for some artificial soil - waste mixtures.

The sensitivity observed in the Collembola reproduction test was higher then that in many other tests (Chap. 1). This was particularly obvious in SOI waste, where *F. candida* was the only test organism which was negatively affected by this substrate. WOO waste affected the reproduction of *F. candida* and of the earthworm *Eisenia fetida* to the same extent (mean EC_{50} values 5.0 and 4.5, respectively).

The EC_{50} values determined from the Collembolan tests were lower than the EC_{50} values from the oat plant and the earthworm acute test. The results of the plant tests with *Brassica* and those from the Collembolan test are very similar (only WOO seems to be more toxic for the plant than for the springtail).

The factor between the highest and the lowest EC_{50} values for each test waste ranged between 4.5 for SOI and 8.9 for WOO. Such differences between tests with the same substrate can be caused by slight changes of waste properties like pH, organic matter and water content in the participating laboratories. According to Crouau et al. (2002), these properties can significantly influence the outcome of Collembolan tests when evaluating waste toxicity. Water content, in particular, influences the EC_{50} values determined in Collembola reproduction tests (Domene et al. 2007). Furthermore, Crommentuijn et al. (1995) found an LC_{50} range of 802 to more than 2,024 mg Cd/kg dry weight for four different clones of *F. candida*, meaning that the min-max factor was 2.5. Test results were less variable where the same clone was used (Jaensch et al. 2005). Taking into consideration that in the present ring test each participating laboratory adjusted the water content of the wastes manually (due to the peculiar properties of each test waste as explained above) and used the corresponding clones of their own *F. candida* laboratory cultures, the min-max factors (4.5–8.9) determined here can be regarded as very low. In fact they are in the same range as those determined for the basic test set (earthworms, plants).

19.5 Recommendations

The data presented in this chapter confirm that the *F. candida* reproduction test has the potential to be used as an important tool in the assessment of waste toxicity. Due to its broad range of ecological requirements and high tolerance in terms of organic matter content and pH (Jaensch et al. 2005), this species is able to survive and inhabit different waste types, detecting and responding to the contaminants present. In comparison with other terrestrial organisms used in this ring test, *F. candida* was among the most sensitive species. Therefore, we recommend the inclusion of the *F. candida* reproduction test in the test set for the assessment of wastes (ISO 2005).

Chapter 20
Earthworm Avoidance Tests

T. Natal-da-Luz, X. Domene, A. Scheffczyk, and J.P. Sousa

Abstract Based on the fact that organisms have the ability to avoid unfavorable conditions, the avoidance tests with earthworms have been used as an important tool to detect contaminants present in soil. Being ecologically relevant, quick and cost-effective, these bioassays have high sensitivity compared to conventional acute and chronic tests. Earthworm avoidance tests were performed in an inter-laboratory ring test of waste in accordance with ISO guideline No 17512-1 to evaluate the usefulness of this type of bioassays in waste characterization. Concentration gradients of three test wastes (INC, SOI and WOO) were tested using the artificial OECD soil as control. The species *Eisenia fetida* and *Eisenia andrei* were used as test organisms. A total of 12 avoidance tests were performed by four laboratories. The results obtained proved the higher sensitivity of this type of bioassay compared to tests using other terrestrial organisms (in particular, the earthworm acute test). Although all three wastes induced an avoidance behavior, the WOO treatments were the most avoided.

Keywords Earthworms, Avoidance tests, Wastes, Avoidance behavior

20.1 Introduction

In recent years, the avoidance behavior of several soil invertebrate taxa has been used as an early screening indicator of soil ecotoxicity. Although until date, most of the tests have been carried out with earthworms (Yeardley et al. 1996; Natal-da-Luz et al. 2004, 2008a,b, Hund-Rinke et al. 2005; Lukkari and Haimi 2005), other taxa, like enchytraeids and collembolans, are also used (e.g. Loureiro et al. 2005; Amorim et al. 2005). This fact contributed to the recent publication of the ISO Guideline No. 17512-1 for the earthworm species *Eisenia fetida* and *Eisenia andrei* (ISO 2007).

T. Natal-da-Luz (✉), X. Domene, A. Scheffczyk, and J.P. Sousa
Instituto do Mar – Coimbra Interdisciplinary Centre, Largo Marquês de Pombal, Department of Zoology, University of Coimbra, 3004 – 517, Coimbra, Portugal

H. Moser and J. Römbke (eds.), *Ecotoxicological Characterization of Waste*, 191
DOI: 10.1007/978-0-387-88959-7_20, © Springer Science+Business Media, LLC 2009

Because of the ability of individual worms to avoid unfavorable conditions, the avoidance tests with earthworms are sensitive compared to acute or reproduction assays, ecologically relevant and quick screening methods to evaluate the influence of contaminant mixtures and single substances in soils (ISO 2007). This type of bioassay can also be used as an early screening tool in site specific risk assessment scenarios. For all these reasons, the earthworm avoidance test is considered useful in the early characterization of wastes. Although avoidance assays had not been included in the list of ecotoxicity tests by the EN 14735 (2005), the advantages of this type of bioassay justified its selection for the ring-test.

20.2 Method

Two-section chamber avoidance tests with the earthworms *E. fetida* and *E. andrei*, based on the ISO guideline No. 17512-1 (2007), were performed for each test waste (INC, SOI and WOO). Artificial soil according to ISO 11268-2 (1998) was used as test substrate in the control and in the waste treatments. These were prepared by mixing each waste with the soil in the following percentages of waste respectively: 3.125, 6.25, 12.5, 25 and 50% DW (INC); 6.25, 12.5, 25, 50, 75 and 100% DW (SOI); 0.78, 1.56, 2.08, 3.125, 3.75, 6.25, 8.33, 12.5, 16.67 and 25% DW (WOO). For each treatment 4–5 replicates were used. Each replicate consisted of a 1 L capacity plastic container, divided into two sections by a cardboard divider. One of the sections was filled with 250 g DW of artificial soil (control section) and the other with the corresponding soil–waste mixture (test section). After the divider's removal, ten adult earthworms (250–600 mg), previously washed and wiped dry, were placed on the middle line of the soil surface. The replicates were covered by a transparent lid and incubated at 20 ± 2°C, under a light intensity between 400 and 800 lux with a photoperiod of 16 h:8 h light: dark for a time period of 48 h. After this period, the number of worms found in each section was recorded. Tests were considered valid when the mortality was lower than or equal to 10%. An additional combination was performed to ensure that no avoidance reaction is detected when control soil is placed on both sides of the test containers (Yeardley et al. 1996; Hund-Rinke and Wiechering 2001). In this case the procedures adopted were the same as those described above with the exception that uncontaminated artificial control soil was placed in both sections of the test chambers.

The individual test results of each participant were analyzed by the Fisher exact test (Zar 1999). A one-tailed test was used, assuming as null hypothesis that half of the total number of earthworms stay in the treatment (test) section, simulating a situation where no avoidance behavior is detected. Null hypothesis was rejected for a probability equal to or lower than 0.05 ($p \leq 0.05$). The NOEC value for a specific waste was defined as the highest soil-waste mixture percentage for which no avoidance behavior was detected when it was combined with the control. The EC_{50} value (and respective 95% upper and lower confidence limits) for each waste was calculated using probit regression.

In this analysis, the percentage of organisms found in the control side along the combinations with the soil-mixture gradient of each waste was taken into account. As a control situation, it was assumed that 50% of organisms were found in each side of the replicates.

The assessment of repeatability and reproducibility was performed for each waste according to ISO 5725 (1994), using the data obtained for each individual test. This approach included the determination of the arithmetical mean of the EC_{50} values for each waste, based on the specific EC_{50} values obtained in each data set. The corresponding standard deviations, coefficients of variation and minimum–maximum factors were also calculated for each test material.

20.3 Results

In all, four participating laboratories performed the two-section earthworm avoidance test with the three test wastes, providing 12 data sets. Since almost no mortality was found in any of the obtained results, all the data were considered valid and taken into account for the analysis (Table 20.1).

The results of two-section earthworm avoidance tests with the test wastes INC, SOI and WOO are presented following the same scheme used in the other chapters.

Three laboratories performed the avoidance tests with earthworms using the INC waste treatments (Table 20.2). The lowest NOEC obtained was 3.125% and the highest NOEC was 12%. Therefore, the NOECs determined for INC waste differed by a factor of 3.84. The NOEC and EC_{50} concentrations of each laboratory differed from each other by a factor of 1.55 to a factor of 3.04.

Table 20.1 Summary of the data obtained in the avoidance tests with earthworms

Substrate	No. labs	No. tests	Invalid	Accepted
INC	3	3	0	3
SOI	4	4	0	4
WOO	4	5	0	5
Sum	n.a.	12	0	12

n.a. Not applicable

Table 20.2 NOEC and EC_{50} values and respective lower (LL) and upper (UL) 95% confidence limits (95% CL) for avoidance behavior of earthworms in the INC waste treatments obtained by each participating laboratory (Lab Code)

Lab Code	NOEC	EC_{50}	95% CL:LL	95% CL:UL
11[a]	3.13	9.5	7.0	12.1
18[b]	12.0	18.6	2.1	37.5
22[a]	6.25	12.9	7.5	16.5

All values are given as % of dry weight of waste in artificial soil

[a]*E. fetida*

[b]*E. andrei*

Four laboratories decided to test the SOI waste, providing a total of four acceptable test data sets (Table 20.3). Two laboratories detected avoidance behavior of the test organisms at the lowest concentration tested (25%). In the other two laboratories, NOEC values of 50 and 25% were determined, differing by a factor of 2. The EC_{50} calculation was possible only in the data sets provided by two laboratories: 31.2% and 48.3%. The NOEC and EC_{50} values determined by Laboratory No 22 differed by a factor of almost 2.

Four laboratories provided five sets of data from avoidance tests with the WOO waste (Table 20.4). In three experiments an avoidance reaction was observed even at the lowest test concentrations used (3.75%, 1.56% and 3.125%); i.e. a NOEC could not be determined. The NOECs of the last two tests, where lower concentrations were used, were very low (0.78% and 2.08%) and differed by a factor of 2.7. EC_{50} values were determined in all tests, but in two of them no confidence limits could be calculated. This occurred in those tests where the most extreme EC_{50} values were found (0.4% and 6.5%). Considering only the other three data sets, the EC_{50} values varied between 1.8 and 3%. In the two cases where the difference between NOEC and EC_{50} values could be defined, they differed by a factor of about 3.

The WOO test waste was most strongly avoided by the earthworms but the EC_{50} values differed considerably. INC and, in particular, the SOI test waste, caused less avoidance, but the EC_{50} values were in a much smaller range differing only by a factor of 1.5–2.0 (Table 20.5).

Table 20.3 NOEC and EC_{50} values and respective lower (LL) and upper (UL) 95% confidence limits (95% CL) for avoidance behavior of earthworms in the SOI waste treatments obtained by each participating laboratory (Lab Code)

Lab Code	NOEC	EC_{50}	95% CL:LL	95% CL:UL
11[a]	<25	31.2	16.0	41.0
18[b]	50	>100	n.d.	n.d.
22[a]	25	48.3	40.8	57.7
54[a]	<25	n.d	n.d.	n.d.

All values are given as % of dry weight of waste in artificial soil; *n.d.* Not determined
[a]*E. fetida*
[b]*E. andrei*

Table 20.4 NOEC and EC_{50} values and respective lower (LL) and upper (UL) 95% confidence limits (95% CL) for avoidance behavior of earthworms in the WOO waste treatments obtained by each participating laboratory (Lab Code)

Lab Code	NOEC	EC_{50}	95% CL:LL	95% CL:UL
11[a]	<3.75	0.4	n.d.	n.d.
18[b]	<1.56	1.8	1.2	2.3
22[a]	<3.125	3.0	2.0	3.8
22[a]	2.08	6.5	n.d.	n.d.
54[a]	0.78	2.8	2.1	3.5

All values are given as % of dry weight of waste in artificial soil; *n.d.* Not determined
[a]*E. fetida*
[b]*E. andrei*

Table 20.5 Summary of the earthworm avoidance test results with the three test wastes (all values are given as % of dry weight of waste in artificial soil)

Substrate	Mean EC_{50}	SD	CV	Factor Min–Max
INC	13.7	4.6	33.6	2.0
SOI	39.8	12.1	30.4	1.5
WOO	2.9	2.3	77.9	16.3

Arithmetic EC_{50} mean values and respective standard deviation (SD), coefficient of variation (CV) and the Min/Max-factor between the lowest and highest EC_{50} values for each waste are given

20.4 Discussion

The methodology adopted in the earthworm avoidance tests was suitable for the test materials, but the adjustment of the moisture of certain soil-waste mixtures was problematic. As already mentioned in previous chapters, the maximum water holding capacity of the WOO waste (451%) provided a lower fraction of water available to the organisms when compared to that in the control soil (63%). In addition, the water holding capacity measured in WOO wastes did not take into account the higher absorption of this test material compared to that in the control soil. Therefore, the moisture was determined and adjusted "manually" so that no free water appeared when the substrate was pressed by hand (see also Chap. 11, Sect. 4).

The results obtained in the earthworm avoidance tests showed that the endpoint used (avoidance behavior) allowed the effects of the test substrate to be detected earlier than in the earthworm acute tests for all three wastes. For example, only the pure SOI waste caused an effect on mortality in the earthworm acute tests, while in the avoidance tests with SOI, the worms reacted even at lower mixtures. However, the coefficient of variation calculated from the avoidance reactions were always higher compared to those obtained in the acute tests. A similar observation was made by Hund-Rinke et al. (2003) who found a higher sensitivity but also higher variability in the avoidance tests than in acute tests, when testing to the reaction of compost worms to different contaminated and remedied natural soils.

The toxicity of SOI on soil invertebrates was observed only in the earthworm avoidance tests and in the Collembola reproduction tests (Chap. 19), but the EC_{50} values of the former were always lower than those calculated for the springtails. In this case, however, the coefficient of variation was lower for the avoidance behavior of earthworms in almost all test wastes. The exception was the WOO waste where a coefficient of variation of 77.9% was found in the avoidance tests vs. a coefficient of variation of 63.2% obtained in Collembola reproduction tests. Finally, the earthworm avoidance behavior was also at least as sensitive as the growth of terrestrial plants, while the CV values of these tests were in the same order of magnitude.

In agreement with the results from a laboratory intercalibration test with contaminated soils (Hund-Rinke et al. 2003), it was shown in the avoidance tests with these wastes that, in about one third of all tests performed, considerable differences were identified between participating laboratories (in this case mainly with the WOO substrate). Furthermore, it has already been shown by Natal-da-Luz et al. (2008b) that the avoidance behavior of *Eisenia andrei* is an endpoint highly influenced by natural soil properties such as organic matter content and texture. This means that the high variability found in the present experiments could be related to the slight differences between the artificial soil constituents used in each laboratory (e.g. sand mineral constitution and peat source). In addition, the use of two very similar yet different test species could have contributed to increase the variability of results, as, despite the ecological and physiological similarities of *E. fetida* and *E. andrei,* their responses to stress can be different (Dominguez et al. 2005).

20.5 Recommendations

The sensitivity of the earthworm avoidance tests was clearly shown in the present ring test, as this test provided lower EC_{50} values than the other terrestrial bioassays. This fact and the short test period are advantages that confirm the high utility of this test in evaluating waste toxicity. For these reasons, we recommend the inclusion of the earthworm avoidance assay in the set of tests for waste characterization. Nevertheless, avoidance tests with other organisms (e.g. collembolans or enchytraeids) can provide complementary results which may also be useful for the evaluation of wastes. Therefore further research evaluating the usefulness of the avoidance behavior of other soil invertebrate species in waste toxicity evaluation is still needed.

Chapter 21
Bacteria Contact Test

H. Neumann-Hensel and C. Roeber

Abstract Bacteria play an important role in all substance cycles in the environment. Therefore, they are an indispensable part of any test battery for the ecotoxicological characterization of waste. In this ring test a bacterial solid contact test with *Arthrobacter globiformis* was used. The test was performed with the three waste types INC, SOI and WOO, according to the draft guideline ISO 10871. A total of ten participating laboratories conducted the contact test and provided 28 data sets. After statistical evaluation, only three test results were rejected. The different levels of toxicity of the waste material are given by the final EC_{50} values: INC = 22.6%, SOI = 12.4% and WOO = 0.6%. The solid contact assay showed a higher sensitivity than the other contact tests. It is easy to handle and is carried out within one day. However, due to the limited experience with the test system, the solid contact assay should be evaluated in detail to assess its suitability as a standard method for the testing of waste.

Keywords Arthrobacter globiformis; Waste; Soil; Copper; PAH

21.1 Introduction

Biological methods in combination with chemical analysis can provide important information for the assessment of pollutants in waste. Microbial activities are used in standardized methods as ecologically relevant endpoints, to measure the activity or structure of the local microbial community (Lachmund et al. 2003; Kostanjšek et al. 2005). In addition to these community-based methods, there is also a need for the development of biotests, which use selected microorganisms for testing the toxicity of pollutants (Brandt et al. 2002; Shaw et al. 2000). In these tests, the

H. Neumann-Hensel
Dr. Fintelmann und Dr. Meyer GmbH, Chemische und biologische Handelslaboratorien, Mendelssohnstr. 15 D, 22761, Hamburg, Germany

H. Moser and J. Römbke (eds.), *Ecotoxicological Characterization of Waste*,

DOI: 10.1007/978-0-387-88959-7_21,

potential hazard of contaminants is evaluated by integrating their intrinsic toxicity and their availability to the test organisms (Brohon & Gourdon 2000). Another advantage of the utilization of cultivated microorganisms is the standardization and comparability of the test results. Usually, such tests with added microorganisms are performed with eluates of materials like soil or waste. In this case, information about the water soluble pollutants is obtained, but no information about the adsorbed subfraction of the pollutants is available (Ahlf 2004). Therefore, a so-called solid contact test is needed, in which the test organisms have direct contact with contaminated material (Feiler et al. 2005; Heise & Ahlf 2005; Hollert et al. 2003; Duft et al. 2002; Neumann-Hensel 2004). Such solid-contact tests may provide a different and maybe more comprehensive assessment of waste than corresponding aqueous extract tests (Shaw et al. 2000; Weber et al. 2006). One standardized solid contact assay is the assay with *Arthrobacter globiformis* (ISO 10871 Draft and DIN 38412 L48). This bioassay has already been recommended as a quick and sensitive method for the risk assessment of waste (LfU 2004). To optimize the test for routine use, it was miniaturized by changing the kind of detection (Neumann-Hensel & Melbye 2006). In addition, the utilization of freeze-dried bacteria was proposed in order to develop a test kit. The advantages of using freeze-dried bacteria include the possibility of long-term storage, easy handling and little effort in cultivation. To assess the ecotoxicological hazard potential of wastes to terrestrial organisms, the earthworm acute test was selected (CEN 2005). However, in order to clarify whether the bacteria contact test using *A. globiformis* may be an alternative or worthwhile addition, several participants performed this test as part of the additional test battery.

21.2 Methods

The test principle is the measurement of the dehydrogenase activity of the test organism *A. globiformis* after an incubation time of 2 h with the solid material. The test uses a redox-active dye (resazurine), which changes colour when reduced by bacterial dehydrogenase. Toxic substances can inhibit the rate of dye reduction. The freeze-dried or fresh sample is weighed into a 24-well microplate, with 0.6 g in each well, with four replicates. After adding 0.6 mL deionised water, the microplates are pasteurized to inactivate microorganisms. Then 0.4 mL test bacteria (which have to be in the logarithmic phase) are added in every well and the microplates are incubated for 2 h at 30°C on a horizontal shaker. After this contact time, the redox-dye resazurin (dissolved in potassium-phosphate-buffer with glucose) is added to each well. The microplates are shaken for 60 min at 30°C. During this time the dehydrogenase is determined by measuring the product resorufin every 15 min in the presence of the soil, using a fluorimeter (em. 535 nm, exc. 590 nm). The slope of the relative fluorescence of the sample in relation to the slope of uncontaminated reference gives the inhibition as percentage. An overview of the test procedure is given in Fig. 21.1. For testing waste, a geometric dilution series between waste and control material (sand W4) is used. The test waste is mixed with the control material using appropriate techniques. Mixture

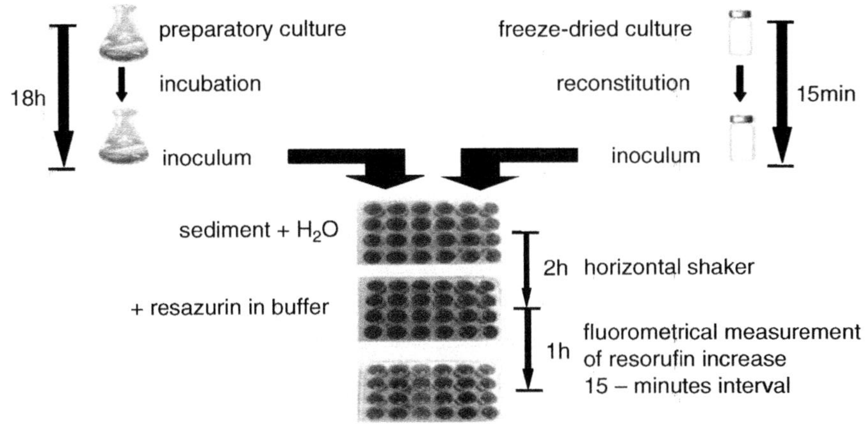

Fig. 21.1 Test procedure of the solid contact assay with *A. globiformis*

rates of 0%, 12.5%, 25%, 50% and 100% are suggested. Benzalkonium chloride (BAC) is recommended as reference substance. One soil was spiked with BAC in a concentration of 600 mg/kg.

The test is valid if the fluorescence of the control increases by a factor of more than 5 (measuring time 0–60 min) and the reference substance effects an inhibition between 30 and 80%.

The EC_{50} value (and respective 95% upper and lower confidence limits) for each waste was calculated by means of probit regression. However, in almost all cases (22 out of 25) the calculation of the confidence limits was not possible. Due to the low number of test results no specific statistical evaluation of the whole data set was performed.

21.3 Results

In total, ten participating laboratories performed the contact assay with *A. globiformis* (ISO 10871 Draft and DIN 38412 L48) and provided 28 data sets (Table 21.1). Because in three cases only a limit test was performed (and thus had to be classified as invalid) only 25 data sets could be evaluated. However, all these 25 tests fulfilled the validity criteria concerning the performance of the test.

The results of the *Arthrobacter* tests with INC, SOI and WOO are presented. In Table 21.2 the main test results, i.e. the EC_{50} values, are shown. Since in many tests the EC_{50} value was outside the tested range, only three INC, five SOI and six WOO data sets could be used for further evaluation.

By far the most toxic waste was WOO, while SOI and in particular INC were clearly less toxic to the bacteria. In those cases where detailed EC_{50} values could be determined, the range of toxicity was usually small: just a factor of three in WOO, six in INC and ten in SOI tests.

Table 21.1 Participating labs and number of tests performed

Substrate	No. labs	No. tests	Invalid tests	Accepted tests
INC	8	8	1	7
SOI	10	10	1	9
WOO	10	10	1	9
Sum	n.a.	28	3	25

n.a. Not applicable

Table 21.2 Compilation of the results (EC_{50} values) of the 28 tests with three wastes (data expressed as % waste in sand W4)

Lab Code	Substrates INC	SOI	WOO
2	43.4	3.2	1.3
12a	<21.9	<5.2	<0.8
12b	<20.1	<8.2	0.4
12c	>25.0	13.1	0.4
12d	n.p.	<16.2	0.4
12e	n.p.	2.7	0.5
29	17.3	27.8	<1.6
31	<12.5	<12.5	<25.0
33	7.6	15.2	0.5
Mean	22.6	12.4	0.6
Range	7.6–43.4	2.7–27.8	0.4–1.3

n.d. Not detectable; *.n.p.* Not performed

21.4 Discussion

Most of the participants had established the test to participate in the ring test. The implementation of the test in their laboratories ran smoothly. Methodologically, almost no problems were reported. The test is easy to handle and replicable. The noted range of the EC_{50} value is broad (see Table 21.2), mainly because several EC_{50} values were reported as "lower than" and not as an exact value.

The order of the toxicity found in this test is remarkable, as in most other tests performed in this ring test, INC was more toxic than SOI and the latter showed rarely any toxicity at all. However, here almost the opposite was found: SOI was more toxic than INC but both caused clear effects, meaning that the sensitivity of the bacteria contact test is high compared to the other additional terrestrial tests. The *Arthrobacter* test shows higher toxicity in WOO and SOI (mean value) compared to the reported effects in the other additional terrestrial tests. Therefore, and because of the toxicity specifics, this test system qualifies as a bioassay for waste testing, and can be integrated in a future test battery, after more experience with different waste materials has been gained.

21.5 Recommendations

Currently, it is not possible to recommend the inclusion of this test in the basic test set for wastes because of the limited experience with this test system. However, due to its high sensitivity, easy handling and performance, the solid contact assay with *A. globiformis* should be evaluated in detail to assess its suitability as a standard method for the testing of wastes.

Part III
Additional Investigations

Chapter 22
Reproducibility and Repeatability of the Results of the European Ring Test on the Ecotoxicological Characterisation of Waste

J. Römbke, C. Van der Wielen, and H. Moser

Abstract The evaluation of the outcome of a ring test has to assess the uncertainty of the results by determining the repeatability and the reproducibility of the data sets. This evaluation is the most important step in defining the validity of an individual test method or even a whole set of tests. The results of the aquatic and the terrestrial basic test set are statistically analyzed by two parameters: reproducibility standard deviation and repeatability standard deviation (ISO 5725-1, 1994).

The discussion of the statistical performance of the ring test is focused on the basic test battery and the waste material WOO and INC, fulfilling the given requirements such as number of data sets, calculable test results and valid test procedure according to the guidelines and SOPs (the test substrate SOI was excluded here due to its low toxicity; very often no EC_{50} could be determined).

The comparison of repeatability standard deviation s_r and reproducibility standard deviation s_R covers a range between 2.16 and 15.0. 42% of the biotest results are within a s_R/s_r ratio of 4 and about 85% are within a range for the s_R/s_r ratio from 2 and 8, taking into account that this range already covers influence factors like eluate preparation or the use of different test species. It can be summarized that the basic test set forms a reliable methodology for waste ecotoxicity tests, which can be optimized with experiences gained in the ring test (see also Chap. 1.5)

Keywords Basic test battery, Laboratory tests, Algae, Daphnia, Daphnia, Repeatability, Reproducibility, Evaluation of test results

---

J. Römbke
ECT Oekotoxikologie GmbH, Boettgerstr. 2-14, 65439, Floersheim, Germany

H. Moser and J. Römbke (eds.), *Ecotoxicological Characterization of Waste*,
DOI: 10.1007/978-0-387-88959-7_22, © Springer Science+Business Media, LLC 2009

22.1 Introduction

When assessing the results of any ecotoxicological test, it is important to have an idea of the uncertainty level of the results. This is especially important when evaluating the outcome of a ring test, which is by definition one of the most important steps in the process of defining the validity of an individual test method or even a whole set of tests. For this purpose, international guidelines (e.g. ISO 1994; ISO 2002) recommend determining the repeatability and the reproducibility of such data sets. These measures of uncertainty are defined as follows:

Reproducibility standard deviation s_R (ISO 5725-1 1994) is a measure of the dispersion of the distribution under reproducibility conditions, i.e. conditions where test results are obtained with the same method on identical test items in different laboratories with different operators using different equipment.

Repeatability standard deviation s_r (ISO 5725-1 1994) is a measure of dispersion of the distribution under repeatability conditions, i.e. conditions where independent test results are obtained with the same method on identical test items in the same laboratory by the same operator using the same equipment within short intervals of time.

In the following sections, reproducibility is briefly discussed, as the respective information has already been presented in the individual chapters describing the ring test results. In contrast, the repeatability of the ring test results has not been discussed so far, so this measure of uncertainty is presented in detail here. In both cases, the focus is on the basic test set (Table 22.1). A similar evaluation of the additional test set is – with a few exceptions – not possible due to the low number of data sets provided (usually < eight per test substrate). All LC/EC_{50} values reported here were re-calculated using the ToxRat® (2003) program.

As discussed in Chap. 4, any statistical evaluation of a ring test must fulfill certain requirements as laid down in ISO guideline 5725-2 (2002). In particular, all participants have to use the same method, preferably as described in a Standard Operation Procedure (SOP). For the calculation of the repeatability standard deviation, every laboratory should report the same number of test results ($n \geq 2$)

Table 22.1 Brief overview on the five tests belonging to the basic test set

Name	Guideline	Species
Eluate (aquatic) waste tests		
Algae	ISO 8692 (2004)	*Desmodesmus subspicatus* or *Pseudokirchneriella subcapitata*
Daphnia	ISO 6341 (1996)	*Daphnia magna*
Luminescent bacteria	ISO 11348-1/2 (2005)	*Vibrio fischeri* (3 sources, alternatively)
Solid (terrestrial) waste tests		
Earthworms (acute)	ISO 11268-1 (1997)	*Eisenia fetida* or *Eisenia andrei*
Plants	ISO 11269-2 (2004b)	*Avena sativa* and *Brassica rapa*

per test and test sample. In addition, there must be enough participants to calculate the reproducibility (the minimum is considered to be eight laboratories). In this context it must be noted that the data presented have to be valid according to the respective test guidelines. In addition, a distinctive test endpoint must be calculable (i.e. results like $EC_{50} > 50\%$ are not suitable). For this reason, in the following sections, no results of tests with SOI will be discussed because of the low toxicity of this waste substrate.

22.2 Reproducibility

In the ring test on the ecotoxicological characterization of waste, the requirements for assessing reproducibility were mostly fulfilled for the tests belonging to the basic test battery (Table 22.1; for details see Chaps. 7–11), as SOPs were provided for them. Tests which violated these SOPs or which did not fulfill the validity criteria or missed the expected range of results of reference tests were not accepted for further statistical evaluation. Also, statistical outliers as defined in the ISO guidelines (1994, 2002) and those data sets outside the warning limits as defined by Environment Canada (2005) were not considered – see Chap. 4 for details of the statistical evaluation. In addition, for each test, far more than eight valid data sets were provided (between 52 (earthworms) and 161 (daphnids)).

In the case of the aquatic tests, the variability of the results presented in this chapter integrates not only the variability of the ecotoxicity tests, but also the variability of the entire procedure: sampling of the test portion, eluate preparation (leaching procedure, separation of the solid and liquid phases) and ecotoxicity test. A preliminary analysis of the raw data provided by the participating labs clearly showed that some of them did not strictly follow EN 14735 (CEN 2005) for eluate preparation (e.g. amount of waste for the leaching test, volume of eluate collected) and for the storage time of eluates before performing aquatic tests (see Chap. 28).

Due to this complexity the reproducibility of the aquatic tests is not as good as that of the terrestrial tests, for instance when looking at the factor between the minimum and maximum EC/LC_{50} values or at the factor between the lower and upper limit of the warning limits (Table 22.2). The highest factors for the terrestrial tests are 19 and 22 for the lower and upper limits and the minimum–maximum values, respectively. For the aquatic tests, the numbers for the same factors are 108 and 138. Due to the low toxicity, the factors are always lowest for SOI. With the exception of Algae tests with the species *Pseudokirchneriella subcapitata,* the factors are particularly high when results of tests with two species (Algae) or tests with different sources (luminescent bacteria) are combined. In fact most of the "non-combined" data sets showed satisfying results, especially in the tests with the medium-toxic INC in contrast to the highly toxic substrate WOO. In addition, it is worthwhile to mention that both ways of expressing the reproducibility of test results did not differ considerably.

Table 22.2 EC_{50} values (% waste) and their minimum–maximum factor based on all accepted data minus those outside of the warning limits

Test system: Species/ source Aquatic tests	Factor between lower und upper limits of the 95% confidence intervals			LC/EC_{50} values: Minimum–maximum factor		
	INC	SOI	WOO	INC	SOI	WOO
Algae: all	85	4	94	59	n.a.	138
D. subspicatus	56	n.a.	3	15	n.a.	2
P. subcapitata	31	n.a.	55	20	n.a.	138
D. magna	7	n.a.	30	10	n.a.	31
Lumi-Bacteria: all	10	2	108	8	2	122
Lumi-Bacteria: freeze	8	2	21	8	n.a.	20
Lumi-Bacteria: liquid/ fresh	n.a.	n.a.	5	n.a.	n.a.	11
Terrestrial tests						
E. fetida/E. andrei	2	n.a.	5	2	n.a.	3
A. sativa	13	n.a.	15	17	n.a.	10
B. rapa	19	4	16	22	3	11

n.a. Not applicable: no EC_{50} determinable due to low toxicity

22.3 Repeatability

During the ring test, some labs repeated the tests with a particular test system on the different waste substrates several times. However, most of the participating labs reported single test results only, because the performance of the tests is tedious and elaborate. This is especially true of the terrestrial tests: due to the effort related to the performance of the long (about two weeks) earthworm and plant tests, respectively, such repetitions were made only with aquatic tests. These results will be presented in the following sections, using Algae and Daphnia tests as examples. In addition, some experiences gained from other tests (luminescent bacteria, Lemna, umu and Arthrobacter; see Chaps. 9, 12, 16, 21) are discussed. Usually they were performed with the most toxic test substrate WOO.

All the requirements which have been listed in Section 22.2 are also valid when evaluating the repeatability of the data sets (such as the use of SOPs etc.). Even those who had reported more than one result might not meet the demand "within short intervals of time", as the time interval between two tests was often two weeks or even one month and more. Therefore, it is doubtful whether these time intervals meet the requirement for repeatability of ISO 5725-2 (2002).

In Table 22.3 the results of 17 INC and 15 WOO Algae tests performed in five and six labs, respectively, are compiled, showing the following tendencies:

- The range of EC_{50} values increases when different test conditions are integrated (e.g. different test vessels).
- The size of the test vessels does not seem to be the major factor of influence concerning intralaboratory repeatability.
- The variability of the INC tests is higher than in the WOO tests.

Table 22.3 Overview on repeated Algae tests with the substrates INC and WOO

Participant (Code No.)	No. of test performed	Mean EC50 (% eluate)	Coefficient of Variance	Range in % of mean EC_{50}	Test conditions
Test substrate: INC					
1a,b	2	10.8	23	33	D.s.: V, M
12a,b	2	3.9	64	91	D.s.: M
12c,d	2	32.5	24	35	D.s.: V
12a,b,c,d	4	3.2	53	113	D.s.: V, M
36a,b	2	1.3	–	19	P.s.: M
49a,b,c	3	26.3	29	50	D.s.: V
61a,b	2	6.8	9	13	P.s.: V, M
Test substrate: WOO					
1a,b	2	1.2	6	9	D.s.: V, M
12c,d	2	1.0	8	12	D.s.: V
12b,c,d	3	1.2	25	87	D.s.: V, M
36a,b	2	0.2	–	25	P.s.: M
49a,b	2	1.4	1	2	D.s.: V
51a,b	2	1.2	20	28	D.s.: V
61a,b	2	0.2	7	10	P.s.: V, M

D.s. D. subspicatus; *P.s. P. subcapitata*; *V* Big test vessels; *M* Microplates

Table 22.4 Overview on repeated *Daphnia* tests with the substrates INC and WOO

Participant (Code No.)	No. of test performed	Mean EC^{50} (% eluate)	Coefficient of Variance	Range in % of mean EC_{50}	Test conditions
Test substrate: INC					
12a,b,c	3	2.2	8	15	V, D
33a,b	2	4.3	11	16	V, S
36a,b	2	1.9	13	18	V, D
49b,c	2	3.0	–	6	V, S
50a,b	2	7.0	–	50	V, S
51a,b	2	3.6	8	12	V, S
Test substrate: WOO					
8a,b	2	0.32	59	83	V, ?
12a,b	2	0.12	1	2	V, D
21a,b	2	0.26	9	13	V, ?
36a,b	2	0.16	37	52	V, D
49a,b,c	3	0.21	11	22	V, S
50a,b	2	0.85	1	1	V, S
51a,b	2	0.23	14	19	V, S

V Big test vessels; *S* Same eluate; *D* Different eluates

The same tendencies were observed in a comparable study in three French labs performed with the same test substrates (Pandard et al. 2007), despite the fact that in the ring test, due to the inclusion of the less sensitive species *Desmodesmus subspicatus,* the overall range of EC_{50} values was clearly broader.

In Table 22.4 the results of 13 INC and 15 WOO Daphnia tests performed in six and seven labs, respectively, are compiled, leading to the following statements:

- The same repeatability is observed using the same or different eluates.
- The variability of the WOO tests is higher than in the INC tests.

The second conclusion is not backed up by the French ring test, maybe because the number of tests performed (5–10) was higher than in this ring test (usually, $n = 2$).

In the case of the tests with luminescent bacteria, many test results (i.e. EC_{50} values) with INC and SOI wastes could not be determined as they were outside the tested range (usually higher than the highest test concentration). Therefore, the following evaluation of repeatability focuses on the tests with WOO (Table 22.5). In this compilation it is obvious that the tests performed with freeze-dried bacteria (lab No. 36) differ clearly in terms of sensitivity from those tests performed with fresh/liquid bacteria (see Chap. 9 for details). However, there is no clear difference between these two sources in terms of variability. Concerning the coefficient of variance as well as the range of the individual test values, expressed in percent of the mean EC_{50}, the numbers determined for the luminescent bacteria tests are even lower than those found in the Daphnia and, in particular, the Algae tests. In the literature, referring to tests repeated 5–10 times, CVs of 12–18% are reported, which is in good agreement with the results presented here (Pandard et al. 2007).

In the following sections, the information about repeated tests performed with three other test systems is summarized (all of them conducted by Laboratory No. 12):

Lemna-test. The test with WOO was repeated five times and with INC and SOI, twice for each. For WOO, the mean EC_{50} was 2.46 ± 0.32% (range: 2.12–2.98%), meaning that there was no significant difference between the individual test runs. Also the EC_{50} values determined in the INC and SOI tests were very similar.

Arthrobacter-test. INC, SOI and WOO substrates were tested three, five and five times, respectively. Since many of these tests (in particular with INC and SOI) did not reveal EC_{50} values (just ranges), no detailed evaluation of the repeatability of these results is possible.

Umu-test. All three substrates were studied two to three times. No differences at all were found in these tests (i.e. at the highest eluate concentration of 66.7% no indication of genotoxicity was found).

Despite the small data base the information revealed from these tests shows an acceptable variability, i.e. the repeatability of the results is high.

Table 22.5 Overview on repeated luminescent bacteria tests with the substrate WOO

Participant (Code No.)	No. of test performed	Mean EC_{50} (% eluate)	Coefficient of Variance	Range in % of mean EC_{50}	Test conditions
Test substrate: WOO					
12a,b,c,d,e	5	7.2	11	29	F/L
13a,b,c,d,e	5	4.8	9	23	F/L
21a,b	2	4.7	26	36	F/L
33a,b	2	5.6	5	7	F/L
36a,b	2	0.6	29	41	Fr
49a,b,c,d,e,f	6	4.6	19	55	F/L
51a,b	2	7.3	10	15	F/L

22.4 Discussion

The first possibility for the assessment of variation is the comparison of repeatability standard deviation s_r and reproducibility standard deviation s_R (see Chap. 4). In chemical and physicochemical ring tests a s_R/s_r ratio of less than 4 is an indication of robustness. In the ecotoxicological ring test with wastes, this ratio covers a range between 2.16 and 15.0 (Table 22.6). About 42% of all ratios are within the "chemical" range, while 83% are located between 2 and 8. In fact, several of the high ratios can be explained either by combining results which are hardly comparable (e.g. different Algae species or different sources of luminescent bacteria) or which may have been influenced by low toxicity (and thus a low number of distinct EC/LC_{50} values) as in the case of the SOI tests. Taking into account also the differences between chemical/physicochemical and ecotoxicological ring tests, this variation is considered acceptable. In other words, for ecotoxicological ring tests with such substrates like wastes these results show that the basic test set forms a reliable methodology for waste ecotoxicity tests.

However, it is worthwhile to mention that despite an unavoidable inhomogenity of the waste substrate and a greater time interval between repeated tests, the EC/LC_{50} values generated with a particular test system for a particular test substrate did not differ considerably.

Finally, some ideas concerning the improvement of test results (i.e. a decrease of uncertainty) are briefly outlined:

- One tool well known in science is the concept based on the causal effect diagram (Fig. 22.1) as proposed by Ishikawa (1963; as cited in Kamiske & Brauer 1999). In a step-wise approach the total variability is divided into its various (potential) causes, i.e. material, method, test substrate, and environmental conditions of the operator.

Table 22.6 Comparison of the ratio between reproducibility and repeatability (Ratio s_R/s_r) to evaluate the overall robustness of the test results

Test system: Species/source Aquatic tests	Ratio between reproducibility and repeatability (Ratio s_R/s_r)		
	INC	SOI	WOO
Algae: all	6.89	4.09	7.47
D. subspicatus	5.46	n.a.	2.16
P. subcapitata	5.90	n.a.	5.97
D. magna	3.75	n.a.	7.38
Lumi-Bacteria: all	4.70	3.24	11.6
Lumi-Bacteria: freeze	3.86	2.32	6.38
Lumi-Bacteria: liquid/fresh	n.a.	n.a.	5.03
Terrestrial tests			
E. fetida/E. andrei	2.26	n.a.	2.73
A. sativa	3.20	15.0	3.19
B. rapa	9.67	11.1	2.59

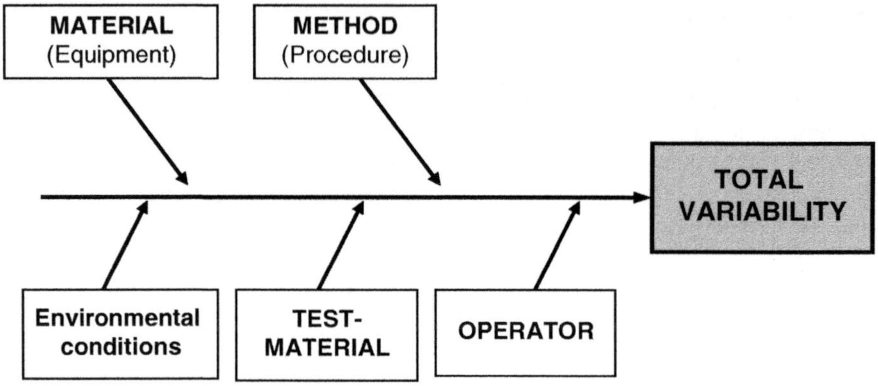

Fig. 22.1 Causal effect diagram (Ishikawa (1963), as cited in Kamiske & Brauer 1999)

- From a scientific point of view the exchange of endpoints may be helpful, but this is very test system-specific and cannot be used for all the aquatic tests discussed here (probably most easily for sublethal tests like the Algae test).
- Finally, and probably most importantly, reproducibility and repeatability can be improved by constant training of the test performers and a well established test quality assessment.

22.5 Recommendations

1. As the ring test was not designed for repeatability evaluations, labs should be asked to provide more than one result per test and test sample (for aquatic tests only) to enable calculation of repeatability according to ISO 5725-2.
2. Such repeatability tests should be performed with ten replicates. A special experimental design is needed to identify and/or confirm factors influencing reproducibility and repeatability.
3. Available information from the ring test (see also Pandard et al. 2007) indicates that the influence of the elution step on repeatability is probably small.
4. Since an enhancement of reproducibility and repeatability is a daily exercise when performing tests, tools such as effect cause diagrams, control chart (=warning limits) or studies to identify causality should (regularly) be undertaken. But most important is to continue training.

Chapter 23
Comparison Between Toxkit Microbiotests and Standard Tests

G. Persoone and K. Wadhia

Abstract In addition to their practicality and user-friendliness, Toxkit microbiotests have the major advantage over "standard" bioassays in that they are all independent of the (costly and time consuming) year-round culturing of live stocks of the test species, which makes them very popular for routine applications in aquatic and terrestrial ecotoxicology. In the framework of the international ring test for the characterization of waste, several labs from eight European countries performed assays with various Toxkits. A detailed discussion of the results of these microbiotests, with specific considerations of their interlaboratory variability, is given in Chap. 13 of this volume. In the present chapter, Toxkit data are compared with the results obtained with "standard tests". The findings confirmed those obtained in recent years from similar intra-lab and inter-lab comparative studies on pure chemicals and on mixed environmental samples, namely that the sensitivity of the Toxkit microbiotests is identical to that of the standard tests with the same test species. The present study furthermore revealed that the precision (in terms of variation coefficients) of the Algaltoxkit and Daphtoxkit microbiotests was better than that of the standard tests. Based on the previous findings and taking into account the advantages of microbiotests, it is recommended that the Algaltoxkit and Daphtoxkit should be taken into consideration as valid alternatives to the standard algal growth inhibition assay with *Pseudokirchneriella subcapitata* and the acute *Daphnia magna* test for the evaluation of the toxicity of solid waste eluates.

Keywords Toxkit, Eluate tests, Interlaboratory variability, Methodological comparison

23.1 Introduction

As mentioned in the Introduction to Chap. 13 of this volume, acute and short chronic Toxkit microbiotests are available for various test species. In addition to being practical and user-friendly, these assays have a major advantage over "conventional"

G. Persoone (✉) and K. Wadhai
MicroBioTests Inc. Kleimoer 15, 9030 Mariakerke-Gent, Belgium

H. Moser and J. Römbke (eds.), *Ecotoxicological Characterization of Waste*,
DOI: 10.1007/978-0-387-88959-7_23,

toxicity tests: they are independent of the (costly and time-consuming) year-round culturing of live stocks of the test organisms. This asset makes these "culture/maintenance free" Toxkit assays particularly attractive for routine toxicity testing of aquatic as well as terrestrial environments (Persoone et al. 2000). As the results of "additional test methods" (beyond the five selected standard bioassays) could also be submitted in the framework of the European ring test on the ecotoxicological characterization of waste and waste eluates, a number of labs from eight European countries performed assays with different Toxkit microbiotests (Algaltoxkit, Daphtoxkit, Thamnotoxkit, chronic Rotoxkit and Phytotoxkit). The results of these microbiotests, and in particular the interlaboratory variability, were analyzed and the findings are discussed in detail in Chap. 13 of this volume. The objective of the present chapter is to compare the results of the Toxkit microbiotests with those of the standard tests performed by the other labs which participated in this international ring test on wastes. The emphasis is on the comparison of the Algaltoxkit data with those of the standard algal growth inhibition assay with *Pseudokirchneriella subcapitata*, and on the Daphtoxkit results with those of the standard acute *Daphnia magna* test, both in terms of sensitivity of the assays and their precision. The results of the chronic Rotoxkit tests have been compared with those of the chronic rotifer tests performed in different test containers and with a different algal species used to feed the rotifers during the exposure time. In order to permit ease of comparison, the rotifer tests performed according to the methodology of the pending ISO norm for chronic rotifer tests has been referred to as "standard" chronic rotifer tests, and the alternative assay as the chronic "Rotoxkit" microbiotests.

23.2 Methods

As mentioned in Chap. 13, Toxkit microbiotests are performed with the materials included in each kit, and by following the specific detailed "Standard Operational Procedure" of the particular bioassay (see references Algaltoxkit F (2004), Daphtoxkit F (2001), Thamnotoxkit F (2001), chronic Rotoxkit F (2006) and Phytotoxkit (2004)). The Algaltoxkit and Daphtoxkit test procedures strictly follow the ISO and OECD Guidelines for the 72 h algal growth inhibition test and the acute *D. magna* test respectively. In the case of the Daphtoxkit assays only the 24 h results were assessed as the 48 h data were not required to be submitted for the standard *D. magna* assay. The only difference with the standard tests and the Toxkit microbiotests is that the micro algae needed for the Algaltoxkit assay originate from "algal beads" and for the Daphtoxkit test the test organisms are hatched from dormant eggs (ephippia). The species used in the Algaltoxkit microbiotest is *P. subcapitata.*

The Thamnotoxkit microbiotest is a 24 h mortality test, performed on larvae hatched from cysts of the anostracan crustacean *Thamnocephalus platyurus*, and the chronic Rotoxkit is a 48 h reproduction inhibition assay performed using the rotifer *Brachionus calyciflorus*, also hatched from cysts. The Rotoxkit essentially follows the French standard (norm) for the determination of chronic toxicity to *B. calyciflorus* in 48 h (AFNOR NF 90-377 2000), but employs slightly different test containers and a different type of algae (stored in algal beads). The AFNOR constitutes the basis of the "48 h population growth inhibition test" and is presently in the final stage of evaluation by ISO for acceptance of this short-chronic assay (ISO/CD 20666 2005).

The Phytotoxkit microbiotest is a 3-day seed germination and early plant growth microbiotest performed in flat transparent test containers, with measurement of root and shoot growth by image analysis. In the present ring test, the assays were not performed with the three plant species employed in the Phytotoxkit, but with the two "imposed" plant species (*Avena sativa* and *Brassica rapa*). Furthermore the test duration of the assay was 4 days and only the root length was measured.

The evaluation of the Toxkit test results reported in Chap. 13 was made by calculating the geometrical means of the EC_{50} values, the corresponding standard deviations and coefficients of variance, based on the Warning Limit Approach (EC 2005). For the standard tests, the 72 h EC_{50}s for the algal assays with *P. subcapitata* and the 24 h EC_{50}s for the *D. magna* tests were kindly provided by the organisers of the ring test. The comparative assessment of the Toxkit data with the results of the standard *P. subcapitata* and *D. magna* tests was implemented by determining the geometrical mean of the EC_{50}s values, with standard deviation and coefficient of variance.

23.3 Results

Table 23.1 shows the total number of results for the standard test and the Toxkit microbiotests for algae, *D. magna* and rotifers, and the relative number of "acceptable" results. It should be emphasized that for the standard tests with algae, the comparison is limited to the tests performed using *P. subcapitata,* and that no distinction has been made between the results obtained in test vessels (of different volumes) and in microplates.

From the table, the deductions that can be made are as follows:

- The number of results for the *Daphnia* tests was twice as many as those of the algal tests. However, for the standard algal tests, the figure only reflects the results obtained with *P. subcapitata*. Many standard algal tests were also performed with *Desmodesmus subspicatus*, the second species of micro-algae indicated in the

Table 23.1 Summary of the number of standard tests and Toxkit microbiotests performed using algae, *Daphnia* and rotifers, and the number of accepted tests (in parenthesis)

Substrate	Standard algal test	Algaltoxkit	Standard Daphnia test	Daphtoxkit	Standard rotifer test	Rotoxkit
INC	20 (17)	7 (5)	43 (39)	11 (8)	3 (3)	4 (4)
SOI	21 (14)	7 (7)	43 (40)	11 (11)	3 (3)	3 (3)
WOO	19 (16)	7 (6)	42 (40)	11 (11)	3 (3)	3 (3)
Sum	60 (47)	21 (18)	128 (119)	33 (30)	9 (9)	10 (10)
Percentage acceptable	78	86	93	90	100	100

() Number of acceptable tests

ISO norm for algal tests. In order to make a strictly fair comparison of the Algaltoxkit test results with those of the standard algal assay, only the data obtained with the same test species (*P. subcapitata*) were examined in the evaluation.

- The number of standard algal tests with *P. subcapitata* was more than twice as many as the number of Algaltoxkit tests and for the acute *Daphnia* assays; there were in fact four times more test results conveyed for the standard test than for the Daphtoxkit microbiotest.
- The percent of "acceptable" results for the algal tests and the *Daphnia* tests was relatively high (from 78 to 92%) for both the standard tests and the microbiotests. The number of chronic rotifer tests performed with the Toxkit method and the (pending) ISO "standard" test procedure were much fewer (only 3–4 results for each waste) in comparison with the algal and *Daphnia* tests, but all the test results were found to be acceptable.

The mean EC_{50} values of the acceptable test results with the corresponding variation coefficients have been calculated for the standard tests and the Toxkit microbiotests for algae, Daphnia and rotifers. They are given in Table 23.2.

The following salient points may be seen from this table:

- The mean EC_{50} values for the standard tests and the Toxkit microbiotests were of relatively the same magnitude for the three test species.
- The assays performed to test SOI waste eluate with all three test species confirmed that this particular waste was not (very) toxic and that in most cases no effects were evident even in the undiluted (100%) waste eluate.
- The mean EC_{50} variation coefficients differed for the three test species and those between the standard tests and the microbiotests exhibited a substantial range.
- The variation coefficients for the INC waste eluate and the WOO eluate for the standard algal and *Daphnia* tests were considerably different.
- Both the standard algal test and the Algaltoxkit microbiotest have very high variation coefficients, but the CVs for both the INC and the WOO waste eluate were lower for the microbiotest than for the standard test.
- The CV of the standard *Daphnia* test was also relatively high for the WOO waste eluate (90%); and for both types of wastes the CVs were lower for the microbiotest than for the standard test.
- For the rotifer tests, the CVs of the (few) EC_{50}s were on the lower side (16–29%), with the exception of the WOO Rotoxkit results (40%).

Table 23.2 Summary of the mean EC_{50}s (in percentage dilution of the eluates) and the corresponding variation coefficients (in parenthesis and in bold font) for the standard tests and the Toxkit microbiotests for algae, *Daphnia* and rotifers

Substrate	Standard algal test	Algaltoxkit	Standard *Daphnia* test	Daphtoxkit	Standard rotifer test	Rotoxkit
INC	3.31 (94%)	4.24 (79%)	3.32 (55%)	2.41 (41%)	4.44 (25%)	5.35 (16%)
SOI	>100 (na)	>100 (na)	>100 (na)	>100 (na)	>100 (na)	>100 (na)
WOO	0.42 (151%)	0.36 (93%)	0.42 (90%)	0.61 (32%)	0.09 (29%)	0.13 (40%)

na Not applicable

In order to provide a better perspective of the individual EC_{50} results of the Algaltoxkit and Daphtoxkit tests in comparison with those of the standard algal and *Daphnia* assay, results have been transposed into graphical format (Figs. 23.1–23.4). This consists of the microbiotest EC_{50} values marked as individual points in relation to the mean EC_{50} of the standard tests with ± 1SD and ± 2SD (warning limits) presented as line graphs.

Figures 23.1–23.4 show that virtually all the Toxkit microbiotest EC_{50}s appear between the mean EC_{50} and ± 1SD (of the corresponding standard tests), and that each Toxkit EC_{50} is within the 2SD warning limit range. The only exception is the result from an Algaltoxkit, which is slightly outside the upper warning limit (Fig. 23.1).

In view of the small dataset of the rotifer results, a similar graphical compilation was not feasible; however, all Rotoxkit EC_{50} data corresponded reasonably well with those of the standard chronic rotifer tests.

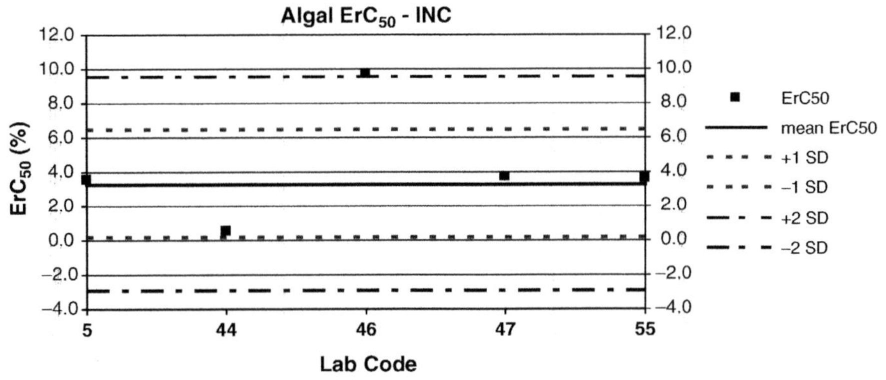

Fig. 23.1 ErC_{50} values (72 h) for the Algaltoxkit tests for the INC waste plotted in relation to the mean ± 1SD and ± 2SD (warning limits) for the standard tests

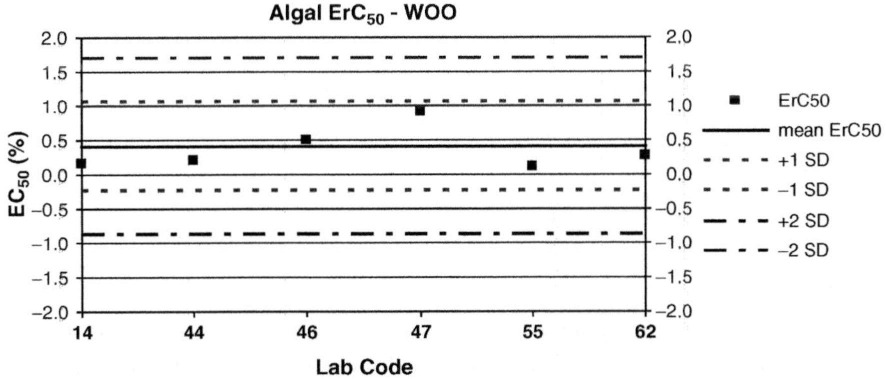

Fig. 23.2 ErC_{50} values (72 h) for the Algaltoxkit tests for the WOO waste plotted in relation to the mean ± 1SD and ± 2SD (warning limits) for the standard tests

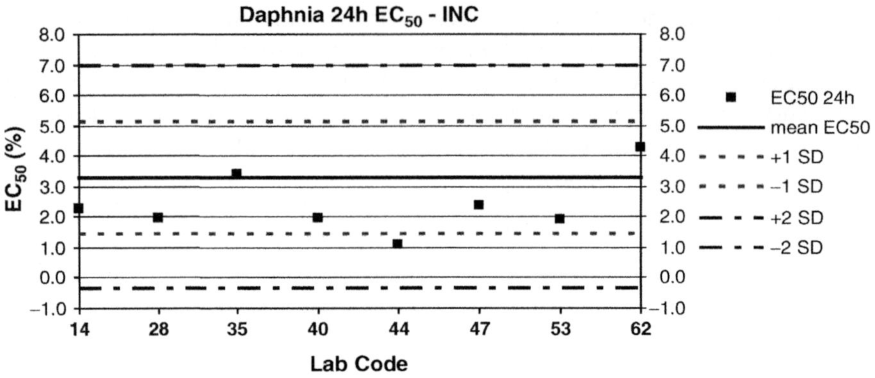

Fig. 23.3 EC_{50} values (24 h) for the Daphtoxkit tests for the INC waste plotted in relation to the mean ± 1SD and ± 2SD (warning limits) for the standard tests

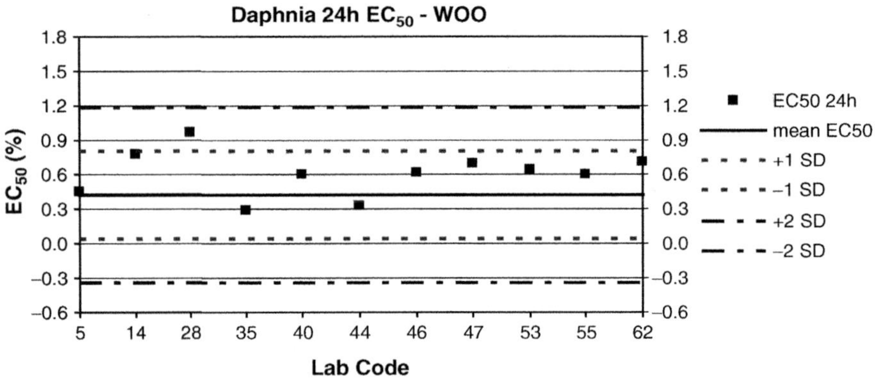

Fig. 23.4 EC_{50} values (24 h) for the Daphtoxkit tests for the WOO waste plotted in relation to the mean ± 1SD and ± 2SD (warning limits) for the standard tests

Table 23.3 Mean EC_{50}s of the standard *Daphnia* test and Daphtoxkit vs. the EC_{50} obtained with the Thamnotoxkit for the three waste eluates

Substrate	Standard *Daphnia* test	Daphtoxkit	Thamnotoxkit
INC	3.32	2.41	13.4
SOI	>100	>100	>100
WOO	0.42	0.61	0.31

With reference to the Thamnotoxkit, of the two results submitted, only one was considered acceptable. The EC_{50} results for this microbiotest, the mean EC_{50}s of the standard *Daphnia* test and the Daphtoxkit for the three waste eluates, are given in Table 23.3.

Interestingly, from this table it can be noted that the EC_{50} of the Thamnotoxkit (the test species for which is an aquatic crustacean phylogenetically closely related to Daphnids) is substantially higher for the INC waste eluate than the mean EC_{50} of the *Daphnia* tests, but for the WOO waste eluate the Thamnotoxkit EC_{50} is in contrast somewhat lower than the mean EC_{50} of the two *Daphnia* assays. This first appraisal, based on a very small number of data sets, gives a good starting point for further investigations on the sensitivity of the Thamnotoxkit vs. that of other toxicity tests.

Table 23.4 gives a comparison of the mean EC_{50}s of the standard plant tests and the EC_{50} (only one test result for each waste) obtained with the Phytotoxkit using the two designated plant species (*A. sativa* and *B. rapa*). The results for the standard plant tests have been taken from Chap. 10 of this volume, which deals specifically with the plant tests.

The test methodologies and the effect criteria of the standard plant test and the Phytotoxkit microbiotest are totally different: the standard test is indeed carried out in pots and measures the shoot biomass after 14 or 21 days, whereas the Phytotoxkit is carried out in flat transparent test containers and determines root length by image analysis after 4 days.

Despite these methodological differences, it is interesting to note that the EC_{50}s of the Phytotoxkit tests were found to be relatively close to the mean EC_{50}s of the standard plant tests (e.g. both indicating no effect of SOI to both plant species). With the exception of the test with *A. sativa* and INC, the EC_{50} values of the other three Phytotoxkits with INC and WOO are, however, higher than those of the respective standard plant tests.

23.4 Discussion

The good correlation between Toxkit data and results obtained with standard tests actually confirms the results of the many studies which have already been implemented on the sensitivity of the Algaltoxkit and the Daphtoxkit vs. that of the "standard" ISO and OECD assays with this test species. Within-lab (intra) and

Table 23.4 Mean EC_{50}s of the standard plant tests and the EC_{50} obtained with the Phytotoxkit for the three wastes

Substrate	Standard test *Avena sativa*	Phytotoxkit *Avena sativa*	Standard test *Brassica rapa*	Phytotoxkit *Brassica rapa*
INC	29.4	24.3	23.9	50
SOI	57.8[a] to >100[b]	>50[c]	63.0	>50[c]
WOO	10.0	32.9	2.64	4.38

[a] Six results
[b] 12 results
[c] Highest concentration tested

between-lab (inter) comparisons have been performed during recent years with the Algaltoxkit on pure chemicals (Persoone 1998a; Van der Wielen & Halleux 2000; Lucivjanska et al. 2000; Persoone & Janssen 2006), on sediment pore waters (Vangheluwe et al. 2000), on waste leachates (Latif & Zach 2000), on fly ash leachate (Lucivjanska et al. 2000) and on industrial effluents (Daniel et al. 2004). For the Daphtoxkit, similar comparative studies with the standard *D. magna* assay have been published for pure chemicals (Persoone 1998b; Fochtman 2000; Lucivjanska et al. 2000; Venturin & Ros 2003; Baudo et al. 2004; Baudo et al. 2008), household products (Ulm et al. 2000), waste leachates (Latif & Zach 2000), fly ash leachate (Lucivjanska et al. 2000) and industrial effluents (Daniel et al. 2004).

Table 23.2 nevertheless shows that variation coefficients of the Toxkit results for the waste eluates were high for the algal bioassay (79–93%), but also relatively high for the Daphtoxkit and the chronic Rotoxkit for some waste eluates. In most cases, the CVs were above the 30% limit indicated by Environment Canada for inter-laboratory repeatability of toxicity tests (EC 2005). However, the data in Table 23.2 also show that the inter-laboratory precision of the microbiotests is at least as good (if not better) than that of the corresponding standard assays. The CVs of the Algaltoxkit and the Daphtoxkit assays are indeed in all cases lower (and in several cases even substantially lower) than those of the standard tests.

As discussed in other chapters of this volume, the high CVs found for the different types of assays performed on the waste eluates are probably related to a great extent to "technical" problems (homogeneity of the wastes and the preparation of the eluates) rather than to the methodology of the bioassays.

As only one test result for each type of waste was available, no firm conclusions can be drawn on the sensitivity of the Thamnotoxkit microbiotest vs. that of the *Daphnia* assay. The EC_{50}s of this alternative crustacean microbiotest show that *T. platyurus* was less sensitive than *D.magna* for the eluate of the INC waste, but was well within the same sensitivity range for the EC_{50}s of the SOI and the WOO waste eluate.

Although one of the four Phytotoxkit EC_{50} values was outside the "warning limits" (mean + 2 SD) of the standard plant assays, the other results for this new assay (which is more rapid and much simpler than the standard tests in pots) are relatively close to the mean EC_{50}s of the standard plant tests.

All the facts and figures given and discussed above again corroborate the results of former studies on pure chemicals and on environmental samples, and reveal the potential of the microbiotests – and in particular that of the Algaltoxkit and the Daphtoxkit – as practical and reliable bioassays for the evaluation of the toxicity of solid wastes. The conclusion reached in a recent paper on the results of "a demonstration project to assess the comparative sensitivity and practicality of new assays in comparison to standard tests for the assessment of the toxicity of industrial effluents" (Daniel et al. 2004) therefore also seems to be valid for "solid waste eluates", namely that "the Algaltoxkit and the Daphtoxkit have the best overall correlations to the standard test and are well-suited for cost-effective monitoring of industrial effluents".

23.5 Recommendations

The facts and figures discussed in Chap. 8 on the Toxkit microbiotest results submitted for the European ring test on wastes confirm the robustness and the precision of these assays as already reported in numerous scientific papers.

The comparison of the Toxkit results with those of the standard tests also confirm (once again) that the "culture-maintenance free" Algaltoxkit and Daphtoxkit microbiotests have the same sensitivity as the standard tests using test species cultured in the lab.

It is therefore recommended that the former two microbiotests should be taken into consideration as suitable alternatives to the standard tests with *D. magna* and micro-algae, for the ecotoxicological characterization of wastes and waste eluates.

In view of the interesting correlation of the Phytotoxkit results with those of the standard plant test, and taking into consideration the shorter duration of the former assay and its practicality, it is also suggested that further research should be performed on toxicity evaluations of solid wastes with this microbiotest.

Although the rotifers were found to be more sensitive than the crustaceans and the micro-algae for WOO waste eluate, it is nevertheless highly probable that if the acute *D. magna* test was prolonged to 48 h (as recommended in the ISO norm and imposed in the OECD norm) this bioassay would probably have resulted in most cases to be as sensitive as the (more complex) 48 h rotifer tests.

Chapter 24
Comparison and Characterization of OECD Artificial Soils

J. Hofman, I. Hovorková, and J. Machát

Abstract OECD artificial soil is a widely used substrate in soil toxicity tests. Despite its apparent necessity as a defined mixture relevant for solid phase exposure, several problematic issues have been revealed recently which must be considered seriously. It is not clear if the OECD artificial soil is really a standardized reference material omitting the influences of varying natural soil properties or if there is still significant variability present, which may influence toxicity results. Under the auspices of the EU ringtest for the ecotoxicity of wastes, a new project has been started with the aim of addressing the variability of nearly 20 artificial soils. Although the collected soils were declared to have been prepared strictly according to OECD guidelines, they were different even at the first look and the organic carbon content in the artificial soils varied from 1.4 to 6.0%. This indicates the variability in the organic carbon content of peat from different sources, producers and countries. The cadmium sorption experiment on selected soils suggests that the cadmium K_d varies among artificial soils over one order of magnitude. It is apparent from our pilot results that there are differences between the OECD artificial soils from various labs in the EU.

Keywords Artificial soil, Cadmium, Soil organic matter, Phenanthrene, Sorption, Standardized tests

24.1 Introduction

Artificial soil was first introduced as a substrate for the earthworm acute toxicity test (OECD 1984; ISO 1997). It has been recommended as a medium for other tests (acute and reproduction tests with enchytraeids, springtails, and nematodes) and it is a "reference soil" in the testing of complex solid samples (e.g. wastes or contaminated soils).

J. Hofman(✉), I. Hovorková, and J. Machát
1RECETOX – Research Centre for Environmental Chemistry and Ecotoxicology, Faculty of Science, Masaryk University, Kamenice 126/3, CZ62500, Brno, Czech Republic

H. Moser and J. Römbke (eds.), *Ecotoxicological Characterization of Waste*,
DOI: 10.1007/978-0-387-88959-7_24,

OECD artificial soil is defined as a mixture of 70% fine quartz sand (50% particles 0.05–0.2 mm), 20% kaolin clay (kaolinite content preferably above 30%), and finely ground *Sphagnum* peat. The composition of the artificial soil was standardized in the early 1980s when the effects of changing peat content, clay type and content, pH, and water content on the toxicity of 2-chloroacetamide and benomyl were tested (Heimbach and Edwards 1983). Artificial soil was developed to achieve a standardized "soil-like" medium, much more representative of natural soils than filter paper, artisol, solutions or agar that were previously used in soil ecotoxicity tests, which can be used to produce reproducible and comparable data (Edwards 1984; Heimbach 1984). Recently, even an artificial soil modified for tropical regions has been proposed (Garcia et al. 2007).

Despite the novelty, undoubted necessity and usefulness of such a "reference material", the frequent and widespread use of OECD artificial soil has revealed several problematic issues which must be considered seriously. The toxic effects and risks of test chemicals are strongly dependant on their behavior, adsorption and bioavailability in soil, and thus are influenced by the soil properties. Because properties of the OECD artificial soil are significantly different from natural soils, the extrapolation possibilities are still under discussion (Amorim et al. 2005a,b; Hofman et al. 2007). Originally, it was claimed that the adsorptive capacity of the artificial soil is the same as a typical loamy soil (Edwards 1984; Heimbach 1984). However, this was solely based on the cation exchange capacity, which is important for the sorption of metals, but not for most organic hydrophobic contaminants (van Gestel 1992). The high sand content and type of clay (kaolin with low binding capacity) of the artificial soil should represent a so called "worst case scenario" for the behaviour of contaminants in soil – they should be readily bioavailable (Edwards 1984; Heimbach 1984). On the other hand, the artificial soil contains a relatively high amount of the organic matter which might be expected to adsorb particularly organic contaminants very strongly.

Another problem is the fact that although the amount of compounds used for artificial soil preparation is strictly defined, specific properties of the compounds are only briefly defined and they probably vary a great deal among the producers and countries. This could be the reason for the high variability of the toxicity results between laboratories, though standardized procedures and reference substances are used. The span of the LC_{50} values was shown to be over 1–2 orders of magnitude for the pentachlorophenol and chloroacetamide toxicity on *Eisenia fetida* in the international ringtest performed when introducing the first OECD soil test (Edwards 1984). Similarly, EC_{10} values of the derosal toxicity on the reproduction of *Enchytraeus albidus* were spread over 2–3 orders of magnitude (Römbke & Moser 2002). When testing wood-chip waste, the span of the LC_{50} for *E. fetida* differed from 8 to 36% (Chap. 11 of this book). Based on these facts, the question arises: Is the OECD artificial soil really a standardized reference material omitting the influences of varying soil properties or is there still significant variability present which influences toxicity results and risk assessment and which should be addressed and then reduced?

In the EU ringtest for ecotoxicity of wastes, artificial soil was used as a control and mixture material in various invertebrate tests. Using this opportunity, a new project has been started with the aim to address the variability related to OECD artificial soils. Here only the pilot results of the project are shown: the comparison of WHC_{max} and C_{org} in 20 artificial soils, and the cadmium sorption in four selected soils. The future plan of the project is to compare (a) properties of artificial soils from EU labs; (b) cadmium and phenanthrene sorption in these soils; (c) cadmium and phenanthrene aging, extractability, and bioavailability in time experiments; (d) habitat function of these soils for model organisms; and (e) toxicity of reference compounds (e.g. cadmium, phenanthrene, chloroacetamide, carbendazim) in these soils.

24.2 Materials and Methods

The study was started with the collection of OECD artificial soils prepared in various soil labs in the EU. The ringtest participants were asked to prepare 2–3 kg of dry artificial soil according to the OECD guideline (OECD 1984) including the addition of $CaCO_3$ for final pH adjustment. The participants were also asked to provide information about their artificial soils and components: how much $CaCO_3$ they added for pH adjustment, actual pH value, maximal water holding capacity WHC_{max}, amount of water to add for 50% WHC, detailed peat, clay and sand descriptions (particle size, mesh size, TOC, pH, producer etc.). In winter 2006/2007, 20 artificial soils were delivered from the labs of several European countries: one from Belgium, three from the Czech Republic, one from Denmark, four from France, five from Germany, two from Netherlands, one from Poland, one from Portugal, and one from Spain. Soils were dry at 98% dry matter content approximately and they were stored at laboratory temperature. Their properties were put into a small database enabling further explanation of the differences.

The maximal water holding capacity of soils was determined by the method described in Annex C of ISO 11268-2 (ISO 1998). Organic carbon content in the soils was measured by high temperature oxidation on LiquiTOC automatic analyzer (Elementar, Germany). Sorption coefficient of cadmium was determined according to the recommendations of OECD Guideline 106 (OECD 2000). The method was performed in 50 mL conical polypropylene tubes which were pre-cleaned with 5% HNO_3. Five grams of the soil (four replicates per soil) were weighted into tubes and 40 mL of 0.01 M $CaCl_2$ solution was added. Tubes were shaken for 12 h (100 rpm) at 25°C. Then, 5 mL of 0.01 M $CaCl_2$ containing 0.2 mg/mL of cadmium was added and the suspensions were shaken again for 72 hrs. Based on the preliminary experiments, this duration was confirmed as adequate equilibrium time. The suspensions were centrifuged (4,000 rpm), 10 mL supernatant was carefully taken from the tubes, filtered through 0.2 μm syringe filter to 22 mL vial, 50 μL of concentrated HNO_3 was added, and the samples were stored at 4°C. In the batch of samples, control for sorption of cadmium on tube walls (without soil) was measured in triplicate. Blank was measured without cadmium addition in duplicate for each soil. Samples were analysed for the cadmium concentration by ICP-MS. The soil–water partition coefficient was calculated as $K_d = C_{soil}/C_{solution}$, where $C_{solution}$ is the measured cadmium concentration and C_{soil} is the concentration of adsorbed cadmium calculated from the difference between the amount added initially and that remaining in the solution divided by the known weight of soil (5 g).

24.3 Results and Discussion

During the collection of soils, information was provided that in two soils (G, K) the amount of peat was decreased to 5% and in one soil (G) the amount of clay was only 10%. However, we decided to include these samples in the final comparison keeping this in mind. Although other soils were declared to be prepared strictly according to the OECD guideline (OECD 1984), they were different even at the first look (Fig. 24.1).

The soils differed in color from white–grey to yellow–brown, but mostly they had grey colour as expected. The structure and particle distribution of the soils were apparently different. Some soils were very fine but some contained larger sand grains or pieces of peat. It should be verified whether all participants used sand with 50% particles between 50 and 200 μm as required by the guidelines. Information provided by the participants revealed that sieves from 1 to 10 mm were used for the

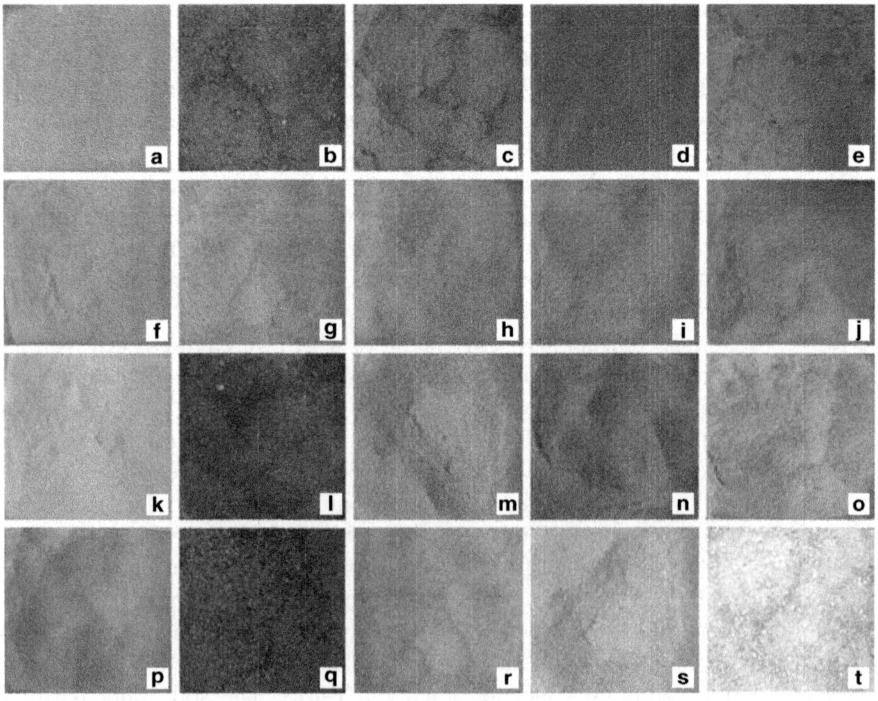

Fig. 24.1 First look comparisons of the collected OECD artificial soils from EU labs

peat sieving in the different labs. OECD and ISO guidelines for *E. fetida* tests (OECD 1984; ISO 1997) recommend finely ground peat and the guidelines for the enchytraeid reproduction test (OECD 2004a; ISO 2004) specify a desired size of 2 ± 1 mm. The peat particle size is likely to affect the sorption of pollutants and should be defined more precisely. Smaller particles should be preferred due to higher homogeneity of the mixed soil. However, if the peat is too fine the oxygen content could become so low that small animals like enchytraeids can be affected (T. Moser, pers. comm.).

The participants provided the information that they usually add 0.2–0.75% $CaCO_3$ to reach the desired pH value of soil and 0.19–0.48 mL water to 1 $g_{dw\ soil}$ to reach 50% WHC. Our results of WHC_{max} (Fig. 24.2) determination showed that there is quite good agreement of the water retention properties in artificial soils with the exception of soils I, M, N, P, and S. No clear explanation of the higher WHC_{max} in these five soils was found in the information provided by the respective participants.

The organic carbon content in the artificial soils varied from 1.4 to 6.0% with a median value of 4.8% (Fig. 24.3). This indicates variability in the organic carbon content of peat from different sources, producers and countries. When the C_{org} content in the soils G and K is multiplied by two and two soils with extremely low C_{org} (E, T) are excluded, the remaining soils have C_{org} values from 3 to 6%.

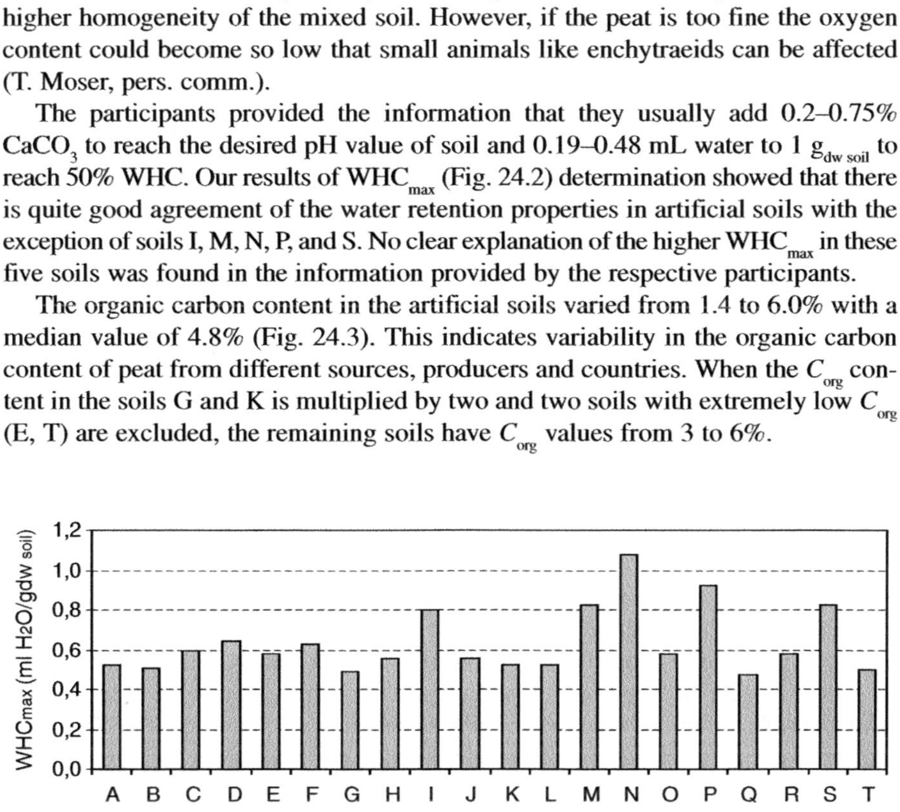

Fig. 24.2 Results of maximal water holding capacity of the OECD artificial soils from EU labs

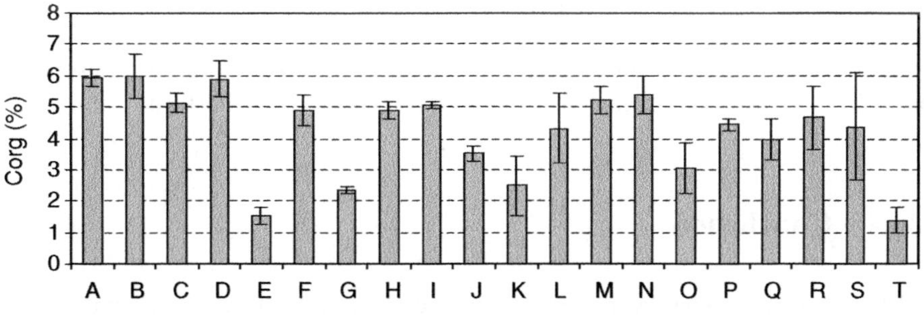

Fig. 24.3 Results of organic carbon content in the OECD artificial soils from EU labs. Whiskers are S.D ($n = 4$)

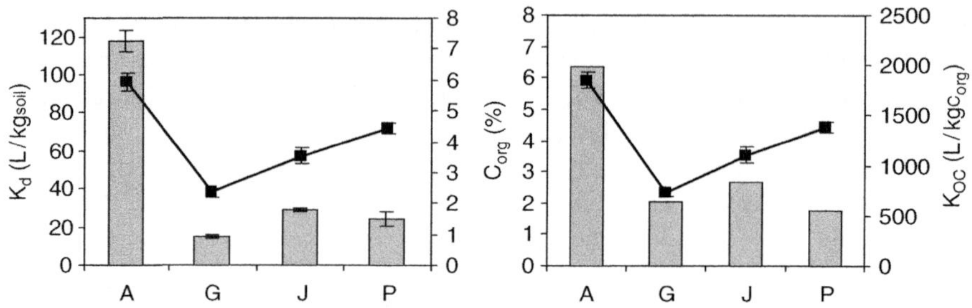

Fig. 24.4 Soil–water (columns in the *left* chart) and organic carbon–water (columns in the *right* chart) sorption coefficients of four selected artificial soils. *Squares* show organic carbon in these soils. Whiskers are S.D ($n = 4$)

Pilot results of the cadmium sorption experiment on four selected soils with different C_{org} contents revealed that the K_d values are comparable to those found in the literature (Anderson & Christensen 1998; Lee et al. 1996). Figure 24.4 shows that the cadmium K_d in artificial soils varies over one order of magnitude.

When K_d values were normalized for C_{org} content (K_{OC}), there were still remarkable differences. This means that the relation of cadmium sorption and C_{org} content does not explain the variability between soils. This might be expected because not only organic matter content but predominantly cation exchange capacity and pH are known to be important for the sorption of metals in soils (van Gestel 1992; Lock and Janssen 2001b; Spurgeon and Hopkin 1996). However, the effect of pH is not responsible in this case, because of the same pH values of all soils (6.0 ± 0.5). The most probable reason is the different cation exchange capacity which is related to the nature of peat and kaoline clay varying in soils from different labs. Similarly, Lock and Janssen (2001a) reported variability in Cd and Zn toxicity on *E. albidus* in artificial soils with three types of clay and seven types of organic matter. The cation exchange capacity was identified as the probable reason because the range was over one order of magnitude among different types of organic matter.

It is apparent that not only the total content of organic matter but also its character (chemical structure, ionic charge, polarity, aromacity etc.) is important for the fate and behavior of contaminants in soils. Our results and the results of Lock and Janssen (2001a) show that varying cation exchange capacity of different peat influence the sorption of metals. In the case of organic chemicals, where the sorption to soil organic matter is the main factor of their soil fate, the effect of the varying peat properties must be expected (Belfroid et al. 1996; Grathwohl 1990; Rutherford et al. 1992).

24.4 Conclusion

OECD artificial soil has been used very frequently as a standardized substrate for soil toxicity tests. It is apparent from our pilot results, that there are differences between the OECD artificial soils from various labs in the EU. Variable soil properties (e.g. C_{org} content and character) cause the different behavior of contaminants, particularly sorption. It is still an unanswered question if this has a significant impact on toxicity evaluation in the framework of ecological risk assessment. It appears that peat is the most problematic component of artificial soil because it is not standardized material and it can differ substantially among countries and suppliers. The way of manipulation, drying, and sieving can also affect peat sorption properties. Finally, microbiological activity may be totally different in different peat sources with strong implications for the fate of test compounds (degradation). Therefore, the peat component of artificial soil needs to be better defined in the test guidelines.

Acknowledgement The authors greatly appreciate all participants who agreed to provide their artificial soils for the purpose of this study (in alphabetic order – does not correspond with the order in Figs): BBA (Germany); ECT Oekotoxikologie GmbH (Germany); ENVISAN-GEM Inc. (Czech Republic); Eurofins-GAB GmbH (Germany); Grontmij/AquaSense (Netherlands); IBACON GmbH (Germany); INERIS (France); INSAVALOR-POLDEN (France); Institute of Chemical Technology (Czech Republic); IRH Environnement (France); Jagellonian University (Poland); MicroBioTests Inc. (Belgium); NERI (Denmark); Paul-Verlaine University in Metz (France); Universitat Autónoma de Barcelona (Spain); Université Paul Verlaine – Metz (France); University of Aveiro (Portugal); VITO (Belgium); Vrije Universiteit Amsterdam (Netherlands). In addition, the authors thank the ringtest organizers for encouragement to conduct this study. Financial support for experimental part of the study is obtained from Ministry of Education of the Czech Republic (project MSM 0021622412 INCHEMBIOL).

Chapter 25
Leaching and Chemical Speciation Modeling of Wastes

H.A. Van der Sloot and A. Van Zomeren

Abstract In this chapter, the wastes to be tested for their ecotoxicological response by exposing them to the eluate as obtained from EN 14735 are characterized in greater detail by the leaching tests developed in CEN/TC292/WG 6 (pH dependence leaching tests TS 14429 and percolation test TS 14405). These methods allow chemical speciation modeling, which provides information on the "free" forms of inorganic and organic substances in solution as opposed to dissolved organic carbon (DOC) associated and colloid bound forms. The unbound form of the substances is considered bioavailable and thus responsible for the observed response. In addition, the repartitioning of substances between bound and unbound forms as a result of the dilution of the eluate with a medium to quantify the ecotoxicological response can be modeled; it has been shown not to be a simple dilution factor as the relative concentration of the "free" forms increased upon dilution. The combination of more detailed leach testing and ecotoxicological studies provided the basis for a more detailed evaluation of ecotoxicological response than was possible earlier.

Keywords Leaching, Chemical speciation, Waste characterization, Bioavailability

25.1 Introduction

In the light of the validation of the EN 14735 (2006), it was considered crucial to expand the range of testing conditions to allow a better insight into the factors controlling leaching of substances and the consequences for responses of organisms (Proceedings "Problems around Soil and Waste III, 2005). In addition, the more elaborate test methods applied are suitable for chemical speciation modeling to see

H.A. Van der Sloot (✉) and A. Van Zomeren
Energy Research Centre of the Netherlands (ECN), Environmental Risk Assessment Group, P.O. Box 11755, ZG Petten, The Netherlands

H. Moser and J. Römbke (eds.), *Ecotoxicological Characterization of Waste*,
DOI: 10.1007/978-0-387-88959-7_25,

what consequences might arise from the choice of experimental conditions within the EN 14735. For the three waste samples, a basic characterization consisting of a combination of pH dependence (CEN/TS 14429) and a percolation test (CEN/TS 14405) was therefore carried out. In combination with chemical speciation modeling, partitioning of constituents between dissolved (free and dissolved organic matter associated substances) and solid phases (solid organic matter, clay surfaces, iron oxides and minerals) can be calculated. This is necessary for understanding the observed ecotoxicity, because ecotoxicological effects can be related to bioavailable concentrations (i.e. "free" ionic forms) instead to total content or total dissolved concentrations in eluates.

25.2 Experimental

25.2.1 Samples

The sampling, homogenization, and distribution of the samples were coordinated by the organizers of the ring test, the German UBA (Chaps. 1 and 2). The test samples – Incinerator ash (INC), Gasworks soil (SOI) and Waste wood (WOO) were shipped to Grontmij AquaSense (where the ecotoxicity tests were carried out) and stored at 4°C in the dark upon arrival. Sub-samples from these batches were sent to ECN for testing of the leaching behavior. The artificial soil used as medium to dilute the waste in the terrestrial ecotoxicity tests was also tested.

25.2.2 Characterization of Wastes

The soil, artificial soil, wood and bottom ash were leached with a pH dependent leaching test (CEN/TS 14429, 2005), which is similar to ISO 21268-4. About 15 g portions of each sample were leached with 150 mL water (*L/S* = 10, corrected for the moisture content of the sample) and equilibrated at pH values between 3 and 12. The pH was continuously monitored and adjusted to the pH set point by dosing 1 M HNO_3 or 1 M NaOH. After 48 h of equilibration the suspensions were filtered (0.45 μm) and analyzed.

Column experiments on the same materials were carried out according to CEN TS 14405 (2004), which is similar to ISO 21268-3. In this column test seven eluate fractions were collected within the range of *L/S* = 0.1–10 L/kg. The total test duration was approximately 21 days. The leachant was dematerialized water (DMW). The test material was applied as received (<4 mm) and up-flow (14 mL/h) was applied through a column waste height of 28 cm and a diameter of 5 cm. The eluates were filtered through 0.45 μm membrane filters and analyzed.

25.2.3 Chemical Analyses

Besides pH and conductivity, the leachates and eluates for bioassays were analyzed for major, minor and trace elements by ICP – AES (Al, As, B, Ba, Ca, Cd, Co, Cr, Cu, Fe, K, Li, Mg, Mn, Mo, Na, Ni, P, Pb, S, Sb, Se, Si, Sn, Sr, TI, V, Zn). DOC (dissolved organic carbon) and TIC (total inorganic carbon) were analyzed by a Shimadzu TOC 5000a analyzer. Cl, F, ammonium and sulphate were analyzed by ion-chromatography. In the gasworks soil, which smelt strongly of organic contamination, 16 PAH were measured in an $L/S = 10$ extract.

25.2.4 Estimation of Model Parameters

The quantities of "reactive" organic carbon in the solid phase (i.e. HA and FA) were estimated by a batch procedure (van Zomeren and Comans 2007), which is derived from the procedure currently recommended by the International Humic Substances Society (IHHS) for solid samples. In short, the procedure is based on the solubility behavior of HA (flocculation at pH < 1) and the adsorption of FA to a polymer resin (DAX-8). The amount of amorphous and crystalline iron (hydr)oxides in the waste mixture was estimated by a dithionite extraction (Kostka and Luther III 1994). The amounts of amorphous aluminum (hydr)oxides were estimated by an oxalate extraction (Blakemore et al. 1987). The extracted amounts of Fe and Al were summed and used as a surrogate for hydrous ferric oxides (HFO) in the model.

25.2.5 Geochemical Speciation and Release Modeling

The database/expert system LeachXS (LeachXS 2005) was used for data management, e.g. pH dependent leaching data, percolation test data, lysimeter, and field leachate data and for visualization of the calculated and measured results (Van der Sloot et al. 2007). The coupled LEACHXS – ORCHESTRA combination allowed very quick data retrieval, automatic input generation for modeling, processing of calculated results and graphical and tabular data presentation. Chemical speciation of the solutions was calculated with the ORCHESTRA modeling framework (Meeussen 2003) embedded in LeachXS. Aqueous speciation reactions and selected mineral precipitates were taken from the MINTEQA2 database. Ion adsorption onto organic matter was calculated with the NICA-Donnan model (Kinniburgh et al. 1999), with generic adsorption reactions as published by Milne et al. (2003). Adsorption of ions onto iron and aluminum oxides was modeled according to the generalized two layer model of Dzombak and Morel (1990).

The input to the model consists of metal availabilities, selected solubility-controlling minerals, active Fe-and Al-oxide sites (Fe- and Al-oxides were summed and used as input for HFO), particulate organic matter (POM) and a description of the DOC concentration as a function of pH (polynomial curve fitting procedure). Basically, the speciation of all elements is calculated in one problem definition in the model with the same parameter settings. This limits the degrees of freedom in selecting parameter settings considerably, as improvement of the model description for one element may deteriorate the outcome for other elements. As a starting point for the model calculations, the maximum value as obtained in the pH dependence leaching test (between pH 3 and 13) was used as the available concentration. It was found that total leachable carbonate concentration plays an important role in the calculations. This parameter was not measured and was therefore estimated on the basis of the model output. The concentration was adjusted until the major (and some minor) elements showed a reasonably good match with the observed leaching data. There is a clear need for more data on total (available) carbonate concentrations in waste materials to enhance model predictions.

The mineral phases that were allowed to precipitate were selected after calculation of their respective Saturation Indices (SI) in the original pH dependence leaching test eluates. Saturation indices were calculated for all the 650 or more minerals in the thermodynamic database and a selection of the most likely and relevant phases was made based on the degree of fit over a wider pH range, the closeness of the SI value to 0, and the expert judgement on the suitability of possible minerals for the waste mixture (e.g. exclusion of high temperature minerals). Generally, minerals were selected if the SI was in the range of –0.2–0.2 for more than two pH data points.

25.3 Results

25.3.1 Characterization of Leaching Behavior of the Wastes

The raw test data generated are given in H-14 navigator (http://ecotoxwasteringtest.uba.de/h14/). In addition, a comparison of the wastes with other similar waste was carried out using the LeachXS database/expert system (www.leachxs.org), to find out how representative the samples are for the type of waste selected. In Fig. 25.1 this is illustrated for Cu in soil, Cu in incinerator bottom ash and Cu in wood. In the case of wood, a natural wood was inserted for comparison. The variability in soil is rather large, which is not surprising given the highly variable nature of soils. The incinerator bottom ash data are very consistent even across different countries. This evaluation, based on a wider range of parameters (H-14 navigator), leads to the conclusion that the samples studied are reasonably representative of the respective types of waste studied. The leaching data were processed in LeachXS-Orchestra to provide the partitioning between free, DOC complexed elements in solution and the partitioning in the solid phase. "Free-concentrations" are in this context the non-DOC bound element concentrations, which include the different ionic complexes (in case of Cu that implies that besides Cu_2^+, $Cu(OH)^+$, $CuCl^+$, $CuCl_3^-$ are included). In Fig. 25.2 the relation between the two main characterization tests (pH dependence and percolation) is illustrated and additional lines in the graphs presented are explained.

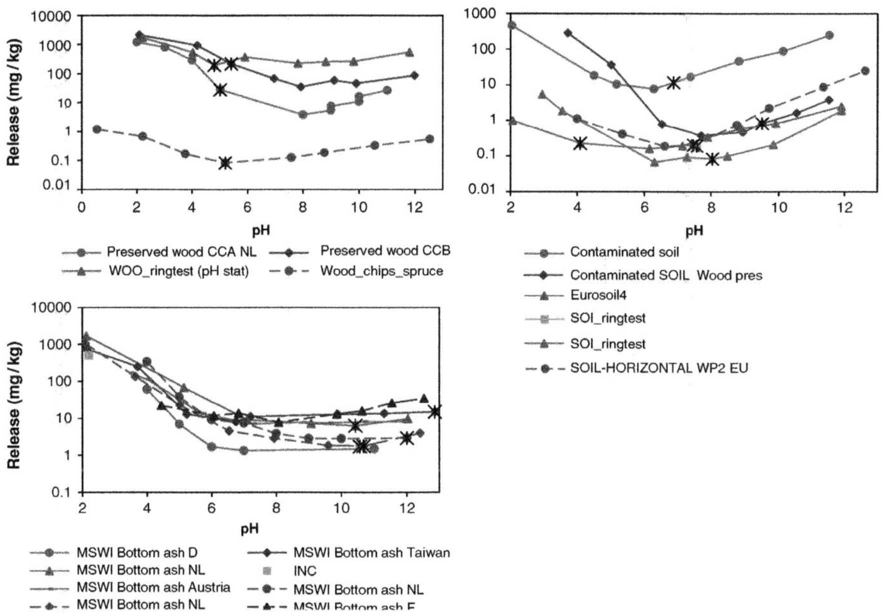

Fig. 25.1 The waste samples used in the study have been compared with other similar wastes to determine the extent to which the samples can be considered reasonably representative for the type of waste. Here data for Cu leaching are given for WOO, SOI and INC in comparison with samples from comparable materials. The asterisk within each dataset represents the native pH of the material

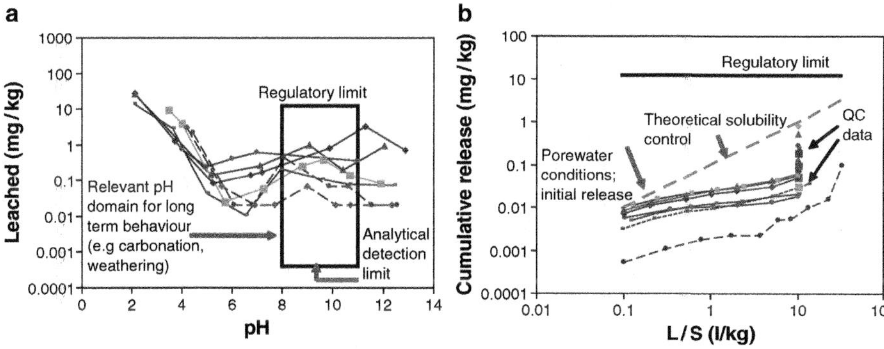

Fig. 25.2 The pH dependent leaching (TS 14429; *L/S* = 10) in relation to percolation testing (TS 14405; *L/S* = 0.1–10) for Cr from several incinerator bottom ashes. (**a**) pH dependent leaching of Cr, (**b**) Cumulative release of Cr

25.3.2 Raw Data

The raw test data for TS14429, TS 14405 and EN 14735 can be obtained from the H-14 navigator. These data make it possible to identify release mechanisms and release controlling factors. As a wide range of pH and *L/S* is covered, a wide range of environmentally relevant exposure conditions can be addressed.

25.3.3 Comparison with Other Similar Materials

The database in LeachXS allows material comparison at different levels. A comparison has been made for the soil, the waste wood and the incinerator ash. Here a comparison to demonstrate the consistency of the data for the test samples with other related samples suffices (H-14 navigator). In Fig. 25.3 an example is given for incinerator ash (for SOI & WOO more data is available in H-14 navigator).

The comparison clearly shows that leaching behavior, both in terms of pH dependence as well as in *L/S* (liquid solid ratio or time dependence) is quite comparable. Most deviations can be traced back. For example in the case of wood, the CCA treated material obviously has a substantially higher Cr and Cu leachability than other wood samples. The soil contaminated with wood preservatives shows a clearly increased dissolved concentration for Cu and Cr (from CCA treated wood) compared to eluate concentrations for these elements in other soils.

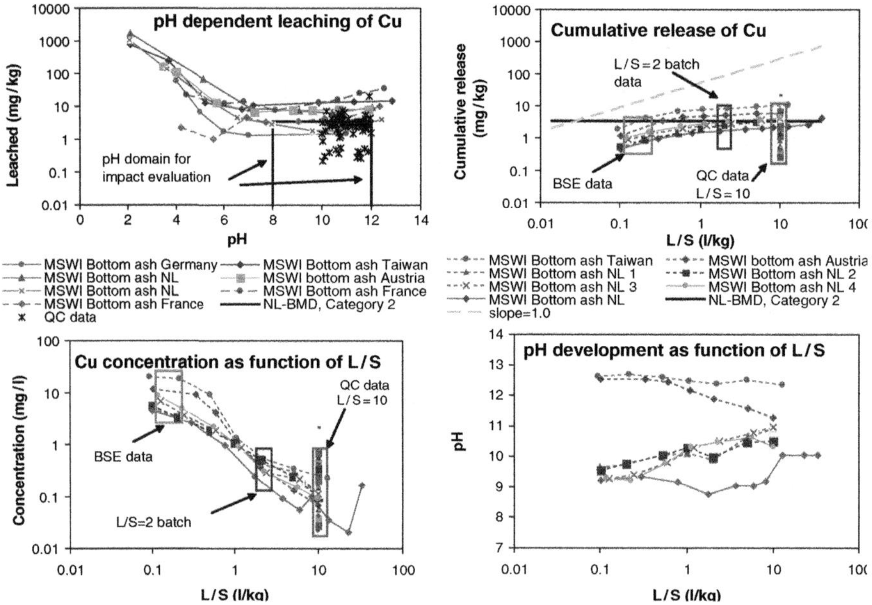

Fig. 25.3 Integrated evaluation of leach test data as function of pH (speciation) and *L/S* (time dependent release) for Cu from MSWI bottom ash. Samples from different sources show similar behavior. Comparison with regulatory criteria is shown. Relation between characterization and compliance test data given. Specific judgment conditions and comparison with other tests indicated (i.e BSE = soil saturation extract). Note: legend with top left plot limited to that graph; legend under top right figure applicable to both bottom graphs. Not all ashes subjected to pH dependence test have been assessed using the percolation test

25.3.4 Partitioning Between Dissolved, Complexed and Bound Forms

By applying chemical speciation modeling using LeachXS-Orchestra, the partitioning between dissolved DOC bound and particulate phases can be estimated. As the ecotoxicity relevant form is mostly the free ion concentration, this is a very relevant subdivision for data interpretation. The availability for leaching of all elements used as input for geochemical speciation modeling is given in H-14 navigator. It should be noted that the availability was determined as the maximum concentration that was obtained in the pH-static leaching test. In addition to the element availabilities, the sorptive surfaces (organic matter, clay, Fe oxide and Al oxide as determined) are also provided.

DOC is a sum parameter for all organic carbon species present in a solution. Not all the compounds present in the DOC are reactive in metal binding. Therefore a fraction of the total DOC is used in the calculation. Both at low and high pH, a larger proportion of the DOC is in the reactive form (van Zomeren and Comans 2004), while at neutral pH the lowest proportion of reactive DOC is found. A varying proportion of reactive DOC is assumed from low to high pH, based on the analysis of sub-fractions of DOC (hydrophilic, fulvic, and humic) using a rapid batch method (van Zomeren and Comans 2007). The concentrations of humic and fulvic acids are assumed to consist of humic acid in the model, for which the interaction parameters according to Nica Donnan (Kinneburgh et al. 1999) are used. The corrected DOC data from the pH dependence test were fitted to a polynomial function to describe the pH dependent leaching of DOC for intermediate pH values. ORCHESTRA calculates the geochemical speciation from pH 1 to 14 with intervals of 0.2 pH values. The calculation is carried out at both $L/S = 10$ as well as at $L/S = 0.3$ under the assumption that the distribution of sorptive phases and the mineral set is the same. This latter condition represents pore water and can be compared with the first fractions of the percolation test, which have been inserted in the graphs. The results of the $L/S = 0.3$ modeling need to be judged at the pH as observed in the percolation test. Initial speciation calculations indicate several possible solubility controlling minerals (SI saturation index close to 0). From this, a set of minerals was selected for the model prediction calculations (see H-14 navigator), of which the minerals listed in Table 25.1 proved relevant for the three specific matrices.

The results of the geochemical model for several major and minor elements in comparison with the measurements in the pH-static leaching test are also given in H-14 navigator. This gives an overall impression of the degree to which prediction and test results for major, minor, and trace elements match in a multi-element model run. For one element, proper stability data may be lacking (Sb, B, Li) while for another the stability data as obtained from the literature may not be entirely adequate ($PbMoO_4$). However, the overall agreement between measured and calculated acid/base neutralization capacity (ANC/BNC) based on the selected minerals and sorption phases is a good indication that the selection of minerals and phases to describe major element behavior will soon be possible (Fig. 25.4).

Table 25.1 Mineral phases relevant for the three matrices apart from DOC, Clay, Fe and Al oxides (detailed information provided in H-14 navigator)

Sample	Mineral phases			
INC	$3CaO\_Al_2O_3\_6H_2O[s]$	Magnesite	$Ni[OH]_2[s]$	P-Wollstanite
	$3CaO\_Fe_2O_3\_6H_2O[s]$	Portlandite	OCP	Tenorite
	Brucite	alpha-TCP	$Pb[OH]_2[C]$	Wairakite
	Calcite	$BaSrSO_4[50\%Ba]$	$Pb_2V_2O_7$	Willemite
	$Fe[OH]_3[microcr]$	$Cd[OH]_2[A]$	$Pb_3[VO_4]_2$	
	Gypsum	Manganite	$PbMoO_4[c]$	
SOI	$AA\_Fe[OH]_3[am]$	Birnessite	$Ni[OH]_2[s]$	$PbMoO_4[c]$
	Albite[low]	Calcite	Ni_2SiO_4	Rhodochrosite
	$Ba[SCr]O_4[96\%SO_4]$	Gibbsite[C]	$Pb[OH]_2[C]$	$ZnSiO_3$
WOO	Manganite	Tenorite	Wairakite	SbO_2

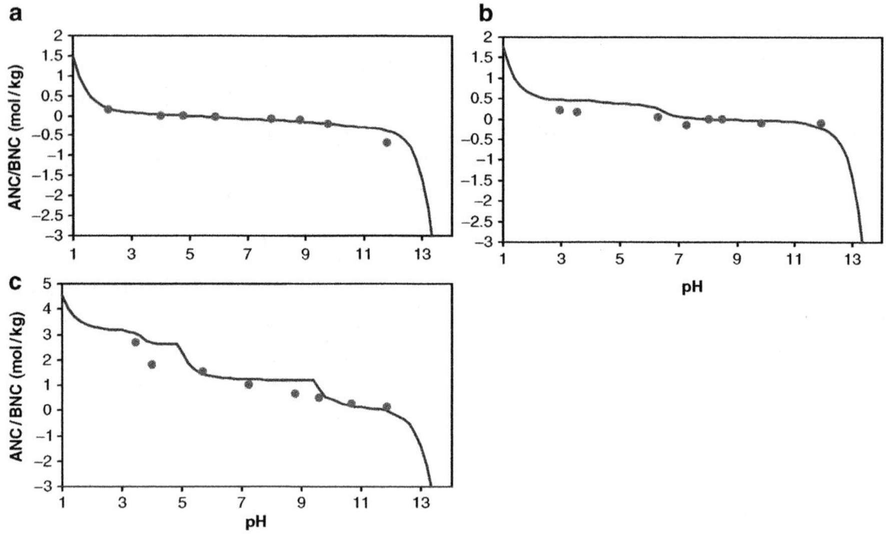

Fig. 25.4 Predicted acid and base neutralisation capacity for the three matrices INC, WOO and SOI as obtained from the chemical speciation modelling. (**a**) Ecotox 2007 Wood, (**b**) Ecotox 2007 Soil, (**c**) Ecotox MSWI Bottom ash

As an illustration Figs. 25.5, 25.6, and 25.7 show the agreement between measurement and prediction for INC, WOO and SOI, respectively, which is further described by the partitioning provided between dissolved and particulate phases.

In general, the model describes the leaching behavior of the wastes quite well, especially considering that changes in input parameters may affect the predicted behavior of several other elements. This implies that the degrees of freedom to vary input parameters are limited dramatically by taking all elements into account simultaneously. As the model assumes equilibrium and it is

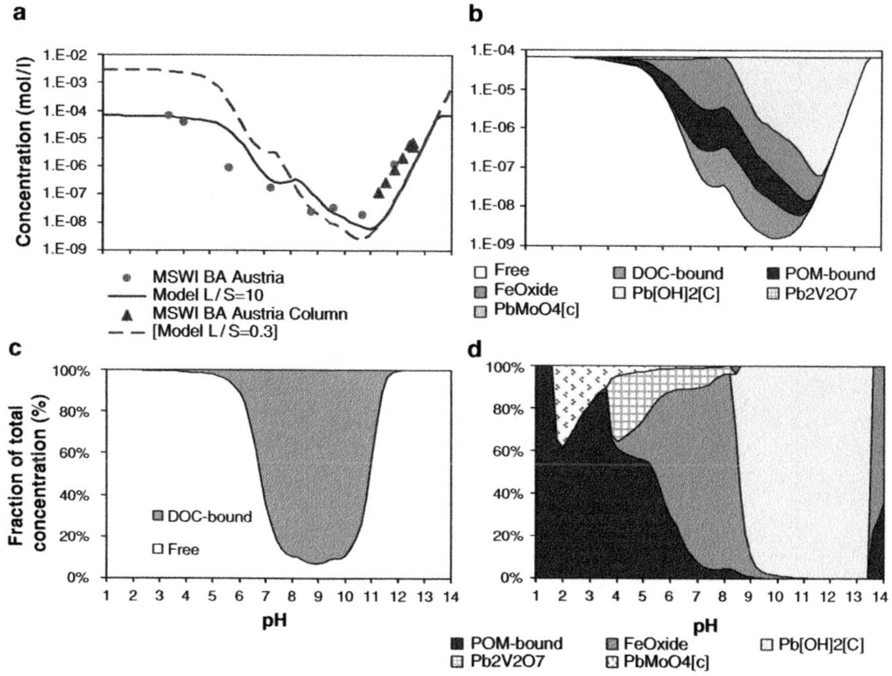

Fig. 25.5 Pb prediction for MSWI bottom ash illustrating the adequate match between test data (*dots*; TS14429) and model results (*drawn line*). For comparison the percolation test data (TS 14405; *triangles*) are given together with the prediction of the Pb concentration in solution for an *L/S* = 0.3 (*broken line*), which corresponds with porewater. Partitioning between dissolved and particulate phases in concentration (*top right*). Fractionation of Pb in dissolved (free and DOC associated) and solid phases (particulate matter: POM, Fe-oxide, clay and specific minerals). (**a**) [Pb + 2] as function of pH, (**b**) Partitioning liquid and solid phase, (**c**) Pb + 2 fractionation in solution, (**d**) Pb + 2 fractionation in the solid phase

known if equilibrium is reached within 48 h contact time, kinetics of dissolution and precipitation will be a factor to reckon with in judging the results. These effects will result in an apparent deviation of the model prediction, whereas the leached concentrations might still increase or decrease due to equilibrium and/or kinetic processes.

Typical examples of such deviations are Mg in INC at pH 9–10, Cr in INC at pH 10–11. Work by Dijkstra et al. (2006) has shown these apparent discrepancies to be minimal regarding the pH of the material, where the system is closest to equilibrium. Another obvious deviation between model measurement and prediction

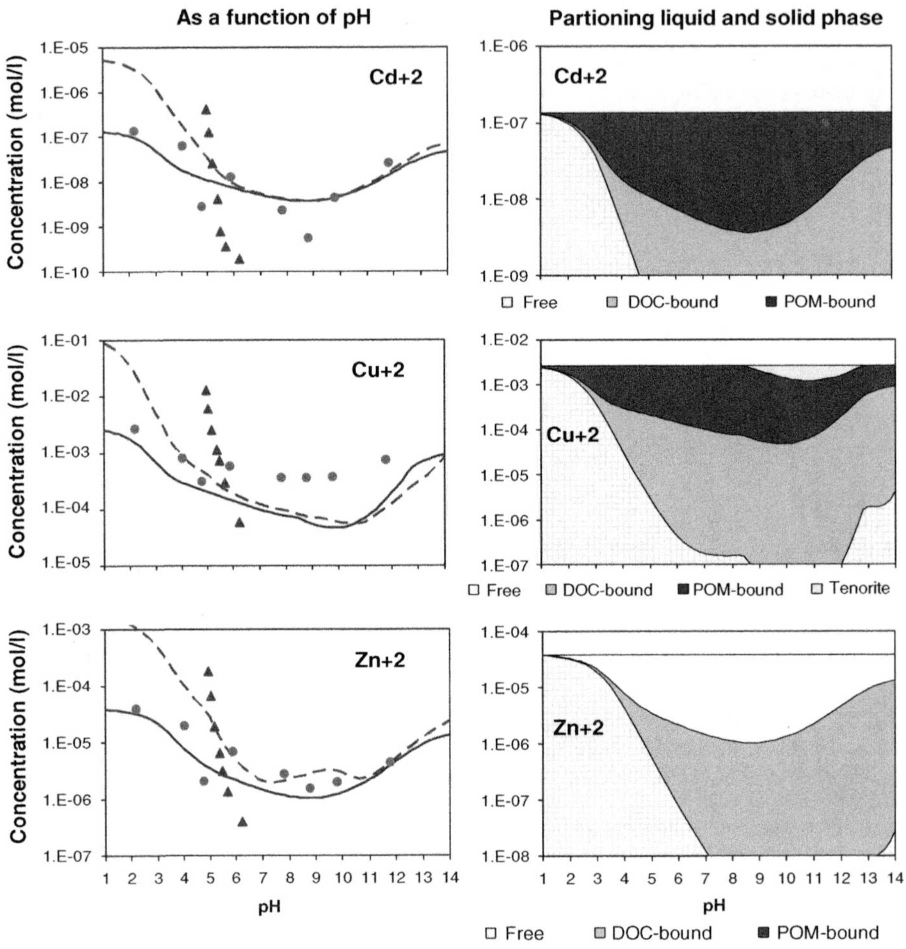

Fig. 25.6 Cd, Cu and Zn prediction for Wood illustrating that particulate and dissolved organic matter is dominant in controlling release from this matrix (*dots*; TS14429) and model results (*drawn line*). For comparison the percolation test data (TS 14405; *triangles*) are given together with the prediction of the Cd, Cu and Zn concentrations in solution for an $L/S = 0.3$ (*broken line*), which corresponds with porewater. Partitioning of the metals between dissolved and particulate organic matter is given in the right hand graphs

is the carbonate in SOI at pH < 6. Under those conditions CO_2 is removed from the solution. The modeling does not take this into account and thus a carbonate level higher than measured is predicted. This implies, however, that there is an adequate understanding of the chemical processes that determine the leaching behavior in these three wastes.

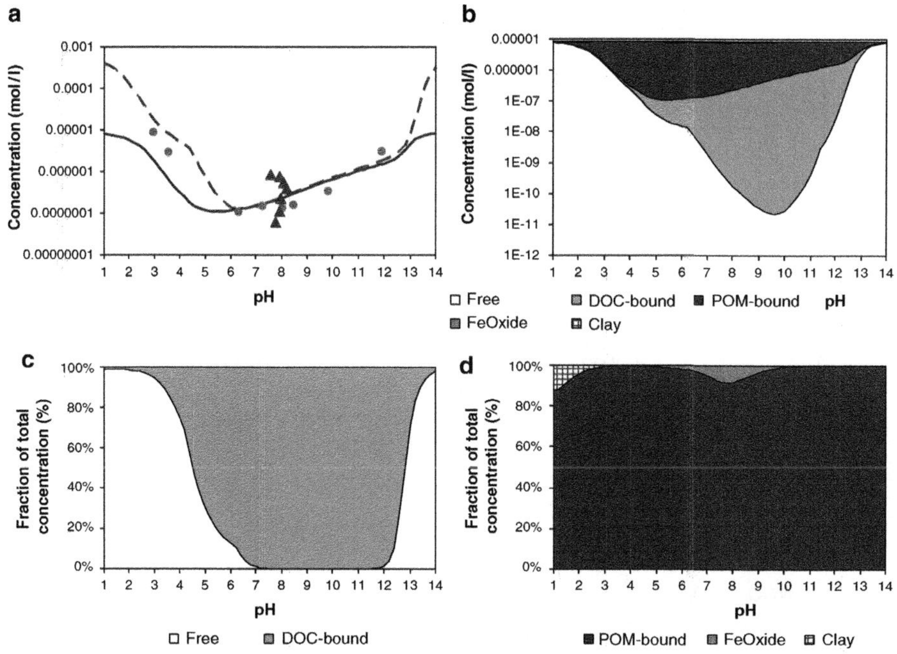

Fig. 25.7 Cu prediction for soil illustrating the dominance of organic matter for Cu leaching from soil by means of test data (*dots*; TS14429) and model results (*drawn line*). For comparison the percolation test data (TS 14405; *triangles*) are given together with the prediction of the Cu concentration in solution for an *L/S* = 0.3 (*broken line*), which corresponds with pore water. Partitioning between dissolved and particulate phases in concentration (*top right*). Fractionation of Cu in dissolved (free and DOC associated) and solid phases (particulate matter: POM, Fe-oxide, clay and specific minerals). (**a**) [Cu + 2] as function of pH, (**b**) Partitioning liquid and solid phase, (**c**) Cu + 2 fractionation in solution, (**d**) Cu + 2 fractionation in the solid phase

25.3.5 Data of EN 14735 in Perspective to Characterization Test Data

The results of EN 14735 eluate analysis performed by the participants in the validation have been studied in relation to the characterization (Fig. 25.8). There is a difference in the eluate pre-treatment. In EN 14735 no filtration or centrifugation is foreseen, whereas filtration or centrifugation is prescribed in the characterization methods. In the case of soil, measured concentrations using EN 14735 are higher than the test data from characterization, which, given the eluate handling, is not surprising, as mobilization of DOC is known to occur in soil samples. The results for wood with the exception of one participant and of bottom ash with the exception of three participants are in better agreement with the characterization data.

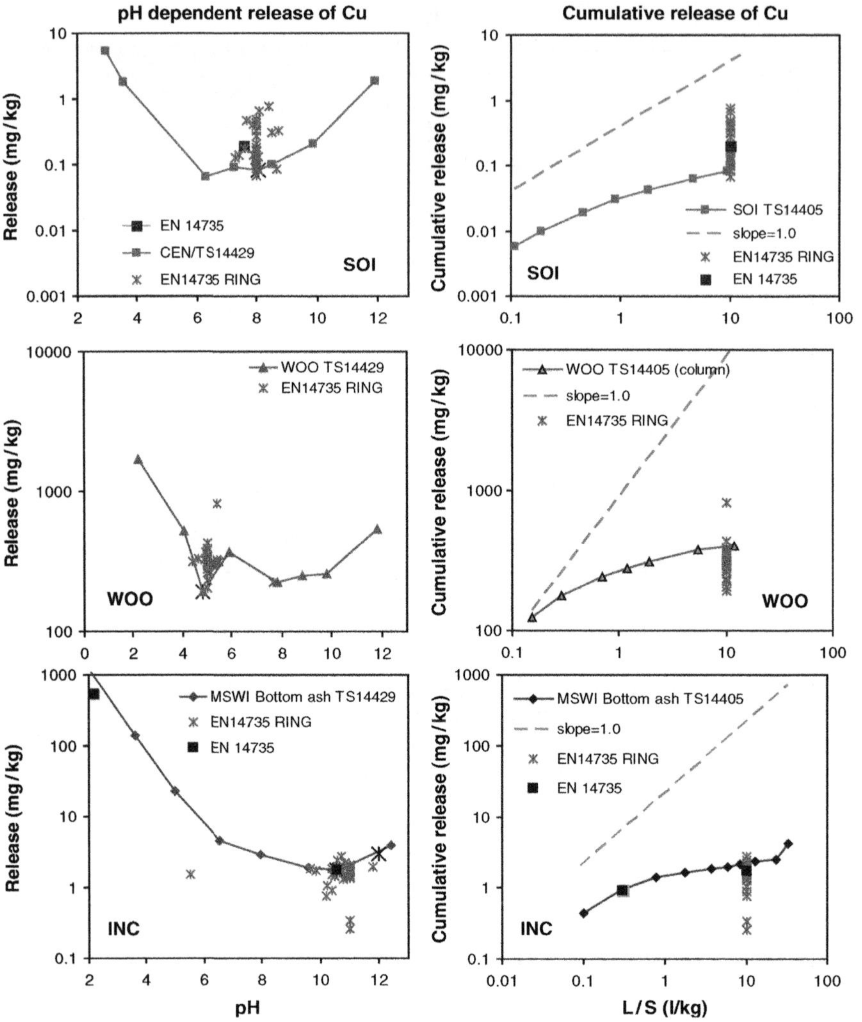

Fig. 25.8 EN 14735 leaching test results of participants in the ecocoxicity validation in relation to the characterization leaching data for Cu from Soil, Incinerator bottom ash and treated wood

25.4 Conclusions

The complete characterization tests (TS14429 and TS14405) allow the test results of EN 14735 to be placed in perspective. In addition, on the basis of these tests, it is possible to identify important chemical processes that determine leaching through geochemical speciation modeling, and also bioavailability and toxicity. pH-static leaching tests, in combination with model predictions of the leaching behavior and the speciation of contaminants in both the solid phase and solution, provide detailed knowledge of chemical processes in these apparent heterogeneous materials and allow the "free" concentrations in eluates to be quantified.

The main conclusions of this study are:

- The combination of chemical analyses and speciation modeling proves to be a powerful tool in being able to assess bioavailable concentrations in eluates ("free" ionic forms). The bioavailable concentrations are considered most relevant in relation to the ecotoxicological response of organisms. The relation between "free" concentrations and the ecotoxicological response is addressed in Chap. 26.
- The methods also provide a basis for the judgment of management options, development of regulations and treatment of materials to improve their environmental compatibility.
- The inclusion or omission of a filtration/centrifugation step in the eluate preparation for analysis may affect the evaluation of ecotoxicological test results, particularly when unfiltered eluates are acidified for chemical analysis of metals. On the other hand, for bioavailability and ecotoxicity, the increased DOC level and higher concentration of colloids in unfiltered samples may not affect the "free" concentrations so much. However, when dilution series are prepared to assess ecotoxicity, the increased DOC and the sorbed quantity of substances on colloids have an effect, because of the repartitioning upon dilution. The question is then reduced to the nature of the DOC and colloid levels in the extract that represent a given situation. It is known that a batch test overestimates the DOC and colloid levels relative to field observations, which implies that a form of solid separation may be advisable.

Acknowledgement The Dutch Ministry of Environmental Affairs is thanked for the financial support provided to complete this work. Mrs J. Leeflang is thanked for her contribution in preparing this chapter.

Chapter 26
Ecotoxicological Response of Three Waste Samples in Relation to Chemical Speciation Modeling of Leachates

J.F. Postma, H.A. van der Sloot, and A. van Zomeren

Abstract Extended leach testing (pH dependence TS 14429 and column TS 14405) provides options for chemical speciation modeling to determine the bioavailable fraction ("free" unbound concentrations of substances in solution) under varying conditions in the test methodology. For aquatic organisms, this has been shown to relate rather well with measured ecotoxicity (Cu in waste wood). Operational aspects of EN 14735 may need modification in the light of the suitability of the method for organic contaminants, the choice of liquid/solid separation in view of the level of dissolved organic carbon and colloids in solution and the relevant liquid to solid ratio for specific questions related to waste use, treatment and disposal.

Keywords Leaching, Chemical speciation, Waste characterization, Bioavailability, Ecotoxicity, Dissolved organic carbon

26.1 Introduction

In the framework of CEN/TC292, work on standardization was initiated to validate methods to characterize the ecotoxicological properties of waste (see Chap. 2). While some choose methods that predominantly use chemical analyses, others are more in favor of direct ecotoxicological measurements. In the evaluation during the workshop "Problems around Soil and Waste III – The H-14 Criterion and (Bio) analytical Approaches for Ecotoxicological Waste Characterization" (Gawlik & Moser 2005) it was clear that either approach on its own cannot resolve the issues. The present work aims to show the benefit of a more integrated evaluation, in which the information derived from elaborate leaching tests and associated chemical speciation modeling is linked with observations obtained in ecotoxicological analyses of eluates resulting from waste.

J.F. Postma (✉), H.A. van der Sloot, and A. van Zomeren
Grontmij|AquaSense, P.O. Box 95125, 1090 HC, Amsterdam, The Netherlands

H. Moser and J. Römbke (eds.), *Ecotoxicological Characterization of Waste*,
DOI: 10.1007/978-0-387-88959-7_26,

Ring tests focus on the inter-laboratory precision of an individual parameter, in this case, toxicity to specific organisms. The strength of the observed toxicity is the final result of a number of processes occurring during all preparation steps, from sampling to the preparation of a concentration series. Critical aspects can (among others) be found in the selection of the leaching conditions (for example pH and solid/liquid ratio) and the choice between batch procedures vs. percolation tests. Within the EN 14735 (CEN 2006) several choices have already been made concerning these aspects. In the light of the present validation of the EN 14735, it was considered crucial to expand the range of testing conditions to allow chemical speciation modeling and to see what consequences might arise from the choices made within the EN 14735. A basic characterization was performed for all three waste samples (Chap. 3). Furthermore, an extended characterization consisting of a combination of a pH dependence and a percolation test was also carried out (Chap. 25). In combination with chemical speciation modeling, partitioning of constituents between dissolved (free and dissolved organic matter associated substances) and solid phases (solid organic matter, clay surfaces, iron-oxides and minerals) can be calculated, which is necessary for understanding the observed ecotoxicity, as ecotoxicological effects are closely related to the chemical form of the constituents.

This chapter presents the results of the ecotoxicological tests in relation to bioavailability, as derived from chemical characterization. Bioavailability in this context is defined as the extent to which constituents are present as "free" ionic (trace elements) and unbound forms (organic pollutants not associated with DOC and/or colloids). The results of the chemical properties, bioavailability, modeling and eluate characteristics as described in Chap. 25 are presented and discussed in the light of the observed toxicity. This knowledge can be used to further our understanding of the ecotoxicological properties of waste, to assess them and try to link speciation to bioavailability, as well as linking the latter to toxicological effects in organisms.

26.2 Material and Methods

Chemical and ecotoxicological analyses were conducted on three different waste samples. The ecotoxicological analyses consisted of three aquatic bioassays using eluates and two terrestrial bioassays performed with the waste sample itself. In this chapter, only the results of the aquatic toxicity tests are discussed. Preparation of the samples was carried out strictly according to EN 14735 and additional information as communicated by the ring test organization. Both the chemical and ecotoxicological analyses are described in detail.

Sampling, homogenization and distribution of the samples are described in Chap. 3. All test samples (incineration ash (INC), waste wood (WOO) and gasworks soil (SOI)) were upon arrival stored at 4°C in the dark and were tested with the following aquatic bioassays on the eluates: Bacterium *Vibrio fischeri* (Microtox), the waterflea *Daphnia magna* and the algae *Pseudokirchneriella subcapitata*. To account for possible interference by confounding factors, several physical-chemical characteristics such as pH, oxygen saturation, conductivity and ammonium contents were measured prior, during and at the end of the different bioassays.

26.2.1 Acute Toxicity Test with the Bacterium *Vibrio fischeri*

Bioassays with the bacterium *Vibrio fischeri* were conducted according to ISO 11348-3. The reduction in the bioluminescence of this bacterium was assessed by measuring the light intensity (using Microtox® equipment) after 5, 15 and 30 min exposure to the following concentrations: 45, 22.5, 11.25 and 5.625 Vol%. All tests were conducted at 15°C and in duplicate. EC_{20}- and EC_{50}-values were estimated, using the statistical programs of the Microtox testsystem.

26.2.2 Acute Toxicity Test with the Waterflea *Daphnia magna*

Bioassays were performed according to ISO 6341, by which the immobility of the organisms after 24 and 48 h was assessed. All tests were performed at 20 ± 2°C and a light regime of 16:8 = light:dark. Tests were started with juveniles <24 h old. The test volume was 50°mL for each vial. All samples were tested using five concentrations and a blank consisting of ISO-medium. Each concentration was tested in four replicates with five individuals per replicate. EC_{50} and NOEC-values were estimated using the statistical package Toxcalc (Tidepool 1995).

26.2.3 Toxicity Test with the Freshwater Algae *Pseudokirchneriella subcapitata*

Bioassays with the algae were performed according to ISO 8692 (2004). An exponentially growing pre-culture was used as an inoculum for the test. According to EN 14735, all eluates were filtered over 0.45 μm before starting the tests. All tests were started with an inoculation concentration of about 10^4 cells/mL. The tests were performed in an incubator, shaking continuously at 200 rpm. Cell densities were measured by means of fluorescence at $t = 0$, 48 and 72 h using a plate-reader (670 nm). Growth rates (μ) were calculated for each test concentration and replicate.

26.3 Results

26.3.1 Eluate Analyses

Results of some basic characteristics of the waste samples and their eluates are given in Table 26.1. A more extended characterization following the methods developed by CEN/TC292/WG6, is given in Chap. 25.

Table 26.1 Sample and analytical characterization of the eluates

Sample	pH	Moisture content (% w/w)	Parameters (mg/L at *L/S* = 10)		
Incineration ash	10.6	2.3	Cu: 0.17	Ni: 0.0015	DOC: 21.7
Waste wood	4.8	9.4	Cu: 19.2	Cr: 0.28	DOC: 1078
Gasworks soil	8.0	0.85	Cu: 0.0083		DOC: 23.3

26.3.2 Ecotoxicological Analyses

The results of the bioassays are summarized in Table 26.2 and discussed in detail. (For all raw data see H-14 navigator).

26.3.2.1 Incineration Ash (INC)

The eluate was clear and colorless, with the following characteristics: pH = 10.6; Conductivity = 1,490 μS/cm; Oxygen sat. = 84% and DOC = 21.7 mg/L (*L/S* = 10).

This pH is rather high and could easily cause toxic effects for several aquatic organisms. Range-finding tests confirmed that toxicity was high and the eluate could best be tested in substantially reduced concentrations. As a result, the following maximum concentrations were tested for INC: Microtox: 45.5 vol%; daphnids: 6.0 vol% and algae 20 vol%. Due to this dilution of the test medium, the maximum pH values in the highest test concentrations were considerably lowered (8.8 for the algae; 8.3 for the daphnids) and direct toxic effects of these lowered pH values are not expected anymore. Microtox was less sensitive, however, and higher test concentrations were consequently used. For this test system, it was concluded that toxicity might be a direct result of increased pH values, as pH in the three highest test concentrations varied between 9.0 and 9.9. Postma et al. (2002) mentioned a safe maximum pH value of 8.5. The three EC_{50}-values varied between 1.0 vol% for the daphnids to 18.1 vol% for Microtox. The algae were characterized by an EC_{50}-value of 9.8 vol%.

26.3.2.2 Gasworks Soil (SOI)

The eluate obtained from the gasworks soil was slightly colored (yellow-brownish) and characterized by the following physical-chemical characteristics: pH = 8.0; Conductivity = 114 μS/cm; Oxygen sat. = 97% and DOC = 23.3 mg/L (*L/S* = 10). In the preparation of the eluates a strong smell typical for gasworks soil was noted, indicating the presence of substantial quantities of oily components. The tests were performed with the following maximum concentrations: Microtox = 81.9 vol%; daphnids = 100 vol% and algae = 50 vol%. Compared with the other two samples, eluates from the gasworks soil proved to be least toxic. For the Microtox assay, an EC_{50}-value was estimated of 54.9 vol%, compared to EC_{50}-values of 0.4 and 18.1 for the other two waste samples. The difference between the daphnids and algae

Table 26.2 Summary of the ecotoxicity tests for the three samples (vol%)

Sample:		Incineration ash	Gasworks soil	Waste wood
Bioassay		INC	SOI	WOO
Bacterium *Vibrio fischeri*	EC_{20}	7.2 (6.2–8.3)	10.2 (8.7–11.9)	0.3 (0.2–0.4)
	EC_{50}	18.1 (15.6–21.0)	54.9 (46.0–65.5)	0.4 (0.3–0.6)
Daphnia	NOEC	0.38	100	0.13
Daphnia magna	EC_{50}	1.0 (0.7–1.3)	>100	0.2 (0.2–0.3)
Algae	NOEC	5	25	<0.025
Pseudokirchneriella subcapitata	EC_{50}	9.8 (8.1–16.4)	>50	0.05 (0.04–0.08)

was even more pronounced, as no EC_{50}-values could be estimated for the gasworks soil. In the bioassay with the daphnids, no toxic effects were observed at all, while the algae showed a significant but small reduction in growth rate (14% reduction) in the highest concentration tested (50 vol%).

26.3.2.3 Waste Wood (WOO)

The eluate obtained from the waste wood sample was yellow colored (probably due to a rather high DOC content) and characterized by the following physical-chemical characteristics: pH = 4.8; Conductivity = 457 µS/cm; Oxygen sat. = 96% and DOC = 1,078 mg/L (*L*/*S* = 10). At a pH of 4.8, toxic effects might be expected on many aquatic organisms. However, in this case, pH was most likely not a direct confounding factor. Due to the high toxicity of the waste wood sample, strong dilutions were applied and a test concentration of 2 vol% was the highest concentration tested (Microtox). For daphnids 0.5 vol% was the highest concentration tested and for the algae 0.4 vol%. At these dilutions, the low pH has been buffered to normal values. For Microtox ten test concentrations were used to enable an accurate estimation of EC_{20}- and EC_{50}-values. Strong toxic effects were observed for all three bioassays and EC_{50}-values varied between 0.05 vol% for the algae up to 0.4 vol% for the Microtox assay (Table 26.2).

26.3.3 Total Dissolved and "Free-concentrations" in Eluates for Aquatic Toxicity Testing

From the speciation modeling shown in Chap. 25, the partitioning in solution (distinction between "free" and bound concentrations of elements and organic micropollutants in solution) can be obtained. Chap. 25 also describes how the inorganic fraction that is not bound to DOC and is necessary for explaining the ecotoxicological response has been derived. For the three matrices a summary of this information is given in Table 26.3 (Free = bioavailable).

Table 26.3 Modeled partitioning between total and free element concentrations at the original pH-values of the different waste samples (Soil pH = 7.6; Incinerator bottom ash pH = 10.6 and Wood pH = 4.8)

	Wood					Incinerator bottom ash					Soil				
	Total concentration		Free	DOC-bound		Total concentration		Free	DOC-bound		Total concentration		Free	DOC-bound	
	Mol/L	mg/L	mg/L	mg/L	%	mol/L	mg/L	mg/L	mg/L	%	mol/L	mg/L	mg/L	mg/L	%
Al	2.00E−05	0.54	0.001	0.54	99.9	2.00E−03	54.0	53.7	0.27	0.5	3.00E−06	0.08	0.036	0.045	55.0
Ba	1.00E−06	0.13	0.107	0.03	19.9	8.00E−06	1.1	1.0	0.03	3.0	1.00E−06	0.13	0.13	0.001	1.0
Ca	8.00E−04	32	27	5.5	17.3	8.00E−03	320	317	3.2	1.0	4.00E−03	160	158	1.6	1.0
Fe	2.00E−06	0.11	0	0.11	100	6.00E−08	0.003	0.0002	0.003	95.0	4.00E−07	0.022	0.0001	0.022	99.5
SO4	1.30E−04	13	13	0	0	2.60E−03	255	255	0	0	3.00E−04	29	29	0	0
Mg	2.00E−04	5.4	4.1	1.3	24.0	1.00E−05	0.27	0.26	0.014	5.0	3.00E−04	8.1	7.9	0.16	2.0
P	1.00E−04	3.2	3.2	0	0	8.00E−06	0.26	0.26	0	0	8.00E−06	0.26	0.26	0	0
Cu	3.50E−04	22	0.45	22	97.9	1.00E−05	0.63	0.019	0.61	96.9	2.00E−07	0.013	0.00003	0.013	99.8
Ni	1.50E−07	0.010	0.001	0.01	92.4	2.00E−08	0.001	0.0006	0.001	50.0	3.70E−08	0.002	0.002	0	2.6
Zn	4.50E−06	0.29	0.063	0.23	78.6	3.00E−08	0.002	0.0004	0.002	78.0	1.50E−07	0.010	0.002	0.007	76.0
Cr	5.00E−06	0.26	2.6E−06	0.25	100	5.00E−06	0.26	0.23	0.03	10.0	1.00E−07	0.005	1E−07	0.005	100
B	1.00E−03	7	7	0	0	1.00E−05	0.070	0.07	0	0	2.00E−05	0.14	0.14	0	0
K	1.70E−03	68	67	0.7	1	1.00E−03	40	40	0	0	6.00E−04	24	24	0	0
Na	7.00E−03	168	166	1.7	1	7.00E−03	168	168	0	0	4.00E−03	96	96	0	0
DOC		1,070										8.5			

Table 26.4 Composition of the media used in the aquatic ecotoxicity tests

	Dilution media		
Element (mg/L)	Microtox	Algae	*Daphnia*
Ca	211	4.9	80
SO_4	1,445	5.9	48
Mg	611	3.2	13
P		1.1	
Cu		3.80 E–06	
Zn		0.001	
B	15		
Mo	0.006	0.003	
K	217	0.46	3.1
Na	5,646	25	18
Cl	10,171	48	145
CO_3	80	36	46
F	0.571		
NH_4		5.0	
EDTA		94	
Co		0.0004	

However, this is not the final partitioning necessary for the ecotoxicity tests, as the eluates are modified considerably before organisms are exposed to the test solutions. This relates to the changes in eluate composition due to the dilution with control medium to obtain the test series. The composition of the dilution media for the three aquatic toxicity tests is given in Table 26.4. Adding control media to the eluates changes the partitioning of elements significantly due to changes in the competition for binding sites on (for instance) the dissolved organic matter. The effects of these processes were calculated for all dilutions (See H-14 navigator). In Table 26.5 three examples are provided, showing the interacting effects of (predominantly) competition with DOC (WOO, *Daphnia* and copper), increased salt levels (WOO, Microtox and copper) and increased Ca levels (INC, algae and copper). These examples illustrate the interacting effects of DOC, salt and Ca on bioavailability. All three examples caused an increase in the percentage of bioavailable copper compared to the undiluted eluate, and as such caused a decrease in the observed EC_{50}-values.

Toxicity in the dilutions is therefore not truly representative of the toxicity in the undiluted sample. In risk assessments and fate calculations, the choice of a representative 'dilution' medium might therefore play a crucial role.

26.4 Discussion

The present research focuses on the combination of ecotoxicological and chemical analyses. The main goals are to see whether chemical speciation and bioavailability modeling can explain observed toxic effects, which toxicants might be the causative factors and what lessons might be learned from this, by looking at the specifications of the EN 14735. In this discussion the focus is on the eluate and the aquatic organisms.

Table 26.5 Copper concentrations obtained after dilution with the media as specified in Table 3.4 and the repartitioning of elements between "free" and DOC bound forms in the waste extracts

	Test solutions (vol%)					
	100%	0.50%	0.25%	0.20%	0.13%	0.10%
(A) Wood sample – *Daphnia* test						
Total Cu in undiluted eluate (mg/L)	22.1					
Free Cu in undiluted eluate (mg/L)	0.5					
Total Cu in test solution (mg/L)		0.110	0.055	0.044	0.029	0.022
"Expected" free-Cu in test solution (mg/L)		0.002	0.001	0.001	0.001	0.000
"Modelled" free-Cu in test solution (mg/L)		0.015	0.007	0.006	0.005	0.004
Percent of total as "free" Cu (%)	2.1	13.6	13.6	13.6	16.6	16.6
	Test solutions (vol%)					
(B) Wood sample – Microtox test	100%	2.08%	1.04%	0.52%	0.26%	0.13%
Total Cu in undiluted eluate (mg/L)	22.1					
Free Cu in undiluted eluate (mg/L)	0.5					
Total Cu in test solution (mg/L)		0.46	0.23	0.12	0.057	0.029
"Expected" free-Cu in test solution (mg/L)		0.009	0.005	0.002	0.001	0.001
"Modelled" free-Cu in test solution (mg/L)		0.25	0.11	0.052	0.024	0.011
Percent of total as "free" Cu (%)	2.1	53.4	48.5	45.2	42.0	38.4
	Test solutions (vol%)					
(C) INC sample – algae test	100%	20.0%	10.0%	5.0%	2.50%	1.25%
Total Cu in undiluted eluate (mg/L)	0.6					
Free Cu in undiluted eluate (mg/L)	0.019					
Total Cu in test solution (mg/L)		0.13	0.063	0.032	0.016	0.008
"Expected" free-Cu in test solution (mg/L)		0.004	0.002	0.001	0.000	0.000
"Modelled" free-Cu in test solution (mg/L)		0.010	0.005	0.003	0.001	0.001
Percent of total as "free" Cu (%)	3.1	8.0	7.3	8.8	8.6	8.5

A comparison is made between the 'expected' free-concentration (assuming a proportional relationship as compared to the undiluted eluate) and the modelled free concentration (in which interacting effects are taken into account)

26.4.1 Contaminated Wood

The chemical analyses in the eluate showed extremely high concentrations of copper (±20 mg/L)[1]. Furthermore, the concentrations of chromium and borate were increased as can be expected in preserved wood samples. For copper, several EC_{50}-values can be found in databases (see below), with which Toxic Units (TU)[2] can be calculated. These TU are based on the copper concentration in pH stat experiments and do not take bioavailability into account.

[1] A Cu conc of 19.2 mg/L was measured in the eluate and 22 in the pH experiments. The modelling is based on the concentration of 22 mg/L.

[2] TU-values are calculated by dividing measured concentrations in a sample by its respective EC_{50}-values. These TU-values can be calculated for all toxicants together and added together to provide an overall insight in the expected toxicity. As a consequence, if a sample exceeds the TU-value of 1, it might be expected that > 50% effect will occur.

	EC_{50}-values for copper (mg/L)	TU-values for copper (in undiluted eluate)
Microtox	0.13	170
Daphnia magna	0.024	919
Algae	0.15	147

As TU values are all well above 1 (and even well above 100!), it is clear that copper might play a dominant role in the toxicity to aquatic organisms, especially in the higher concentrations. Here the focus is on copper and *Daphnia*, which is the most sensitive organism to copper. The actual EC_{50}-value for *D. magna* as measured in the eluate was 0.2 vol%. At this concentration, the total copper concentration of 22 mg/L in the undiluted eluate dropped to 0.044 mg/L. Consequently, even in this very low test concentration, 2 TU for copper are present and the observed toxic effects for daphnids seem understandable.

However, only the bioavailable copper concentration, not the total concentration, will be the causative factor. Using the chemical speciation models as discussed above, free copper concentrations were estimated. In the test concentration of 0.2 vol% and a pH of 7.7 it is estimated that 0.006 mg/L copper is bioavailable, corresponding with a TU-value of 0.25. In the highest concentration (0.5 vol%) the bioavailable copper concentration is 0.015 mg/L and the TU-value 0.6. This is somewhat lower than hoped for, but within a factor of 2–4. Looking at all uncertainties surrounding bioavailability estimates, this seems still quite reasonable and it can therefore be concluded that the copper concentration is the main causative factor in the observed toxicity for *Daphnia*.

Part of this small difference between expected and measured toxicity might have been caused by the fact that toxicity of copper to aquatic organisms (such as *D. magna*) is influenced not only by the hardness of the test medium but also the calcium–magnesium ratios within the medium. Naddy et al. (2002) for example demonstrated that EC_{50}-values for *D. magna* might vary between 12 and 57 μg/L in total copper, depending on both total hardness as well as Ca:Mg ratios. This aspect of Ca:Mg ratios presents a new aspect in the toxicity of this waste wood material, but the EC_{50}-value of 0.024 mg/L mentioned above seems to be representative of the conditions in the actual test (hardness around 250–300 and Ca:Mg of 6). The overall conclusion is that copper most likely causes the toxicity for *D. magna*. The next step is to see whether the other two aquatic organisms are also likely to respond to copper. This seems less likely because both the Microtox assay as well as the algae are less sensitive to copper, while simultaneously the observed EC_{50}-values in the eluate are comparable or even somewhat lower (Microtox: 0.4 vol%; algae: 0.05 vol%).

	Total copper concentrations		Bioavailable copper concentrations	
	Conc. at the EC_{50} (μg/L)	TU-values at the EC_{50}	Conc. at the EC_{50} (μg/L)	TU-values at the EC_{50}
Microtox	88	0.68	40	0.31
Algae	11	0.07	1.4	0.01

From this overview it can be concluded that a remarkable difference exists between the estimated bioavailable concentrations between the Microtox and Algae tests. The % bioavailable copper in the Microtox is much higher (around 40% in this test concentration as compared to 15% in the Algae). The reason for this is the fact that Microtox is performed in a saline environment and the increased salinity causes an increased availability of copper. This effect is further illustrated in Table 26.5. The bottom line of these observations is that for the Microtox it can be concluded that toxicity is mainly caused by copper, while the toxicity in the Algae is still largely unexplained.

26.4.2 Incineration Ash

The chemical analyses of the eluates show that several elements are increased, including several toxic metals such as aluminum and copper. The strongest differences however occur within major elements like Ca and SO_4 (Table 26.6). Ratios of major elements such as Ca:Mg and Na:K might strongly influence basic cell functioning.

Based on this overview the following conclusions can be made:

26.4.2.1 Daphnia

The toxicity of the eluate is so high that direct effects of major elements can be ruled out. At a test concentration of 1 vol% concentrations of elements like Ca,

Table 26.6 Overview of the composition of the eluate, control medium as well as the estimated composition at observed EC_{50}-dilutions (both for daphnids and algae; mg/L)

	Undiluted eluate	Control medium *Daphnia*	Control medium Algea	Concentration of 1 vol%[a]	Concentration of 10 vol%[b]
Al	54.0				
Ca	320	80.1	4.9	82.4	36.4
Mg	0.27	13.3	3.2	13.1	2.9
SO_4	254.8	48.1	5.9	50.2	30.8
Na	168	17.7	25.3	19.2	39.6
K	40	3.1	0.46	3.4	4.4
Cr	0.255				
Cu	0.63				
Ratios					
Ca/Mg	1,185	6.0	1.5	6.3	12.6
Na/K	4.2	5.8	55.4	5.6	9.0

[a]Based upon the control medium for daphnids; EC_{50}-value for *D. magna* was 1 vol%
[b]Based upon the control medium for algae; EC_{50}-value for *P. subcapitata* was 10 vol%

Mg etc., as well as crucial ratios, are not significantly different from the control medium. However, these differences in concentrations might have indirect side effects, as the bioavailability of metals like copper is strongly dependent on hardness. By diluting the eluate, not only the total concentrations but also metal availability changes. These effects are illustrated in detail in Table 26.5. While only 3% of the copper is in its bioavailable form in the undiluted eluate, around 15% of the copper is bioavailable in the test concentrations varying between 0.2 and 6 vol%. Due to this difference of a factor of 5, estimated EC_{50}-values tend to decrease and choices for the dilution medium (and its composition) might significantly affect toxicity. With TU-values around 0.03–0.13, based on bioavailable copper concentrations at the EC_{50} concentration of 1 vol% and one concentration higher (3 vol%), explainability is clearly less as compared to the wood sample, but still it can be concluded that copper plays a role (especially if the uncertainties in the applied modeling and organism sensitivity are taken into account). In this light it is important to reckon that unexpected side effects, such as the observed strong changes in Ca-concentrations as well as Ca:Mg ratios, affect the bioavailability of copper.

26.4.2.2 Algae

For the Algae, major elements might have a more direct effect as the Ca:Mg ratio at the observed EC_{50}-value is a factor of 10 higher as compared to the control medium. Unfortunately, no comparable data could be found in the literature to describe the growth rate of algae at different Ca:Mg ratios. It is therefore concluded that by performing toxicity tests on eluates of waste material, special attention should be paid to major elements.

26.4.3 Gasworks Soil

According to the chemical analyses of the eluates, this waste material is not strongly polluted by inorganic pollutants. Concentrations of most metals are generally low. A low level of pollution (in comparison with the other two samples) is also concluded on the basis of the bioassays, as only slight, or even no toxicity was observed. On the other hand, the gasworks soil itself did show signs of severe contamination and the smell alone should be enough to indicate the presence of increased mineral oil and PAH concentrations. In Table 26.7 the total content of PAH (as provided in Chap. 3) and the leaching of two pH steps according to TS14429 (alternative solids separation by centrifugation) at own pH and pH 12 are given.

The expected presence of organic pollutants in this sample sheds a different view on the EN 14735, as this method seems not to be specifically developed to correctly represent the leaching of organic contaminants. For example, the procedure differs in several essential aspects with the ISO/TS 21268, which was specifically designed

Table 26.7 Total PAH in relation to "available" and leachable PAH at own pH of SOI sample

Parameter	Total (mg/kg)	Concentration at $L/S = 10$ (μg/L) pH 12	Own pH	Leached in mg/kg pH 12[a]	Own pH	Available for leaching (%)	Leached at own pH (%)
Anthracene	23	5.2	1.5	0.052	0.015	0.22	0.065
Benzo(a)anthracene	87	17.6	1.7	0.18	0.017	0.20	0.019
Benzo(a)pyrene	59	9.9	0.66	0.10	0.0066	0.17	0.011
Benzo(b)fluoran-thene	79	15.1	1.05	0.15	0.010	0.19	0.013
Benzo(ghi)perylene	35	1.78	0.43	0.018	0.0043	0.051	0.013
Chrysene	69	17.5	1.70	0.17	0.017	0.25	0.025
Fluoranthene	182	37	10	0.37	0.10	0.20	0.055
Phenanthrene	69	25	0.47	0.25	0.0047	0.37	0.007
Pyrene	146	28	6.9	0.28	0.069	0.19	0.047

[a]Considered to represent the "available" fraction for leaching

Table 26.8 TU-values for Microtox and *Daphnia* based on PAH-concentrations found in the leaching test according to TS14429 at own pH

Parameter	Concentration $L/S = 10$; own pH; μg/L	Microtox EC_{50} (μg/L)	TU-value	*Daphnia* EC_{50} (μg/L)	TU-value
Anthracene	1.52	580	0.0026	268	0.0057
Benzo(a) anthracene	1.66	260	0.0064	10	0.17
Benzo(a)pyrene	0.66	8,100	0.0001	5	0.13
Benzo(b) fluoranthene	1.05				
Benzo(ghi) perylene	0.43				
Chrysene	1.70	1,490	0.0011	0.7	2.4
Fluoranthene	10.0	630	0.0159	106	0.094
Phenanthrene	0.469	49	0.0096	667	0.0007
Pyrene	6.89	1,160	0.0059		

to assess organic parameters in relation to ecotoxicity testing. One key issue is that a batch test does not seem to be a very good representative for organic contaminant leaching as the agitation in a batch method leads to substantially greater release as compared to a column test (ISO/TS 21268-2 and -3, 2006). This was found in the "Sickerwasser prognose" project as well (2007).

From Table 26.8 it can be concluded that in spite of the substantial contamination of the SOI material with organic contaminants (PAH's, Table 26.8), no significant toxicity is found among the aquatic organisms, since all PAHs are likely bound to either particulate or dissolved organic matter in the eluate and thus not present in free (unbound) form for uptake by organisms. In Table 26.8 the sensitivity of organisms towards PAHs is calculated. For Microtox no toxic effect is noted, while for *Daphnia* a response may be found.

The question is whether the concentrations in solution corresponding to EC_{50} are total concentrations, or whether these values relate to the "free "organic contaminant.

Table 26.9 Partitioning of PAH in "free" and DOC-bound phases (LeachXS-Orchestra)

Parameter	DOC-bound (%)	Parameter	DOC-bound (%)	Parameter	DOC-bound (%)
Anthracene	10	Benzo(b) fluoranthene	95	Fluoranthene	90
Benzo(a) anthracene	60	Benzo(ghi) perylene	25	Phenanthrene	70
Benzo(a)pyrene	20	Chrysene	95	Pyrene	40

In case of the latter, it may be expected that part of the PAHs are bound to DOC (or better sub-fractions of DOC). In Table 26.9 the partitioning of the different PAHs is given between dissolved (free and DOC bound) and particulate phases (particulate organic matter). When the above concentrations are based on free PAH, then the "no-effect" on organisms is fully explained.

From Table 26.9 it follows that chrysene (the only one with a high TU-value; see Table 26.8) is not likely to be free in solution in concentrations exceeding the sensitivity and thus is the possible toxic effect considered of limited relevance. The partitioning of organic contaminants in free, DOC and POM bound is of high importance for assessing the long-term effects of organic contaminants. These results clearly show the limitations of judgment regrading organic contaminants based on total composition and the current possibilities to judge based on unbound dissolved contaminants.

26.5 Overall Conclusions

The pH-static leaching tests in combination with model predictions of the leaching behavior, resulting in the speciation of contaminants in both the solid phase and solution, provide detailed knowledge of chemical processes in these apparent heterogeneous materials, which can be linked to an ecotoxicological response. The main conclusions are:

- The combination of chemical analyses and speciation modeling proved to be a powerful tool in finding causative relations between toxicants and observed effects in organisms. In the present waste samples, copper seemed to be of prime importance for the toxicity as demonstrated in (especially) the wood sample and also the incinerator ash sample. In the soil sample, organic pollutants are the most likely cause of toxicity; however, the toxicity level is relatively low, as predicted, due to a strong association of PAH with DOC. For some of the bioassays and sample combinations, this match is rather good and observed effects are deemed understandable. For other combinations, the match is less pronounced and other factors (not yet identified) might have influenced toxicity as well.
- The speciation modeling demonstrated changes in bioavailability as a result of the applied test methods, including the dilution of the eluates with control medium. Interacting effects due to (predominantly) competition with DOC (e.g. WOO, *Daphnia* and copper), increased salt levels (e.g. WOO, Microtox and

copper) and increased Ca levels (e.g. INC, algae and copper) are highlighted. All three examples caused an increase in the bioavailability of copper compared to the undiluted eluate and therefore caused a decrease in the observed EC_{50}-values. Toxicity in the dilutions might therefore not be truly representative of the toxicity in the undiluted sample. In risk assessments and fate calculations the choice for a representative 'dilution' medium will therefore play a crucial role.

- The EN 14735 method was not developed for organic micro pollutants. Further developments may help to resolve this limitation. In addition, solid/liquid separation has a major influence on DOC and colloid levels in solution. The decision to filter or centrifuge the eluate or to abstain from liquid/solid separation cannot be taken lightly, as the present results show the major influence of these parameters on the outcome of an ecotoxicity test. For many decisions on waste use, treatment and disposal, a low liquid to solid ratio is of importance. Translation of results from $L/S = 10$ to a low L/S is impossible, given the influencing factors identified here.

Acknowledgement The Dutch Ministry of Environmental Affairs is thanked for the financial support provided to complete this work. Mrs J. Leeflang is thanked for her help in preparing this contribution.

Chapter 27
H14-Navigator Uses Topic Maps as Application Data Model

G.E. Weber, R. Eilbracht, and S. Kesberg

Abstract The web application H14-Navigator provides metadata and the results of the H14 European ring test on ecotoxicity of waste from an entirely Topic Maps-based backend. It uses ontology for physical, chemical and biological properties, and knowledge models contained in the H14 topic map to generate interface structures, knowledge-oriented access to and navigation paths through a highly networked information space. The frontend generates tables as if it were based on an application-specific relational data model. Content can also be rendered in the so-called topic view, reducing the amount of information displayed, and focusing on relationships and relevant knowledge models for anchoring the focused topic in the framework of the ecotoxicological domain. Where appropriate, the topic view is complemented by Tables providing more detailed information. An intelligent search algorithm completes frontend functionalities. The H14-Navigator is an example of a new kind of frontend, radically doing away with application-specific relational data models, as well as with the distinction between data-oriented and knowledge-oriented storage layers. It uses only one model, which is the Topic Maps data model standardized in ISO/IEC 13250-2 (2006), thus providing evidence of the hitherto underestimated power of this graph-based model. With the availability of the standardized XML syntax ISO/IEC 13250-3 (2007), and the possibility of defining a globally accessible identity for each subject represented by a topic, Topic Maps-based semantic application systems are here to stay.

Keywords Ecotoxicity of waste, Ecotoxicological ontology, ISO 13250-2, Ontology, Ontology-based application, Published subject identifiers, Subject-centric computing, Topic Maps, Topic Maps-browser, Topic Maps-frontend

G.E. Weber (✉), R. Eilbracht, and S. Kesberg
Stuttgart, Germany

H. Moser and J. Römbke (eds.), *Ecotoxicological Characterization of Waste*,
DOI: 10.1007/978-0-387-88959-7_27,

Glossary

Association An association is a "representation of a relationship between one or more subjects"(ISO/IEC 13250-2 2006). In the H14-Navigator, associations are referred to as "relations", this being the more common term.

Backend A backend is storage of information. Backends may use a file-system but are usually residing in a database which is managed by a database management system (DBMS) such as the open source DBMS MySQL.

Data model A data model is a structure which organizes information, especially information and the relationbetween information items. Applications usually depend on an application-specific data model, to which the frontend is closely coupled.

Frontend A frontend is a kind of software which makes information stored in a backend accessible in a user-friendly way. H14-Navigator is a web-application which renders the information in the H14 topic map into web pages, accessible through any computer with access to the internet.

Ontology According to Gruber (1993) an ontology is an explicit specification of a conceptualization. However, the ontology of the H14 topic map aims to go beyond a mere conceptualization and rather subscribes to the realistic approach advocated by Fielding et al. (2004) who stated: In his *Physics*, Aristotle writes, 'When the objects of an inquiry, in any department, have principles, conditions, or elements, it is through acquaintance with these that knowledge, that is to say scientific knowledge, is attained,' and we would do well to keep such words in mind today when we seek to design an adequate inventory of ontological elements for database integration and navigation.

Relation See association.

Topic A topic is a "symbol used within a topic map to represent one, and only one, subject, in order to allow statements to be made about the subject" (ISO/IEC 13250-2 2006). In the H14 topic map, this subject-centric approach is applied in a very far reaching way, and entities such as samples, methods, processes, organizations, but also properties and property values are modeled as subjects, represented by topics.

Table view The table view is one of two visualization patterns of the H14-Navigator. It renders information in the form of tables, as if it was based on an application specific relational data model. However, the H14-Navigator derives all content from a topic map, which does not have a data model designed according to the needs of the H14 ring test, but adheres to the data model defined by ISO/IEC 13250-2 (2006).

Topic Maps Topic Maps, written with capital initials, is "a technology for encoding knowledge and connecting this encoded knowledge to relevant information resources" (ISO/IEC 13250-2 2006). Considering the use of its data model as a very powerful application data model, the standard's self description seems heavily inclined to understatement.

Topic map A topic map is "a set of topics and associations" (ISO/IEC 13250-2 2006), adhering to the definitions given in ISO 13250-2 (2006). All content provided by the H14-Navigator is stored in a single topic map.

Topic view The topic view is one of two visualization patterns of the H14-Navigator. It is focused on one topic, its relations to other topics and information directly linked to the focused topic. For particular topics, it may be complemented by one or more tables.

27.1 Introduction

In the European Union, so far no methodological recommendations for the assessment of the ecotoxicity of waste have been provided. Therefore, the German Federal Environment Agency (UBA) coordinated a European ring test aimed at evaluating a set of methods for assessing the ecotoxicity of wastes and waste eluates described as hazard criterion H14 in the European waste list (see Chapter 3). The results of this ring test are intended to support the drafting of binding European recommendations for the assessment of the ecotoxicity of waste. Given the scientific value of the ring test results, and their intended use as expert inputs strengthening the basis of European environmental policy, all results were published online. However, due to the high complexity of the ecotoxicological domain, project and data structures, a web-frontend, based on an application- specific relational data model, seemed inappropriate to assure easy accessibility of the expert results also for non-experts. Instead, UBA opted for an approach of an entirely Topic Maps-based application for all information about, and for all results of, the ring test. Easy accessibility of the complex content was a prime requirement, intended to improve not only expert usage of the results but also the impact achievable with the outcome of the project.

The ring test focused on three waste substrates which were evaluated by laboratories all over Europe. All in all, 67 laboratories participated in the ring test (60 delivered data), and 17 different ecotoxicological methods using 16 different biological species were applied. About 200 different properties were assessed, including the reference substances tested. The examinations were subject to quality control, and the numerical results were subject to statistical analysis, the results of which provided the basis for the evaluations of the ecotoxicological methods employed, and for the methodological recommendations for assessing the ecotoxicity of waste.

The H14-Navigator[1] is a new kind of frontend operating on a single topic map which contains information about and the results of the H14 ring test. It uses the domain ontology, and the ecotoxicological knowledge models contained in the H14 topic map to render interface structures and access paths to all content.

[1]The H14-Navigator is a commercial customer specific web application created by Hoelle & Huettner AG for the German Federal Environment Agency; to be accessed at http://EcotoxWasteRingtest.uba.de/h14

27.2 ISO Standardized XML-Based Backend

Contrary to today's dominating architectural structure of information systems, which rests on application-specific relational data models for the backend layer, the H14-Navigator does not make use of a data model and data base tables catered to its application-specific needs. Rather it uses the ISO standardized Topic Maps data model (ISO/IEC 13250-2 2006) as its application data model. Considering that all backend content of the H14-Navigator can be exported using the XML syntax of Topic Maps (ISO/IEC 13250-3 2007), the application draws from two open standards: one defining the standardized data model, and one defining the standardized syntax.

27.2.1 Application Specific Ontology

"Ontology as a branch of philosophy is the science of what is, of the kinds and structures of the objects, properties, events, processes and relations in every area of reality" (Smith 2003). In recent years it has been realized that an ontological approach can help to overcome what Smith (2003) called the "Tower of Babel problem" in the field of information processing, where information has hitherto been mostly characterized by application-specific idiosyncratic terms and concepts of dubious, and frequently undocumented, semantics, and application-specific data models. This current situation causes tremendous problems when integration over different information sources is required and usually case-by-case resolutions involve costly mappings at different layers such as models and semantics.

But problems with idiosyncratic solutions are not restricted to integration projects, as users have to be familiar – at least to some degree – with the structure of an information system to use it efficiently. Given that most users rely on a host of different information systems, they are faced with the need to learn about the idiosyncratic structures of each of the systems they are dealing with. Instead of focusing on what is relevant – domain-specific knowledge structures for solving problems – users are forced to deal with a host of alien and rather technical informational constructs.

The use of common backbone taxonomies of relevant entities and relations of an application domain is one part of the solution to these problems. In information science, this part is also called "ontology". However, contrary to philosophy, information science also knows the plural of "ontology", as each ontology is specific to its application domain.

The core of the H14-Navigator backend content is an application-specific ontology. Although chemical and physical properties of the ring test samples have been excluded from the online publication so far, it seemed appropriate to shape the core of the ontology in a way that would allow the mapping of both additional property domains. Therefore, we adapted an ontology of physical, chemical and biological properties published by Dybkaer (2004) to the needs of the ring test project and its results.

We aimed for a realistic ontological approach as advocated by Fielding et al. (2004), rooted in high level universals such as object, process, method, and property. For linking these high level universals to the low level universals characterizing the entities dealt with at the laboratory level, we used the superclass–subclass relation. The adapted ontology on property (Dybkaer 2004), as well as the project ontology and the ontological components for the biological entities to be modeled, were mapped into Topic Maps constructs.

27.2.2 Integrated Knowledge Models

Given that the sole examination principle used by the ecotoxicological methods studied in the ring test was the response of living organisms, taxonomic relations between the species used are highly relevant knowledge structures for the test results. Therefore, a phylogenetic tree was mapped into the Topic Map. It is comprised of the 16 species used in the ring test, and can be considered one of its core knowledge structures. The phylogenetic tree also served to structure the specialization hierarchies of the top universals such as method, process, and property.

27.3 Topic Maps Frontend

27.3.1 H14-Navigator as Application Specific Topic Maps Browser

The H14-Navigator is the frontend for the H14 topic map, and renders the information contained in the topic map into web pages. It uses the domain ontology, and the knowledge models in the topic map for rendering access, and navigation paths through the highly networked space of information about and the results of the ring test, including the project's structure and participants, its subjects, the methods, and the biological species involved, the processes performed, the samples and their kinds of properties examined, and the project's results at five different levels.

The Topic Maps data model is graph-based. A graph is a set of nodes connected by links called lines or edges. In Topic Maps the nodes are called "topics" and the edges are called "associations". Topics represent subjects about which statements may be made. A topic is of one or more particular types, which are topics themselves. Also, an association (or relation) is of a particular type, which is itself a topic. **Figure 27.1** shows a graph structure representing real structures in the H14 ring test project, as well as structures in the H14 topic map. Consider the topic

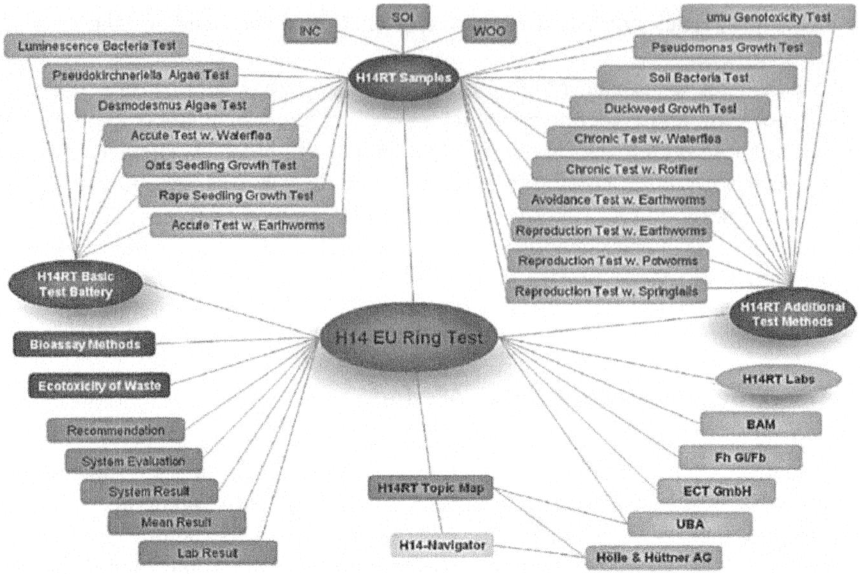

Fig. 27.1 A graph representing structures in the H14 ring test project, as well as in the H14 topic map; each node represents a topic, and each edge represents a typed relation between the involved topics; thus there is a one to one match between structural project components in reality and in the informational structures of the H14-Navigator backend which makes the use of the frontend rather intuitive

"H14 EU Ring Test" in the core of Fig. 27.1, which is of type "project". It is linked to another topic called "H14RT Basic Test Battery" via a relation of type "study-subject", in which the ring test plays the role "study" and the test battery plays the role "subject".

When a relation is rendered by the H14-Navigator, the name displayed for it depends on the direction from which it is viewed. For instance, the home page of the H14-Navigator is a so-called "topic view" of the topic "H14EU Ring Test". Hence, the focus of this page is the ring test itself, and the names of all relations of this topic are rendered according to this perspective. Thus, easily readable interface structures are generated such as: *H14 EU Ring Test – had subject – H14RT Basic Test Battery*; or *H14 EU Ring Test – has results of type – recommendation.* When one of these binary relations is viewed from the side of the other participant, its name is rendered differently, and thus the readability of the interface structures is maintained. For instance: *H14RT Basic Test Battery – studied in – H14 EU Ring Test.* This perspective-dependent naming of relation types was first described by Pepper (2000) whose introduction to Topic Maps is somewhat dated but for the non-technical reader is still a prime reference.

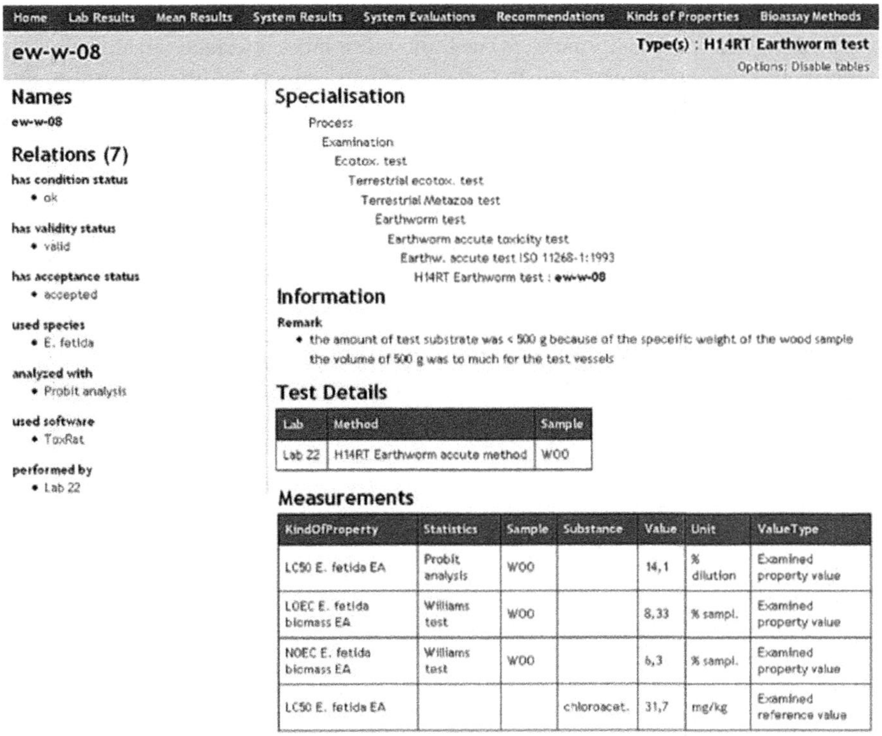

Fig. 27.2 Partial screen shot of the H14-Navigator displaying a topic view of "ew-w-08" characterized by a knowledge model as a special type of an ecotoxicological test process on terrestrial animals; relations of this process are displayed on the left-hand side; further details on the process are rendered as tables on the right-hand side

Figure 27.2 gives an example of a topic view on a particular ecotoxicological test process rendered by the H14-Navigator. Relations of this process are listed on the left-hand side, a knowledge model displayed in the centre helps to allocate the process in its domain, and further details are rendered as tables on the right-hand side. Most components displayed in the interface are topics themselves, and thus with a mouse-click, more information linked to their subjects can be retrieved.

Next to the topic view, the H14-Navigator can render a second visualization pattern which is called the "table view". The tables look as if they originated from an application based on a domain-specific relational data model, but are indeed only particular views of the topic map content. An interactive functionality enables the showing of additional columns or hiding columns irrelevant for a particular user perspective. Figure 27.3 displays a partial screen shot of the table of the mean results of the H14 ring test. Just as in the topic view, most interface objects rendered in the table view are topics, and thus the user can request additional information linked to their subject by a single mouse-click.

An intelligent search algorithm provides additional access paths to relevant content (Fig. 27.4). Matches are qualified as names, or as texts as well as by their type such as

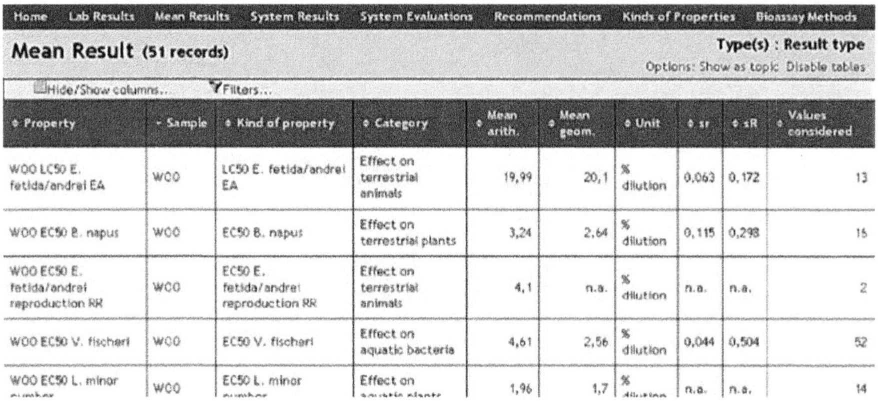

Fig. 27.3 Partial screen shot of the H14-Navigator displaying a table mean results; functionalities for sorting or filtering are available, as well as for showing additional columns or hiding columns irrelevant for a particular user perspective

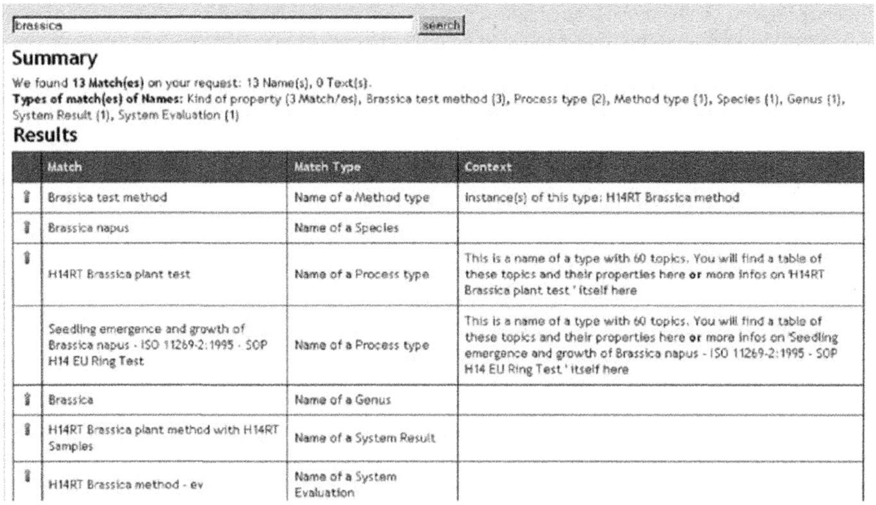

Fig. 27.4 Partial screen shot H14-Navigator search results for "daphnia"; the 22 matches are qualified as names, or as texts as well as by their type such as "Process type", "Species", or "System Evaluation", "Remark" or "Description"

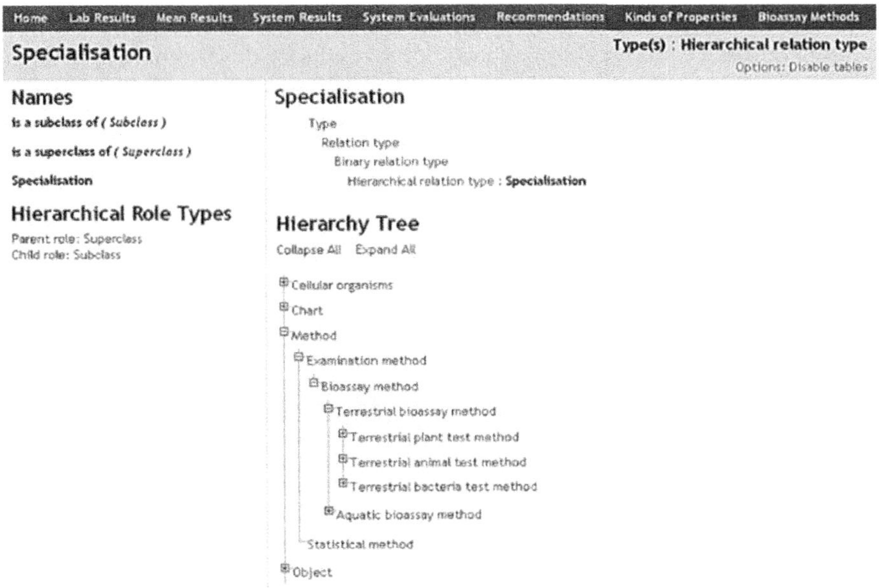

Fig. 27.5 Partial screen shot of the H14-Navigator displaying a hierarchical knowledge model; methods are modelled as a specialisation hierarchy which can be explored interactively and thus provides additional knowledge-oriented access paths to content

"Process type", "Species", "System Evaluation", "Remark" or "Description". This qualification can further reduce the length of access paths to the information sought.

The H14 topic map holds a number of specialization hierarchies containing core knowledge models of ecotoxicology such as a phylogenetic tree of the species used in the ecotoxicological methods investigated, a method-, a process-, and a property-tree. But other entity types of the application such as organizations, results, or objects were also integrated in the specialization hierarchies. The H14-Navigator therefore also renders a particular topic view of the specialization relation, in which an interactively explorable specialization tree offers yet another access path to relevant content (Fig. 27.5).

27.3.2 Multiple Paths to Topics of Interest

The H14 topic map holds about 4,000 topics linked by over 15,000 relations. Hence, any topic can be reached over a multitude of different access paths. To illustrate the variety of access paths, a number of different paths to the laboratory level results of the Desmodesmus algae test are described in detail.

The frontends main menu may be considered a list of shortcut paths to the most sought-after topics. To access laboratory level results of the Desmodesmus algae test, the most direct access path is provided by the menu item "Lab Results", and the user action of activating a filter for the process type of interest.

Another access path to the results of this test is provided by the search functionality. Searching for "Desmodesmus" retrieves the process type "H14 Desmodesmus algae test" and offers direct links to both the topic view, and the table view of this topic. After accessing the table view, the user may apply the hide/show functionality to complement the displayed process attributes by the additional attributes "value" and "unit". The highly networked character of the H14 topic map is demonstrated by a third – much longer – access path[2] to the laboratory level results of the Desmodesmus algae test represented in Fig. 27.6.

In the navigation path depicted in Fig. 27.6, steps 1–5 are rendered as topic-views, whereas the topic of step 6 is by default rendered as a table view listing all processes of type "H14 Desmodesmus algae test" (Fig. 27.7). Whenever a table view can be rendered, a switch to the topic view is offered to reduce the amount of information displayed. The topic view for step 6 is depicted in Fig. 27.8.

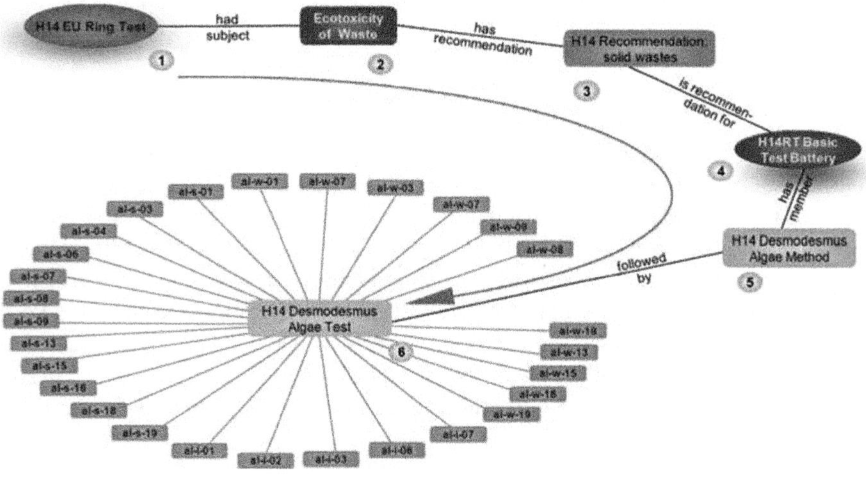

Fig. 27.6 Graph representation of a user navigation path traversed in the H14-Navigator; each numbered symbol represents a topic selected by the user; each named edge represents a relationship between two topics (name given according to the perspective of the topic with the lower number); unnamed edges represent instance-of relations (e.g. "al-w-08" is a process of type "H14 Desmodesmus Algae Test")

[2]This path may be followed by clicking the respective topics at http://EcotoxWasteRingtest.uba.de/h14

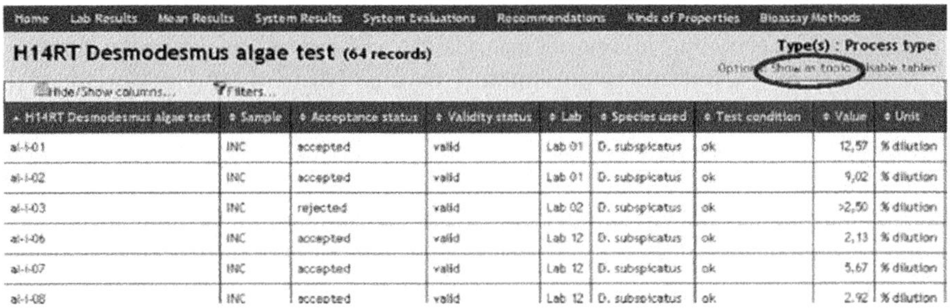

Fig. 27.7 Partial screenshot of H14-Navigator rendering the table view of step 6 of the navigation path shown in Fig. 27.6; next user action "show as topic" encircled

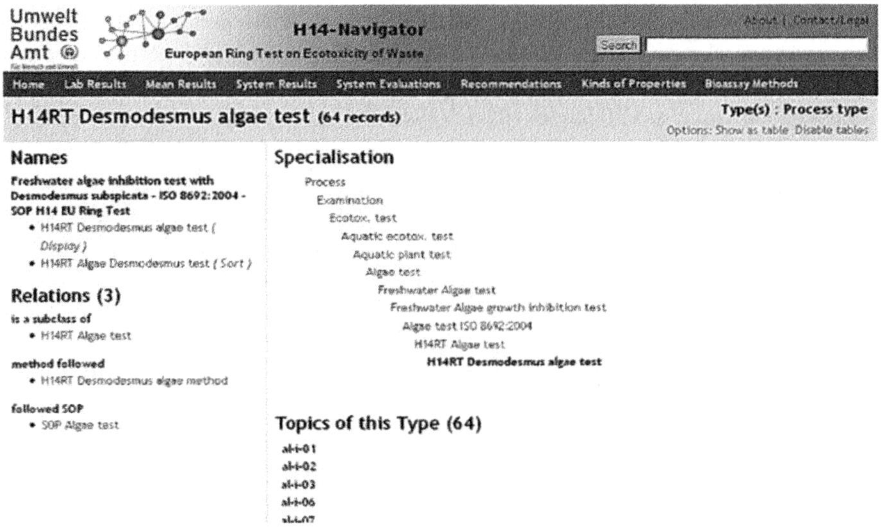

Fig. 27.8 Partial screenshot of H14-Navigator rendering the topic view of step 6 of the navigation path shown in Fig. 27.6; compared to the table view of the same topic given in Fig. 27.7. The topic view reduces the amount of information displayed

27.4 Discussion

The H14-Navigator operates on a backend which is not structured by an application specific data model, but rather by the standardized knowledge-oriented Topic Maps data model (ISO/IEC 13250-2 2006). It uses the domain ontology and the domain specific knowledge models in its backend topic map for rendering knowledge-oriented interface structures, navigation paths through and access paths to all content. Thus, unlike the predominantly data-oriented user interfaces of conventional applications based on idiosyncratic data models, the user interface of the H14-Navigator is knowledge-oriented. As a result of this approach, accessing content changes from navigating a data-centric, application-specific user interface to navigating a domain-specific knowledge map.

As an interdisciplinary science, ecotoxicology draws from concepts in a variety of domains such as ecology, biology, chemistry and toxicology. Considering the complexity of these domains, it would be presumptuous to claim that the ontology developed for the H14-Navigator, can even be close to something like "ecotoxicological ontology". However, its immediate use for structuring ecotoxicological content and knowledge in the H14 topic map, and enabling knowledge-oriented interface structures rendered by the H14-Navigator should be sufficient proof for the usefulness of this approach. Going beyond the framework of the ring test, and thanks to the work of Dybkaer (2004), which inspired some of the core constructs of its ontology, we do believe that it holds the potential to serve as a stimulus for future work for ontology-based application systems focused on properties.

Today, most application systems operate on information in backends with application-specific data models, with functionalities and frontend structures closely coupled to the relational models of their backends. Unfortunately requirements are not static – not even over the development phase of the H14-Navigator – and often data models need to be adjusted, and so do the backends, the closely coupled functionalities, and frontends. However, modifications of such systems are hard and costly, and – possibly an even heavier burden – integrating content in different systems of this kind is a challenge which is even more difficult to master due to model differences, and the usually lacking semantics on their structures and contents. Given today's easily achievable physical connectivity of information systems, and given the ubiquitous need for integrating content from different sources, applications based on application-specific data models without explicitly accessible semantics should be considered as what they are: a costly rather than a precious legacy of the past millennium.

The application-specific approach to modeling can be overcome, and Ahmed (2002) noted that the Topic Maps data model offers one way of doing so. He rightly claimed that the Topic Maps data model matches the decomposition of application design into a set of interacting objects. He further stated that using Topic Maps as application data model would allow modifications of application model structures simply by altering the data which provides the application schema, thus removing the need to re-compile or re-populate database tables. In short, this comes down to adjusting an application model by changing the content of its backend but not the structures of its backend and its data model. As a further aspect of considerable advantage Ahmed (2002) stated that a single application programming interface would enable accessing the data of any such application. Although Ahmed (2005) elaborated on these ideas, so far Topic Maps has not played a very visible role as application data model, which might – to some degree – be due to the slow progress of the Topic Maps standardization process. However, with stable standards for the data model (ISO/IEC 13250-2 2006), and the XML syntax (ISO/IEC 13250-3 2007), with a functional query language (Garshol 2006), and the ISO standardization process for a host of other Topic Maps related standards well under way,[3] it is time to speed up the exploration of the full potential of Topic Maps technology. This is the venture to which the H14-Navigator is meant to contribute.

[3] http://www.itscj.ipsj.or.jp/sc34/open/1025.htm

Part IV
Further Development and Future Application of Biotests in Waste Characterization

Chapter 28
Test Recommendations

H. Moser, J. Römbke, and T. Moser

Abstract The European ring test for the classification of waste was set up as an evaluation study of the guideline EN 14735. A very good data base on the applicability and performance of the tests listed in the Annexure of EN 14735 was compiled and the ecotoxicity tests can now be revised or supplemented with more information regarding their use in waste tests.

The data gathered for waste tests and the results of the reference tests can be used for a further development of the test guidelines, especially for the re-evaluation of the toxic ranges for reference substances.

The reference substance chosen for the aquatic test set was potassium dichromate, but for the Algae test, the current effect range has to be adapted for the broader data base. For the terrestrial test set, boric acid is recommended as the reference substance suitable for most of the tests. In addition to the ecotoxicity tests, other influence factors such as the variability of standard soils and the influence of the sample preparation on the bioavailability and ecotoxicity of the waste samples have been tested. A comparison of a large data set of two aquatic test systems was used to estimate the influence of the eluate preparation on the test results. Finally, the overall appraisal of the test-specific experience showed clearly that the ring test results enable a definition of a standard test set for waste classification.

Keywords Standard test battery, Influence factors, Test comparison, Assessment, Evaluation of results

28.1 Introduction

The data base of biotests suitable for waste ecotoxicity tests was developed by the experts of CENT TC 292 WG7 and laid down in the information annexure of EN 14735. The basic test set was determined during an expert workshop in September

H. Moser (✉), J. Römbke, and T. Moser
Federal Environment Agency, Woerlitzer Platz 1, 06844, Dessau, Germany

H. Moser and J. Römbke (eds.), *Ecotoxicological Characterization of Waste*,
DOI: 10.1007/978-0-387-88959-7_28, © Springer Science+Business Media, LLC 2009

2005 and the selection of the tests was based on the available results and experiences of scientists from several European Member States. As a very good data base on the applicability and performance of the tests has been acquired, the test systems can be revised or supplemented with information regarding their use in waste tests.

28.2 Aquatic Test Set

In the ring test, a large number of aquatic tests were performed for the basic test set (Algae, *Daphnia* and Luminescent bacteria) as well as for the additional tests (*Lemna*, *Brachyonus*, *Ceriodaphnia* and *Pseudomonas*). The gathered data, especially the results of the reference tests, can be used for the further development of the test guidelines, especially for the re-evaluation of the data ranges for reference substances.

Good examples of the additional benefit of the ring test are the results of the reference tests with potassium dichromate. The suitability of the substance for the basic aquatic tests was confirmed. For the *Daphnia*, *Lemna* and *Pseudomonas* tests, the reference tests met the toxic effect range in most cases, whereas for the *Brachyonus* and *Ceriodaphnia* tests, the reference data sets can be used for a revision of the given range, based on the new data.

The results of the ring test can also be used to suggest a final set of waste tests (see Chap. 29) and to give specific information for the selected biotests. It is recommended that the Algae test should be part of the test set, because the test results for waste, waste water and chemicals showed good sensitivity and validity (Pandard et al. 2007). The economic aspects of the application of the ecotoxicity test for waste or waste eluates have to be considered and therefore the use of miniaturized systems or even automated devices is a clear advantage for the selection of a biotest. For the Algae test, microplates or other miniaturized systems are already established in many labs and they seem to be equal to the classic setup. Based on the reference test data collected during the ring test, no suggestion for a preference species was elaborated, but the effect range given in the current Algae guideline for the reference tests was questioned and a new range has been proposed and applied in the selection of valid data sets (see Chap. 5). For the Algae test the different exposure systems (vessel or microplate) did not seem to be the major factor of influence in the intralaboratory reproducibility, but it was clearly shown that the range increases with the integration of various test conditions. In a future revision of the test guideline it should also be taken into account that different endpoints could lead to different results in reference tests (Nyholm 1990).

The *Daphnia* test was highly sensitive to the wastes tested in the ring test, but due to the experience in waste test studies with a broader range of waste types (see Chap. 8), the decision to integrate the test in a final test set is still pending. On the basis of this result, more research is needed to establish a higher tier aquatic biotest for waste ecotoxicity. In addition, more experience in testing other waste types is also needed. Until more elaborate data are available, the *Daphnia* test should be retained in the test set and replaced later if necessary.

The Luminescent bacteria test is perhaps the most common biological test system and is applied in many matrices. The test gives a rapid answer, but it is necessary to give clear definitions for the Luminescent bacteria sources to be used because of the different reactions, especially with heavy metals.

For the Luminescent bacteria test, it was proposed to modify the current guideline in order to establish dose-response testing as in other aquatic tests. 3,5-Dichlorphenol or potassium dichromate was suggested as reference substance. In future, a range for the toxic effects of the reference substance should necessarily be developed for all sources of Luminescent bacteria, because the influence of the culture source (liquid- or freeze-dried or freshly prepared) on the sensitivity of the test organisms was again observed in the ring test (see Chap. 9).

The *Lemna* test was the only biotest from the additional tests to have a sufficient number of participants for statistical assessment. It helps to test the potential phytotoxicity of the waste eluate, using a higher plant organism, but its sensitivity was only moderate. In some labs, the *Lemna* test is used in automated systems, but the methodological effort, e.g. the large volume of eluate needed, is still appreciable. Nevertheless, the *Lemna* test can substitute for the Algae test for colored or turbid eluates and in eluates with a non-neutral pH-value.

Despite the fact that the ring test samples did not show genotoxic potential in the umu test, it is recommended that such tests be included in the basic test set at least for the time being. As there was no indication that the three ring test wastes would display genotoxic potential, not enough data were gathered on positive signals in the genotoxicity tests.

The *Ceriodaphnia* test is a rapid chronic toxicity test on crustaceans, which can be managed methodologically and has good reproducibility. Nevertheless, the test was applied only by a few participants and therefore the data base is too small to suggest that it be included in the final test set.

The *Pseudomonas* test was also part of the additional test set, but was applied by only a limited number of labs. Therefore, the assessment of the test results and its applicability in waste tests are yet to be determined. The availability of a miniaturized microplate version and the ecological relevance of the bacteria for mineral or soil-related waste material point to its value as a good application for waste ecotoxicity tests. On the other hand, the test showed medium sensitivity with only minor shifts compared to Algae and *Lemna* tests.

28.3 Terrestrial Test Set

In contrast to waste eluate testing, the knowledge base for solid waste testing was limited before the ring test started. The definition of the basic test set was based mainly on experiences from the ecotoxicological characterization of soil (ISO 1579 2003). The participants of the ring test applied two standard (earthworm acute and plant growth) and five additional (earthworm repro, earthworm avoidance, enchytraeid repro, *Arthrobacter* contact and collembolan repro) test systems. Both test systems are biotests that require methodological effort. Compared to most aquatic tests, they require long-term exposures, but both test organisms are indicator species for important functions in the ecosystem. Plants form the basis of the terrestrial food webs and their use in biotests covers a close interaction between roots, the pore water and the substrate itself. For a comprehensive ecotoxicological characterization of waste, it is recommended that a higher plant test with two species, differing considerably in taxonomy and physiology (one monocotyledonous and one dicotyledonous), should be used as recommended in the technical standard. The use of biotests for the determination of the environmental hazard of waste has to take the costs of the test set into account. Therefore it might be helpful to set up a limited test system for the classification of waste, in which only one plant species is applied, focusing mainly on the most sensitive species, such as *Brassica rapa* in the ring test. As reference substance for the plant test, boric acidic was suggested (see Chap. 10).

The second higher organism for the waste related test set was *Eisenia fetida* in the earthworm acute test. The integration of a worm test in the set is recommended because in many soils worms are the most important invertebrates. They are often quite sensitive due to their close interaction with the pore water and the substrate via dermal contact and feeding. Furthermore, worms are the most common soil invertebrates and they represent effects on other soft-bodied taxa in the soil invertebrate community (Kupermann et al. 2006c).

The earthworm acute test with endpoint mortality showed a low sensitivity in the ring test. Therefore, more sensitive endpoints should be checked. On the basis of the ring test results, the earthworm acute test is not recommended for the final test set for waste classification. The most obvious result derived from the solid waste tests in the ring test was that more guidelines are necessary regarding the determination and adjustment of the water holding capacity and the moisture regime of the mixtures.

Besides the use of acute toxicity tests, the use of reproduction tests can also be an important instrument for the risk assessment of waste. The earthworm reproduction test is a reliable test system for the ecotoxicological characterization of soils, showing a high sensitivity, sometimes higher than in tests based on behavioral endpoints (ISO 2007). In the ring test, the earthworm reproduction test was among the most sensitive terrestrial tests. Therefore its usefulness as an alternative to the earthworm acute test, the enchytraeid reproduction test, or the Collembolan reproduction test should be evaluated.

Another promising terrestrial test was the earthworm avoidance test. Its application in the ring test clearly showed a high sensitivity by providing lower EC_{50} values than the other terrestrial bioassays. The short test period, the exposure scenario that is similar to field conditions, and the use of a higher test organism are advantages that confirm the high utility of this test. The earthworm avoidance test (ISO 2006) could be a good alternative using the indicator species, and also a faster and more sensitive test system. The use of other organisms (e.g. collembolans or enchytraeids) in avoidance tests can be useful for waste material with specific requirements, e.g. sharp-edged particles or non-neutral pH values. So the further development and evaluation of avoidance behavior tests for waste characterization is needed.

The Enchytraeid reproduction test showed a higher sensitivity in the ring than the acute earthworm test, but a lower sensitivity compared with the earthworm reproduction test and the earthworm avoidance test.

Although the oligochaete test was finally selected, more guidelines are necessary regarding the moisture regime and the pH of the mixtures. The development of an alternative for the currently used reference substance chloroacetomide (boric acid) should be encouraged. However, based on the recommendation of EN 14735 (2005) a worm test should be included in the final test set and a chronic test should be combined with an acute test, at best with different soil-related test organisms. The use of tests with other soil-dwelling invertebrates like collembolans could be an alternative to avoid problems caused by moisture or pH of the waste material (see Chap. 18). The *Folsomia candida* reproduction test has proved to be methodologically suitable for waste tests. At the same time *F. candida* is able to inhabit different waste materials and tolerates a broad range of organic matter content and pH (Jaensch et al. 2005). Compared with the other terrestrial test organisms in the ring test, *F. candida* was among the most sensitive species.

The *Arthrobacter* contact test is a bacterial solid contact test and was not part of the basic terrestrial test set in the ring test, but it is increasingly applied in waste tests (Römbke & Moser 2007). With the growing experience with this test system and due to its high sensitivity and performance, the *Arthrobacter* test could be part of a future standard test set, standing as indicator species for soil dwelling bacteria. The use of a miniaturized microplate version supports the requirements of a routine application in waste characterization. The development of biotests, the standardization of methodology, and the quality assessment of test results require the use of reference substances to test the sensitivity of the test organisms on a reliable base. For these reasons chemical substances, soils from contaminated sites, or waste material was used. The substances and ranges of reference tests for the aquatic test systems are broadly accepted, but the reference substances currently used in terrestrial test systems are often problematic for reasons of availability (pesticides) and human toxicology (chloroacetamide). Therefore, boric acid is recommended as the suitable reference substance for all terrestrial biotests. The reference test results for boric acid from the ring test are not sufficient to draw final conclusions, but they support parallel activities in the normalization working groups such as CEN and ISO (see Chap. 6).

28.4 Microtox Test Systems

One additional activity in the ring test on waste ecotoxicity was the comparison of standard biotest systems with Toxkit microbiotests regarding their sensitivity and performance in waste eluates and waste (see Chap. 13). The results of the Toxkits “Algaltoxkit” and “Daphtoxkit” with the standard tests have shown that there was no difference in sensitivity for the waste materials. The statistical assessment was based on a smaller number of participants, but it did not refer to any major deviations from the results of the standard test set. The use of Toxkit microbiotests could be a suitable and cost-effective alternative for the ecotoxicological characterization of waste, but to receive reliable results, the use of Toxkits also requires knowledge and experience about biotests in the lab. With a sufficient number of parallels and the use of reference substances, the Toxkit microbiotests with *Daphnia magna* and micro-algae can also provide an ecotoxicological characterization of waste.

The experiences and results with the Phytotoxkit, which combines shorter test duration with smaller amounts of waste to an economically interesting alternative, are a good starting point for the further development of the plant test guideline. As the plant test should be part of the final test set for waste classification, the improvement of the test procedure and the adoption of effort-saving new methods (LUBW 2008) are welcomed.

28.5 Variability of Standard Soils

In most of the terrestrial tests, OECD artificial soil is frequently used as standardized substrate for the dilution of the waste material. One participant (see Chap. 24) characterized the various artificial soil materials from the ring test and found significant differences among the soils. Variable soil properties (e.g. C_{org} content) were monitored and the influence of sorption on the toxicity of the waste compounds was analysed. It appears that besides the soil preparation itself, peat as non-standardized material was the most problematic component. In addition, the microbiological activity may influence the availability (or even degradation) of waste compounds. It was recommended that the peat component of artificial soil should be better defined in the test guidelines.

28.6 Influence of Eluate Preparation on the Results of the Aquatic Basic Test Set

The eluate preparation of waste material is an important step in the classification of waste. The particle size, the ratio of liquid to solid phase, and the contact time between eluent and waste material, are important factors. The ring test was set up as an evaluation study of the guideline EN 14735 and therefore followed the guidelines regarding leaching procedure. With the 24 h-batch test and a liquid/solid ratio of 10, the most common leaching method was applied.

Based on the results of the aquatic standard test battery for the waste materials WOO and INC in repeated tests (see Chap. 22), the aim was to estimate the potential influence of the leaching process on the deviation of the results. The comparison of these results with the one obtained on the reference chemical (potassium, dichromate) for *D. magna* and *P. subcapitata* indicated that the steps of eluate preparation do not significantly increase the dispersion of the data, compared to the historical data of the laboratory for the last 5 years (Pandard et al. 2007).

28.7 Influence of Waste Preparation on the Results of the Terrestrial Basic Test Set

The waste samples have been diluted with soil or adapted to the requirements of the test organisms according to the test protocols. One participant of the ring test (see Chaps. 25 and 26) investigated the effects of soil properties and the sample treatment on the bioavailability and ecotoxicological results of the wastes. The exposure conditions in the substrates can differ greatly from the field conditions. If the results of the biotest set are transferred into an ecological risk assessment, natural soil types, which have properties that support relatively high bioavailability of the relevant waste compounds, should be used. With this approach, the results of the biotest set can be used not only for waste classification, but also for the assessment of the risks of a potential release of toxic compounds from waste materials into the environment.

Chapter 29
Ecotoxicological Characterisation of Waste as an Instrument in Waste Classification and Risk Assessment

H. Moser and H. Kessler

Abstract The most important result of the European ring test is the clear evidence that biological test systems are an implementable and reliable instrument for the ecotoxicological characterization of wastes. The experience gained from ecotoxicity waste tests provides the possibility of comprehensively determining potential hazardous waste types with a combination of biological test results and chemical residue analysis on a routine base. The intrinsic properties of waste material can be assessed in a so-called Extended Limit Test design, in which a minimum test set is applied in given dilution steps of the waste to be tested. Based on the present scientific knowledge and the experience of the ring test, toxicity criteria for each test system of the basic test set were developed.
In addition to the determination of intrinsic waste properties, waste ecotoxicity test systems can also be an important instrument for waste risk assessments, e.g., the reuse of waste, by adapting the leaching method and the test battery to the waste management scenario.

Keywords European waste list, Waste classification, Extended limit test, Test battery, Test strategy, Toxicity criteria, Risk assessment, Waste regulation.

29.1 Introduction

The results of the ring tests have clearly shown that biological test systems are an implementable and reliable instrument for the ecotoxicological characterization of wastes. The experiences gained provide the possibility to comprehensively determining potential hazardous waste types with a combination of biological test results and chemical residue analysis. With growing knowledge and experience in waste testing, the use of biological test systems will become accepted as an appropriate instrument of waste analysis (Kostka-Rick 2004).

H. Moser (✉) and H. Kessler
Federal Environment Agency, Woerlitzer Platz 1, 06844 Dessau, Germany

H. Moser and J. Römbke (eds.), *Ecotoxicological Characterization of Waste*,
DOI: 10.1007/978-0-387-88959-7_29,

The significance of the evaluated test systems depends on their practicability (which has been proved in the ring test) and the sensitivity of the test organisms against the waste compounds. Therefore only the use of a test set with different test systems (aquatic and terrestrial tests with organisms from different trophic levels), different effect endpoints (e.g. physiological, behavioral, genotoxic) and different exposure scenarios (e.g. short and long-term) can provide sufficient information on the environmental hazard of the tested waste. In addition to the essential sensitivity of the test organisms, biological test systems to be used in waste enforcement have to be standardized and economically justifiable.

The hazard characterization of waste is closely related to the Dangerous Substance Directive (67/548/EEC). In this respect, waste can be classified by a correlation of chemical compounds and their R-phrases. Each waste type, the composition of which is known, based on analytical data or on its origin, and which can be classified according to the Dangerous Substance Directive, has to be classified as hazardous waste.

For waste types of unknown and/ or complex composition, biological tests are a helpful instrument for:

- The identification of hazardous waste, listed in mirror entries of the European waste list
- The risk assessment of waste in reuse scenarios

29.2 Classification of Waste in Mirror Entries

The mirror entries of the European Waste List often refer to the indefinite description of "hazardous substances", without any further description of what component or which hazard criteria have to be checked. Among the other hazard criteria, the H14-crition "ecotoxic" ensures that the ecotoxic properties of the waste material are determined, especially in waste of unknown origin or composition.

29.2.1 Extended Limit Test Design

The intrinsic properties of waste material can be assessed in a so-called Extended Limit Test design, in which a minimum test set is applied in given dilution steps of the waste to be tested (see Table 29.1). Based on the present scientific knowledge and the experience of the ring test, toxicity criteria for each test system of the basic test set were developed (see Table 29.2). These criteria help to distinguish between ecotoxicologically relevant test results and the results that cannot be differentiated from the test specific reactions of the test species as defined in the respective guidelines (i.e. control validity criteria). Compared to the toxicity criteria of biotests used for substance screening or soil characterization, the toxicity criteria for eluate and

Table 29.1 Test set for the ecotoxicological assessment of waste using the Extended Limit design

Aquatic biotests for eluate testing	
Determination of the inhibition of the mobility of *Daphnia magna* Straus (Cladocera, Crustacea) – Acute toxicity test	DIN EN ISO 6341
Freshwater algal growth inhibition test with *Scenedesmus subspicatus* and *Pseudokirchneriella subcapitata*	DIN EN ISO 8692
Determination of the genotoxicity of water and waste water using the umu test	ISO 13829
Terrestrial tests for solid waste testing	
Soil quality – Determination of the effects of pollutants on soil flora – Part 2: Effects of chemicals on the emergence and growth of higher plants	ISO 11269-2
Soil quality – Avoidance test for determining the quality of soils and effects of chemicals on behavior – Part 1: Test with earthworms (*Eisenia fetida* and *Eisenia andrei*)	ISO 17512-1 (2007)
Solid contact test with *Arthrobacter globiformis*	DIN 38412-48 (2002)/ISO 10871 (2008)

Table 29.2 Limit values recommended for those test methods belonging to the minimum test battery for the ecotoxicological assessment of wastes

	Test organism	Reference	Endpoint	Toxicity criteria[a]
Eluate testing	Algae	DIN EN ISO 8692	Growth	20%
	Daphnids	DIN ISO 6341	Immobilisation	10%
	Salmonella choleraesius	ISO 13829	Gen induction	$D_{min} \geq 2$
Solid waste testing	*Brassica rapa*	ISO 11269-2	Growth	30%
	Eisenia fetida/ Eisenia andrei	ISO 17512-1 (2006)	Behaviour	20%
	Arthrobacter globiformis	DIN 38412-48 (2002)/ ISO 10871 (2008)	Dehydrogenase activity	20%

[a]Toxicity criteria established on the basis of testing a wide variety of wastes or contaminated soil materials. It should be guaranteed that no false positive assessment is obtained.

waste testing are adapted to the specific requirements of waste testing as a more complex field of application for biotests. Due to the higher heterogeneity or more complex composition of waste and waste eluates, the range of deviation of the test results might be larger than in single-substance testing.

The Extended Limit test design is a fast and effort-limited means to combine the chemical and biological classification of waste in mirror entries on a routine base. The Extended Limit design can also be used as a pre-screening method for a comprehensive ecotoxicological characterization of waste.

29.2.2 Test Strategy

The ecotoxicological classification of waste in mirror entries comprises the determination of the aquatic ecotoxicity of the waste eluate with two short-term tests and the assessment of a genotoxicity test. If the waste eluates show no effect in all aquatic tests the waste material itself has to be tested in two terrestrial biotests (see Fig.29.1). Only if there is no effect in all terrestrial tests, the waste can be classified as non-hazardous.

29.2.3 Test Battery

The test set (see Table 29.1) suitable for the determination of ecotoxic properties for most of the waste types in mirror entries was defined based on the present scientific knowledge and the experiences derived from the European ring test. It describes a minimum test battery, which can detect waste compounds with environmental hazard. With the exception of the *Arthrobacter* test (which is still under standardization in ISO) all methods have been validated internationally.

The application of an Extended Limit test design for wastes assumes that the physical and chemical properties of the waste eluate do not exclude the use of test organisms. Specific requirements, e.g. waste eluates with high amounts of nutrients or of dark color or turbidity, require an adaption of the test selected as well as an extension of the suggested test battery.

In order to evaluate the reaction of the test organisms in the chosen waste concentrations, the test result of each dilution step is compared with the test-specific toxicity criteria (see Table 29.2).

If the effect of the waste exceeds the Extended Limit test criteria in the test battery, the waste is regarded as ecotoxic. In order to classify the waste as hazardous according to H14, "ecotoxic" limit values have to be established. These limit values will describe the specific waste concentration, in which the waste may not exceed the toxicity criteria in the biotests and is given as LID (=Lowest Ineffective Dilution) (ISO 17616) If the waste exceeds the toxicity criteria in this dilution step or even in a higher dilution than the ones applied in the limit test, the waste has to be classified as hazardous regarding H14 (see example in Fig. 29.2). Exceeding in one test of the aquatic or the terrestrial set, or the detection of a genotoxic signal leads to the classification as hazardous waste.

The application of the Extended Limit test design is not a comprehensive ecotoxicological assessment of the waste, but by testing the given waste concentration it is possible to distinguish between hazardous and non-hazardous waste with an economically justifiable and time adequate test effort. As such, this approach follows the recommendations already given for contaminated soils (ISO 2007b).

The classification as non hazardous regarding the H14 criterion does not automatically mean that the waste can be classified as non-hazardous at large. The waste has to be assessed regarding all H-criteria given in the European Waste list (H1 to H14, see Chap. 2), before it can finally be classified as non-hazardous.

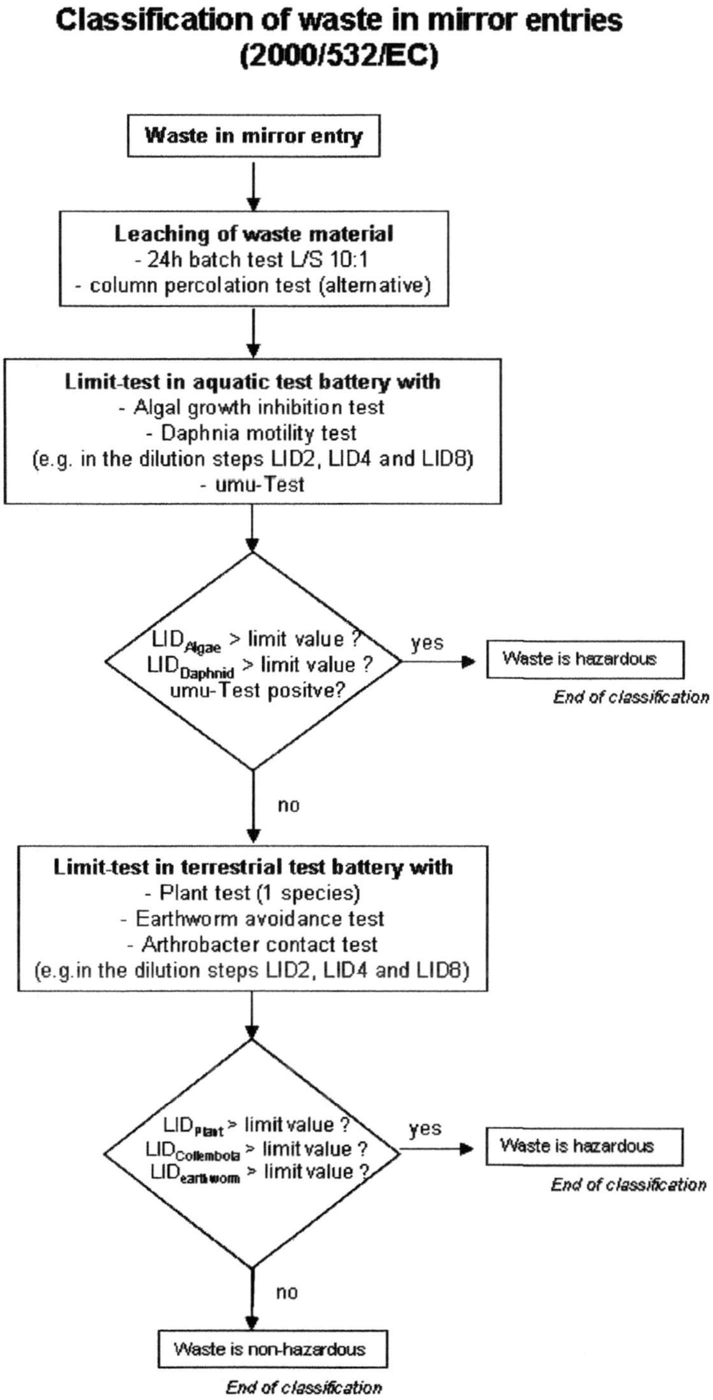

Fig. 29.1 Test strategy for the H14-assessment of waste listed in mirror entries of the European waste list

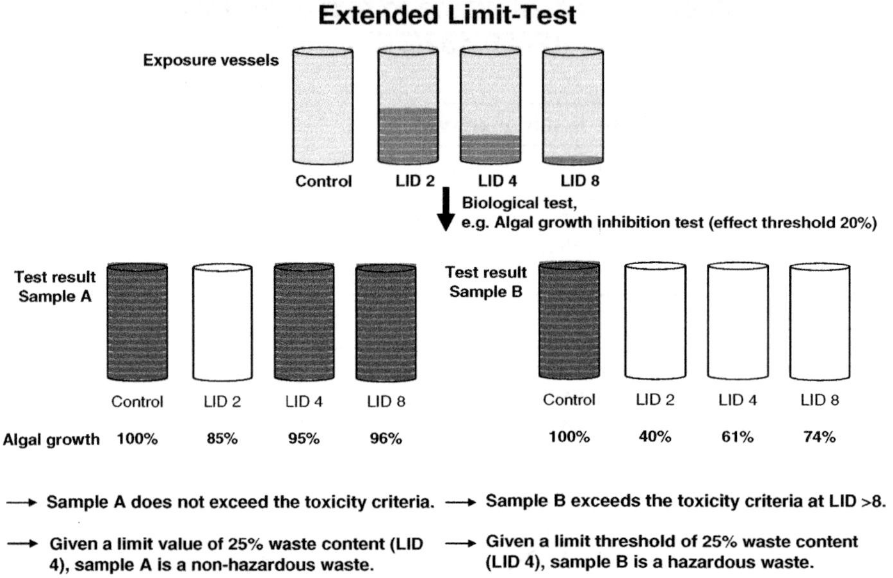

Fig. 29.2 Exemplary classification of two waste samples, tested in the algal test using the Extended Limit test design and applying the toxicity criteria of the algal test (20% growth inhibition) and a limit threshold of 25% waste

The ecotoxicological characterization stands equal to the other H criteria. If a waste can be classified as hazardous according to another H criterion, which can be tested with less effort than H14, the ecotoxicological assessment is not mandatory.

29.3 Ecotoxicological Characterization of Waste in Risk Assessments

The use of biotests for the determination of waste properties potentially hazardous to the environment can also be applied in waste risk assessments. In addition to the determination of waste properties as potentially hazardous to the environment, their ecotoxicological characterization gives information which can be used for their environmental risk assessment (ERA). Such an assessment is needed to clarify whether the waste can be reused without harm to the environment or, equally important, to decide how the waste treatment has to be adapted to minimize the environmentally negative properties of the waste itself. Finally, information is gained for the sustainable redefinition of the planned reuse scenario, e.g. reuse or landfilling of mineral waste. Most of the biological test systems are performed under highly standardized test conditions. However, using accepted ERA rules, including the application of safety factors, their results can be transferred to in-situ conditions (e.g. in environmentally open reuse scenarios), thus providing an estimation of the negative impact on terrestrial (mainly soil) and aquatic ecosystems.

For this kind of risk assessment, the test strategy and the composition of the test set need to be adapted to the anticipatory conditions of the waste treatment or recovery process, especially the environmental compartment which will be exposed to the waste material. The aim of the test strategy is to comprehend the main contamination pathways and assess their environmental impact using suitable leaching and biotest methods (see Fig.29.3). This approach requires experience in waste testing and knowledge of ERA, to define the test conditions and to interpret the

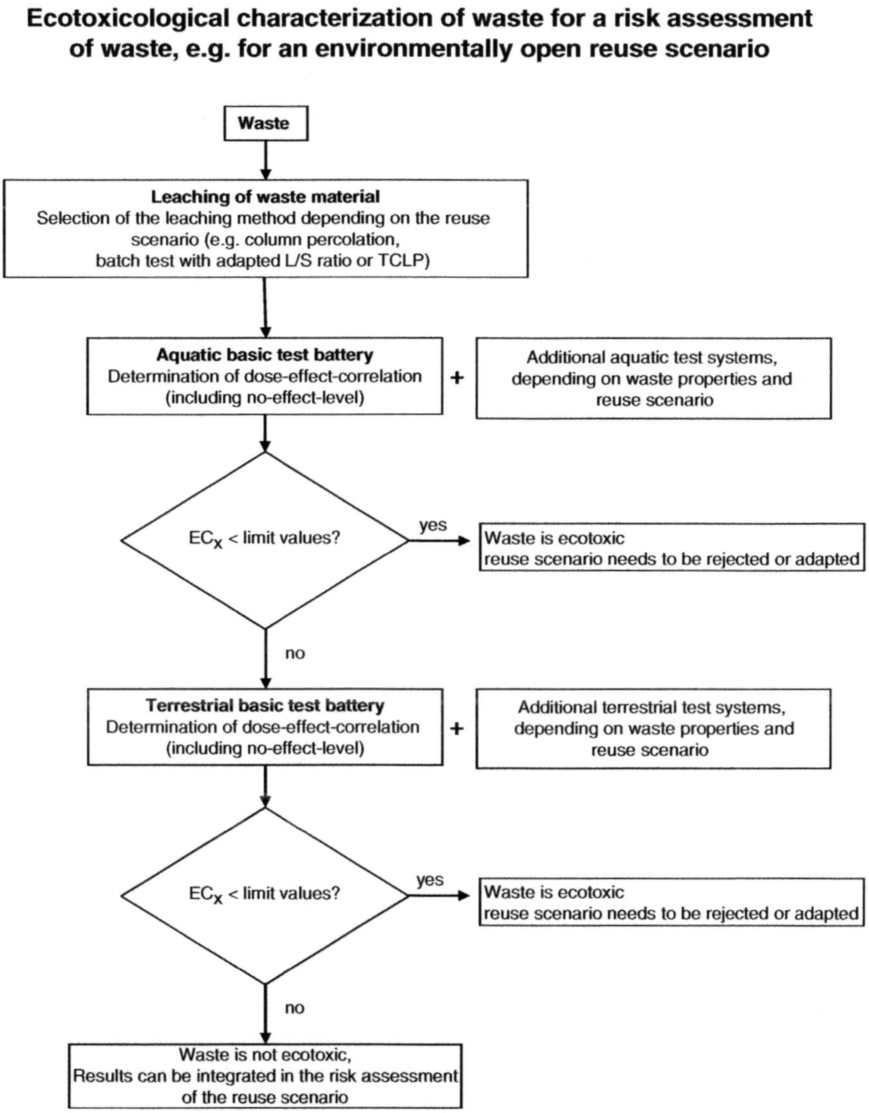

Fig. 29.3 Test strategy for the ecotoxicological characterisation of wastes to be used in an Environmental Risk Assessment (ERA) of waste, e.g. in an open reuse scenario

results. The development of the test strategy will be done case-by-case, because the reuse scenario is often site- and waste-specific. Such an ERA and the respective testing cannot be performed as a simple routine measure like the limit test for waste characterization, but it will lead to a more environmentally sound reuse of waste.

29.3.1 Test Strategy

The complexity of the test strategy is determined by the waste type and the planned reuse scenario. If the waste has already been characterized by biotests, their results can be integrated in the more detailed investigation of the waste management scenario. The main components and the properties of the waste materials influence the selection and number of necessary biotests (see Sect 29.3.2). The risk assessment of waste cannot be based exclusively on the results of a biological Extended Limit test, because the determination of a perhaps complex dose-effect-correlation can only be covered by more differentiated dilution schemes.

29.3.2 Selection of Leaching Methods

The selection of a suitable leaching method is very important to represent the conditions of the waste reuse situation and the modification of the waste material, e.g. aging or generation of in-situ leachate. Especially the particle size, the ration of liquid to solid phase and the contact time between eluent and waste material are important conditions in leaching the waste material (see Table 29.3), which can influence the result of a waste risk assessment.

29.3.3 Selection of Biological Test Systems

The main contamination pathways of the reuse scenario have to be covered in the test set to be selected to address properly the risk of the waste in the environment. Therefore, depending on the exposure pathway(s), aquatic and terrestrial test methods with organisms from three trophic levels (decomposers, consumers and producers) are needed. For the contamination pathway, water aquatic organisms have to be tested in eluates and/or eluate mixtures. The direct or intrinsic toxicity of the waste material needs to be tested in contact tests with the original waste material, using terrestrial (mainly soil) organisms. Furthermore it is important to use test systems with long-term exposure to assess potential chronic biological effects. Biotests suitable for waste testing are listed in the information Annexure B of EN 14735.

Table 29.3 Leaching methods for waste eluates

Characterization of waste – Leaching – Compliance test for leaching of granular waste materials and sludges – Part 1: One stage batch test at a liquid to solid ratio of 2 L/kg for materials with high solid content and with particle size below 4 mm (without or with size reduction)	EN 12457-1:2002
Characterisation of waste– Leaching – Compliance test for leaching of granular waste materials and sludges – Part 2: One stage batch test at a liquid to solid ratio of 10 L/kg for materials with particle size below 4 mm (without or with size reduction)	EN 12457–2:2002
Characterisation of waste – Leaching – Compliance test for leaching of granular waste materials and sludges – Part 3: Two stage batch test at a liquid to solid ratio of 2 L/kg and 8 L/kg for materials with high solid content and with particle size below 4 mm (without or with size reduction)	EN 12457–3:2002
Characterization of waste – Leaching – Compliance test for leaching of granular waste materials and sludges – Part 4: One stage batch test at a liquid to solid ratio of 10 L/kg for materials with particle size below 10 mm (without or with size reduction)	EN 12457-4:2002
Characterization of waste – Leaching behavior tests – Influence of pH on leaching with continuous pH-control	CEN/TS 14997:2006
Characterization of waste – Leaching behavior tests – Up-flow percolation test (under specified conditions)	CEN/TS 14405:2004
Characterization of waste – Leaching behavior tests – Influence of pH on leaching with initial acid/base addition	CEN/TS 14429:2005
Characterization of waste – Leaching behavior tests – Acid and base neutralization capacity test	CEN/TS 15364:2006
Characterization of waste – Methodology for the determination of the leaching behavior of waste under specified conditions	EN 12920:2006

29.3.4 Evaluation of the Test Results

The toxicity criteria of the biotests as defined for the distinction between hazardous and non-hazardous waste also apply for the results of the biotests to be assessed in the context of the planned reuse scenario. No general limit values can be applied, because the risk assessment of waste in reuse scenarios needs to integrate the risk of exposure, the uncertainty regarding biological long-term effects, the results of the chemical analysis of the eluates and the waste material and the additional aspects like long-term stability or the possibility of monitoring or maintenance of the recovered waste.

29.4 Summary and Recommendations

The European ring test for the ecotoxicological characterization of waste produced several results and much experience in ecotoxicity, through the successful cooperation and combined knowledge of the many participants from the European Member States. From the scientific point of view, the implementation of the hazard criterion H14 can now proceed. The basis for common knowledge on waste testing and an EU-harmonized approach for enhancing the European Waste list were made possible by the collaboration of the ring test partners. These experiences can be used not only for the classification of wastes according to the H14 criteria, but also for the identification of a test strategy to be used in an Environmental Risk Assessment (ERA) of waste, e.g., in an open reuse scenario. In this context also, proposals for the evaluation of the results from such tests have been made. Thus, the ring test has proved that biological tests can be successfully applied in waste and waste eluates and that they deliver reliable information for a more environmentally sound and sustainable waste management.

Annex

Table A.1 Summary of the results for the three waste materials in the basic test battery

Sample	Test	N	N-Ak	N-Stat	EC50	UL	OL	U-O	M–M
INC	AL-ges.	48	35	32	4.08	0.44	37.6	85	59
	AL-Ds	21	14	13	8.80	1.18	65.7	56	15
	AL-Ps	27	21	19	2.42	0.44	13.4	31	20
	DA	54	47	38	2.71	1.06	7.33	7	10
	LB-ges	49	45	21	35.4	10.7	111.9	10	8
	LB-freeze	25	23	19	30.8	11.2	84.9	8	8
	RW	18	14	8	45.5	30.6	67.7	2	2
	PL-As	22	19	17	29.4	7.25	93.3	13	17
	PL-Bn	21	18	17	23.9	4.82	91.3	19	22
SOI	AL-ges	48	35	n.a.	56.5	29.20	109.0	4	n.a.
	DA	54	51	2	n.a.	n.a.	n.a.	n.a.	n.a.
	LB-ges	48	42	15	65.5	41.1	97.7	2	2
	LB-freeze	25	21	n.a.	65.8	41.1	97.7	2	n.a.
	RW	17	15	1	n.a.	n.a.	n.a.	n.a.	n.a.
	PL-As	21	18	n.a.	57.8[a]	n.a.	n.a.	n.a.	n.a.
	PL-Bn	20	17	10	63.0	30.9	128.2	4	3
WOO	AL-ges	47	36	35	0.50	0.05	4.84	94	138
	AL-Ds	21	15	14	1.34	0.75	2.39	3	2
	AL-Ps	26	21	19	0.22	0.04	1.93	55	138
	DA	53	51	47	0.38	0.06	1.84	30	31
	LB-ges	57	53	52	2.56	0.26	27.9	108	122
	LB-freeze	24	20	18	0.73	0.16	3.38	21	20
	LB-fresh/liq.	33	33	32	5.60	2.43	12.9	5	11
	RW	17	15	13	20.1	8.47	41.2	5	3
	PL-As	21	18	16	10.0	2.79	43.0	15	10
	PL-Br	19	17	16	2.64	0.67	10.4	16	11

INC Municipal waste incineration ash; *SOI* PAH contaminated soil; *WOO* Waste wood; *n.a.* Not applicable; *N* Number of reported data; *N-Ak* Number of accepted tests; *N-Stat* Number of test data available for statistics; *EC50* Effect value in % dilution (medium or OECD (1984)/LUFA soil); *UL/OL* Upper/lower limit of the 95% confidence interval of the respective EC50-value; *U-O* Factor between UL and OL; *M-M* Factor between minimum and maximum values; *AL* Algae; *Ds Desmodesmus subspicatus*; *Ps Pseudokirchneriella subcapitata*; *DA* Daphnids; *LB* Luminescent bacteria; *ges* In total; *Freeze* freeze-dried; *Fresh/Liq.* Liquid culture; *RW* Earthworms; *PL* Plants; *As Avena sativa*; *Bn Brassica rapa*

[a]Based on just six values

References

2000/532/EC: Commission Decision of 3 May 2000 replacing Decision 94/3/EC establishing a list of wastes pursuant to Article 1(a) of Council Directive 75/442/EEC on waste and Council Decision 94/904/EC establishing a list of hazardous waste pursuant to Article 1(4) of Council Directive 91/689/EEC on hazardous waste (notified under document number C (2000) 1147)

91/689/EEC Council Directive of 12 December (1991) on hazardous waste

94/3/EC: Commission Decision of 20 December (1993) establishing a list of wastes pursuant to Article 1a of Council Directive 75/442/EEC on waste

Achazi, R.K., Roembke, J. & Riepert, F. (2000): Collembolen als Testorganismen. In Toxikologische Beurteilung von Boeden. Edited by S. Heiden, R. Erb, W. Dott, and A. Eisentraeger, Spektrum Verl., Heidelberg. pp. 83–103.

AFNOR (2000) Water quality—Determination of chronic toxicity to *Ceriodaphnia dubia* in 7 days—Population growth inhibition test. NF T 90-376.

AFNOR (2000): Détermination de la toxicité chronique vis-à-vis de *Brachionus calyciflorus* en 48h. Essai d'inhibition de la croissance de la population. T90-377.

Ahlf, W. (2004): Sedimentkontakttests im Überblick. In: Veranstaltungen Statusseminar – Sedimentkontakttests. Bundesanstalt fuer Gewaesserkunde, pp. 5–12.

Ahmed, K. (2002): Topic Maps - A Practical Introduction With Case Studies. In: *Proceedings of XML Europe,* Barcelona 20-23 May 2002; http:/www.indelliance.org/papers/xmle02/dx_xmle02/papers/03-05-01/03-05-01.pdf.

Ahmed, K. (2005): Topic Maps for (Open) Source Developers. In: *Proceedings of XTech 2005: XML, the Web and beyond;* Amsterdam 25-27 May 2005; http:/www.indelliance.org/proceedings/xtech05/papers/04-03-04/.

Ahtiainen, J., Nakari, T., Silvonen, J. (1994): Toxicity of TCF and ECF pulp bleaching effluents assessed by biological toxicity tests. Presentation at the 2nd International Conference on Environmental Fate and Effects of Bleached Pulp Mill Effluents. November 6-10, 1994, Vancouver, Canada.

Al-Ahmad, A., Daschner, F.D. & Kuemmerer, K. (1999): Biodegradability of cefotiam, ciprofloxacin, meropenem, penicillin G, and sulfamethoxazole and inhibition of waste water bacteria. *Arch. Environ. Contam. Toxicol.* 37, 158–163.

Algaltoxkit F. (2004): Freshwater Toxicity Test with Microalgae. Standard Operational Procedure. MicroBioTests Inc. 36 pp.

AMINAL (2000) Handbook for the characterization of the sediments of Flemish surface waters via Triade. Report for the DG Environment of the Flemish Government.

Amorim, M.J.B., Rombke, J., Scheffczyk, A. & Soares, A.M.V.M. (2005a): Effect of different soil types on the enchytraeids *Enchytraeus albidus* and *Enchytraeus luxuriosus* using the herbicide phenmedipham. *Chemosphere* 61, 1102–1114.

Amorim, M.J.B., Roembke, J., Scheffczyk, A., Nogueira, A.J.A. & Soares, A.M.V.M. (2005b): Effects of different soil types on the collembolans *Folsomia candida* and *Hypogastrura assimilis* using the herbicide phenmedipham. *Arch. Environ. Cont. Toxicol.* 49, 343–352.

Amorim, M.J.B., Roembke, J., Schallnass, H. & Soares, A.M.V.M. (2005c): Effect of soil properties and aging on the toxicity of copper for *Enchytraeus albidus*, *Enchytraeus luxuriosus* and *Folsomia candida*. *Env. Toxicol. Chem.* 24, 1875–1885.

Amorim, M.J.B., Roembke, J. & Soares, A.M.V.M. (2005d): Avoidance behaviour of *Enchytraeus albidus*: Effects of benomyl, carbendazim, phenmedipham and different soil types. *Chemosphere* 4, 501–510.

Anderson, P.R. & Christensen, T.H.J. (1998): Distribution coefficients of Cd, Co, Ni, and Zn in soils. *J. Soil Sci.* 39, 15–22.

Bailey, H.C., Krassoi, R., Elphick, J.R., Mulhall, A.-M., Hunt, P., Tedmanson, L. & Lovell, A. (2000): Application of *Ceriodaphnia dubia* for whole effluent toxicity tests in the Hawkesbury-Nepean watershed, New South Wales, Australia: Method development and validation. *Environ. Toxicol. Chem.* 19, 88–93.

Bakker, F.M., Feije, R., Grove, A.J., Hoogendorn, G., Jacobs, G., Loose, E.D. & van Stratum, P. (2003): A laboratory test protocol to evaluate effects of plant protection products on mortality and reproduction of the predatory mite *Hypoaspis aculeifer Canestrini* (Acari: Laelapidae) in standard soil. *J. Soils Sediments* 3, 73–77.

Baudo, R., Sbalchiero, A. & Beltrami, M. (2004): Test di Tossicita acuta con *Daphnia magna*. *Biologi Italiani* 6, 62–69.

Bayly, R.C. & Wigmore, G.J. (1973): Metabolism of phenol and cresols by mutants of *Pseudomonas putida*. *J. Bacteriol.* 113, 1112–1120.

BBA (Biologische Bundesanstalt fuer Land- und Forstwirtschaft) (1994): Richtlinien fuer die Pruefung von Pflanzenschutzmitteln im Zulassungsverfahren, Teil VI 2-2, Auswirkungen von Pflanzenschutzmitteln auf die Reproduktion und das Wachstum von *Eisenia fetida/Eisenia andrei*. Braunschweig, Germany.

Becker R., Donnevert G. & Roembke J. (2008): Biologische Testverfahren zur oekotoxikologischen Charakterisierung von Abfaellen. Abschlussbericht. Umweltbundesamt, Germany.

Becker-van Slooten, K., Rossel, D. & Tarradellas, J. (1999): Gefaehrdungsabschaetzung – Anwendung oekotoxikologischer Testverfahren auf Sickerwasser und Eluate von belasteten Standorten. Bundesamt fuer Umwelt, Wald und Landwirtschaft (BUWAL), 45 pp.

Belfroid, A.C., Sijm, D. & Van Gestel, C.A.M. (1996): Bioavailability and toxicokinetics of hydrophobic aromatic compounds in benthic and terrestrial invertebrates. *Environ. Rev.* 4, 276–299.

Belotti, E. (1998): Assessment of soil quality criterion by means of a soil survey. *Appl. Soil Ecol.* 10, 51–63.

Berger, W., Kalbe, U., Eckardt, J., Fischer, H. & Jansky, H.J. (2004): *Aufbereitung von Referenzmaterialien zur Untersuchung der Eluierbarkeit von Schadstoffen – Processing of reference materials for examination of contaminant leaching behaviour, Aufbereitungstechnik – Mineral Process.* 45, 37–43.

Bernard, C., Persoone, G., Colin, J. & Le Dû-Delepierre, A. (1996): Estimation of the hazard of landfills through toxicity testing of leachates – determination of leachate toxicity with a battery of acute tests. *Chemosphere* 33(11), 2303–2320

Blaise, C., Gagné, F., Férard, J.F., Eullaffroy, P. (2008). Ecotoxicity of selected nano-materials to aquatic organisms. Environ. Toxicol. Chemistry 23: 617-625.

Blakemore, R.J. (1997): Agronomic potential of earthworms in brigalow soils of south-east Queensland. *Soil Biol. Biochem.* 29, 603–608.

Brandt, K.K., Pedersen, A. & Sorensen, J. (2002): Solid-phase contact assay that uses a *lux*-marked *Nitrosomonas europaea* reporter strain to estimate toxicity of bioavailable linear alkylbenzene sulfonate in soil. *Appl. Environ. Microbiol.* 68, 3502–3508.

Brinkmann, C. & Eisentraeger, A. (2008): Completely automated short-term genotoxicity testing for the assessment of chemicals and characterisation of contaminated soils and waste waters. *ESPR – Envir. Sci. Pollut. Res.* 15, 211–217.

Brinkmann, G. & Kuehn, R. (1977): Limiting values for the damaging action of water pollutants to bacteria *Pseudomonas putida* and green algae *Scenedesmus quadricauda* in cell multiplication inhibition test. *Z. f. Wasser- und Abwasserforschung* 10, 338–347.

Brohon, B. & Gourdon, R. (2000): Influence of soil microbial activity level on the determination of contaminated soil toxicity using Lumistox and MetPlate bioassays. *Soil Biol. Biochem.* 32, 853–857.

CEN (2004): Characterization of waste - Leaching behaviour tests - Up-flow percolation test (under specified conditions). CEN/TS 14405. European Standard, Brussels, Belgium.

CEN (2005a): Characterization of waste – Preparation of waste samples for ecotoxicity tests. CEN 14735 45pp. European Standard, Brussels, Belgium.

CEN (2005b): Characterization of waste - Leaching behaviour tests - Influence of pH on leaching with initial acid/base addition. CEN/TS 14429. European Standard, Brussels, Belgium.

CEN (2006a): Characterization of waste - Leaching behaviour tests - Influence of pH on leaching with continuous pH-control. CEN/TS 14997. European Standard, Brussels, Belgium.

CEN (2006b): Characterization of waste - Leaching behaviour tests - Acid and base neutralization capacity test. CEN/TS 15364. European Standard, Brussels, Belgium.

Chao, M.-R. & Chen, C.-Y. (2000): No-observed-effect concentrations in batch and continuous algal toxicity tests. *Environmental Toxicol. Chem.*, 19: 1589–1596.

Chapman, P.M. (1995): Ecotoxicology and pollution – key issues. *Mar. Pollut. Bull.* 31: 167–177.

Christen, O. & Theuer, C. (1996): Sensitivity of *Lemna* bioassay interacts with stock-culture period. *J. Chem. Ecol.* 22, 1177–1186.

Clement, B. & Merlin, G. (1995): The contribution of ammonia and alkalinity to landfill leachate toxicity to duckweed. *Sci. Total Environ.* 170, 71–79.

Clément, B., Colin, J.R. & Le Dû-Delepierre, A. (1997): Estimation of the hazard of landfills through toxicity testing of leachates: 2. *Comparison of physico-chemical characteristics of landfill leachates with their toxicity determined with a battery of tests. Chemosphere* 35, 2783–2796.

Cleuvers, M. & Ratte, T. (2002): Phytotoxicity of coloured substances: is *Lemna* (duckweed) an alternative to the algal growth inhibition test? *Chemosphere* 49, 9–15.

Costan, G., Bermingham, N., Blaise, C. & Férard, J-F. (1993): Potential ecotoxic probe (PEEP): A novel index to assess and compare the toxic potential of industrial effluents. *Environ. Toxicol. Water Qual.* 8, 115–140.

Crommentuijn, T., Stab, J.A., Doornekamp, A., Estoppey, O. & Van Gestel, C.A.M. (1995): Comparative ecotoxicity of cadmium, chlorpyrifos and triphenyltin hydroxide to four clones of the parthenogenetic Collembolan *Folsomia candida* in an artificial soil. *Funct Ecol* 9, 734–742.

Crouau, Y., Gisclard, C. & Perotti, P. (2002): The use of *Folsomia candida* (Collembola Isotomidae) in bioassays of waste. *Appl. Soil Ecol.* 19: 65–70

Daniel, M., Sharpe, A., Driver, J., Knight, A.W., Keenan P.O., Walsley, R.M., Robinson, A., Zhang, T. & Rawson, D. (2004): Results of a technology demonstration project to compare rapid aquatic toxicity screening tests in the analysis of industrial effluents. *Environ. Monit.* 6: 855–865.

Daphtoxkit F. (2001): Crustacean Toxicity Screening Test for Freshwater. Standard Operational Procedure. MicroBioTests Inc. 27 pp.

Deventer, K. (2006): Ecotoxicological characterization of waste - minimum test battery and limit test. Proceedings Workshop: Problems around soil and Waste III – the H-14 criterion and (Bio) analytical Approachtes for Ecotoxicological Waste Characterisation. Gawlik & Moser (eds.), Italy, 2006.

Deventer, K., Zipperle, J. & Kostka-Rick, R. (2004): Oekotoxikologische Charakterisierung von Abfall – Verfahrensentwicklung fuer die Festlegung des Gefaehrlichkeitskriteriums "oekotoxisch" (H-14). Oekologische Umweltbeobachtung 2. LfU Baden-Wuerttemberg, Karlsruche. 123 S.

Devillers, Pandard, Poulsen, Charissou, Jourdain, Férard, Bispo, Grand (2006) Test battery selection for assessment of waste toxicity. Proceedings Workshop: Problems around soil and Waste III – the H-14 criterion and (Bio)analytical Approaches for Ecotoxicological Waste Characterisation. Gawlik & Moser (eds.), Italy, 2006

Didden, W.A.M. (1993): Ecology of terrestrial Enchytraeidae. *Pedobiologia* 37, 2–29.

Dijkstra, J.J., van der Sloot, H.A. & Comans, R.N.J. (2006): The leaching of major and trace elements from MSWI bottom ash as function of pH and time. *Appl. Geochem.* 21, 335–351

DIN (1998): Deutsche Einheitsverfahren zur Wasser-, Abwasser- und Schlammuntersuchung - Suborganismische Testverfahren (Gruppe T) - Teil 4: Bestimmung des erbgutveraendernden Potentials mit dem Salmonella-Mikrosomen-Test (Ames-Test) (T 4). DIN 38415 T4. Weinheim, Germany.

DIN (2002): Testverfahren mit Wasserorganismen (Gruppe *L) – Arthrobacter globiformis*-Kontakttest fuer kontaminierte Feststoffe. DIN 38412 L48. Weinheim, Germany.

Dodard, S., Sunahara, G.I., Sarrazin, M., Gong, P., Kuperman, R.G., Ampleman, G., Thiboutot, S. & Hawari, J. (2005): Survival and reproduction of enchytraeid worms (Oligochaeta) in different soil types amended with cyclic nitramine explosives. *Environ. Toxicol. Chem.* 24, 2579–2587.

Domene, X., Alcaniz, J.M. & Andres, P. (2007): Ecotoxicological assessment of organic wastes using the soil Collembolan *Folsomia candida*. Appl. *Soil Ecol.* 35, 461–472.

Dominguez, J., Velando, A. & Ferreiro, A. (2005): Are *Eisenia fetida* (Savigny, 1826) and *Eisenia andrei* Bouché (1972) (*Oligochaeta, Lumbricidae*) different biological species? *Pedobiologia* 49, 81–97.

Doube, B.M. & Brown, G.C. (1998): Life in a complex community: Functional interactions between earthworms, organic matter, microorganisms and plants. In: Earthworm ecology. Edwards, C.A. (ed.), CRC Press, Boca Raton, FL. pp. 179–212.

Duft, M., Tillmann, M., Schulte, U., Markert, B. & Oehlmann, J. (2002): Development of a new sediment bioassay with the freshwater mudsnail *Potamopyrgus antipodarum* (Gastropoda: Prosobranchia). *UWSF – Z Umweltchem Oekotox* 14, 12–17.

Dybkaer, R. (2004): An ontology on property for physical, chemical, and biological systems. *APMIS Suppl* 117, 1–210.

Dzombak, D.A. & Morel, F.M.M. (1990): Surface complexation modeling: hydrous ferric oxide, Wiley, New York, NY.

EC (Environment Canada) (1990): Biological test method: acute lethality test using *Daphnia* spp. EPS. Report EPS 1/RM/11 July 1990 with May 1996 amendments, 55pp.

EC (Environment Canada) (1999): Guidance Document on Application and Interpretation of Single-species Tests in Environmental Toxicology. EPS 1/RM/34, 203 pp.

EC (Environment Canada) (2004): Biological test method: Tests for toxicity of contaminated soil to earthworms (*Eisenia andrei, Eisenia fetida*, or *Lumbricus terrestris*). Ottawa, Canada. EPS 1/RM/43.

EC (Environment Canada) (2005): Guidance Document on Statistical methods for in Environmental Toxicity Test. Report EPS 1/RM/46, 241 pp.

Edwards, C.A. & Bohlen, P.R. (1996): Biology of earthworms. Chapman and Hall, London, UK. p 276.

Edwards, W.M. & Shipitalo, M.J. (1998): Consequences of earthworms in agricultural soils: aggregation and porosity. In: Earthworm ecology. Edwards, C.A. (ed.), CRC Press, Boca Raton, FL. pp. 147–161.

Eisentraeger, A., Rila, J.-P., Hund-Rinke, K. & Roembke, J. (2004): Proposal of a testing strategy and assessment criteria for the ecotoxicological assessment of soil or soil materials. *J Soil Sediment* 4, 123–128.

Eisentraeger, A., Hund-Rinke, K. & Roembke J. (2005): Assessment of ecotoxicity using bioassays. In: Manual for Soil Analysis – Monitoring and Assessing Soil Bioremediation. Margesin, R. & Schinner, F. (eds.), Soil Biology (Vol. 5). Springer, Berlin, pp. 321–359.

Eisentraeger, A., Reifferscheid, G., Dardenne, F., Blust, R. & Schofer, A. (2007): Hazard characterization and identification of a former ammunition site using microarrays, bioassays and chemical analysis. *Environ. Toxicol. Chem.*, 26, 634–646.

EN (2002a): Characterisation of waste - Leaching - Compliance test for leaching of granular waste materials and sludges - Part 1: One stage batch test at a liquid to solid ratio of 2 l/kg for materials with high solid content and with particle size below 4 mm (without or with size reduction). EN 12457-1. European Standard, Brussels, Belgium.

EN (2002b): Characterisation of waste - Leaching - Compliance test for leaching of granular waste materials and sludges - Part 2: One stage batch test at a liquid to solid ratio of 10 l/kg for materials with particle size below 4 mm (without or with size reduction). EN 12457-2. European Standard, Brussels, Belgium.

EN (2002c): Characterisation of waste - Leaching - Compliance test for leaching of granular waste materials and sludges - Part 3: Two stage batch test at a liquid to solid ratio of 2 l/kg and 8 l/kg for materials with high solid content and with particle size below 4 mm (without or with size reduction). EN 12457-3. European Standard, Brussels, Belgium.

EN (2002d): Characterisation of waste - Leaching - Compliance test for leaching of granular waste materials and sludges - Part 4: One stage batch test at a liquid to solid ratio of 10 l/kg for materials with particle size below 10 mm (without or with size reduction). EN 12457-4. European Standard, Brussels, Belgium.

EN (2006): Characterization of waste - Methodology for the Determination of the Leaching Behaviour of Waste under Specified Conditions. EN 12920. European Standard, Brussels, Belgium.

Erséus, C. & Kaellersjoe, M. (2004): 18S rDNA phylogeny of Clitellata (Annelida). *Zoologica Scripta* 33, 187–196.

European Economic Community (1991): Directive 91/689/EEC on hazardous waste. Official Journal of the European Community, L377, Brussels, Belgium.

Farré, M., Martinez, E., Hernando, M.-D., Fernandez-Alba, A., Fritz, J., Unruh, E., Mihail, O., Sakkas, V., Morbey, A., Albanis, T., Brito, F., Hansen, P.D. & Barcelo, D. (2006): European ring exercise on water toxicity using different bioluminescence inhibition tests based on *Vibrio fischeri*. *Talanta* 69, 323–333.

Feiler, U., Ahlf, W., Hoess, S., Hollert, H., Neumann-Hensel, H., Meller, M. Weber, J. & Heininger, P. (2005): The SeKT Joint Research Project: Definition of reference conditions, control sediments and toxicity thresholds for limnic sediment contact tests. *Environ. Sci. Pollution Res.* 12, 257–258

Ferrari, B. & Férard, J-F. (1999) : Evaluation de l'écotoxicité intrinsèque aà l'aide d'une batterie de tests d'écotoxicité realises après une lixiviation X30-402. Proceedings of Stab&Env99 conference, Lyon, pp 188-191

Ferrari, B., Radetski, C.M., Veber, A-M. & Ferard, J-F. (1999): Ecotoxicological assessment of solid wastes: A combined liquid- and solid-phase testing approach using a battery of bioassays and biomarkers. *Environ. Toxicol. Chem.* 18: 1195–1202.

Fielding, J.M., Simon, J., Ceusters, W. & Smith, B. (2004): Ontological theory for ontological engineering - biomedical systems information integration. In: *Proceedings Ninth International Conference on the Principles of Knowledge Representation and Reasoning,* Whistler, BC, 2-5 June 2004; http://ontology.buffalo.edu/bio/KR_2004.pdf.

Fochtman, P. (2000): Acute toxicity of nine pesticides as determined with conventional assays and alternative microbiotests. In: New microbiotests for routine toxicity screening and biomonitoring. Persoone, G., Janssen, C. & De Coen, W. (eds.), Kluwer/Plenum, New York, NY, pp. 233–242.

Folker-Hansen, P., Krogh, P.H. & Holmstrup, M. (1996): Effect of dimethoate on body growth of representatives of the soil living mesofauna. *Ecotoxicol. Environ. Saf.* 33, 207–216.

Fomin, A., Moser, H. & Pickl, C. (2000): Ecotoxicological investigations of extremely acidic mining lakes using bioassays suitable for testing at low pH. *Toxicol. Environ. Chem.* 76, 237–254

Fritz-Langen, H., Blessing, B. & Krebs, F. (1994): Bakterientoximeter mit Reinkulturen (*Photobacterium phosphoreum* und *Pseudomonas* putida) in der Gewaesserueberwachung (Automated bacterial toxicity test systems operating with pure cultures (*Photobacterium phosphoreum* und *Pseudomonas* putida) for monitoring water quality). *Schriftenreihe des Vereins fuer Wasser-, Boden- und Lufthygiene* 93, 249–273.

Gaines C.S. (2000): Collaborative study of methods for solvent retention capacity profiles (AACC method 56–11). *Cereal Foods World* 45, 303–306.

Garcia, M.V. (2004): Effects of pesticides on soil fauna: Development of ecotoxicological test methods for tropical regions. Ecology and Development Series No. 19. Zentrum fuer Entwicklungsforschung, University of Bonn, Germany. p 281.

Garcia, M.V., Roembke, J., De Brito, M.V.T. & Scheffczyk, A. (2008): Effects of three pesticides on the avoidance behaviour of earthworms in laboratory tests performed under tropical conditions. *Environ. Pollut.* 153, 450–456.

Garshol, L.M. (2006): tolog – A Topic Maps Query Language. In: TMRA05 International Workshop on Topic Map Research and Applications. Maicher, L. & Park J. (eds.), Springer, Berlin, pp. 183–196.

Gawlik, B.M. & Moser, H. (2005): Problems around soil and waste III – The H14 criterion and (bio)analytical approaches for ecotoxicological waste characterization. Workshop proceedings. EUR 22152, Ispra, Italy.

Gellert, G., Stommel, A. & Paulus, D. (1996): Eignung der Mikrotiterplattenmethode fuer aquatische Toxizitaetstests mit Bakterien. *Vom Wasser* 87, 145–152.

Gomes, N.C.M., Kosheleva, I.A., Abraham, W.-R. & Smalla, K. (2005): Effects of the inoculant strain *Pseudomonas putida* KT2442 (pNF142) and of naphthalene contamination on the soil bacterial community. *FEMS Microbiol. Ecol.* 54, 21–33.

Gourmelon, A & Ahtiainen, J. (2007): Developing test guidelines on invertebrate development and reproduction for the assessment of chemicals, including potential endocrine active substances – The OECD perspective. *Ecotoxicology* 16, 161–167.

Grathwohl, P. (1990): Influence of organic matter from soils and sediments from various origins on the sorption of some chlorinated aliphatic hydrocarbons – implications on Koc correlations. *Environ. Sci. Technol.* 24, 1687–1693.

Gruber, T.R. (1993): A translation approach to portable ontology specifications. *Knowl. Acquisition* 2:199–220.

Gurevitch, J., Scheiner, S.M. & Fox, G.A. (2002): Plant ecology. Sinauer Associates, Sunderland, MA. p 523.

Heiden, S.; Erb, R., Dott, W. & Eisentraeger A. (eds.) (2000): Toxikologische Beurteilung von Boeden – Leistungsfaehigkeit biologischer Testverfahren. Spektrum Akademischer Verlag, Heidelberg, Berlin, ISBN 3-8274-1095-9.

Heimbach, F. (1984): Correlations between 3 methods for determining the toxicity of chemicals to earthworms. *Pesticide Sci.* 15, 605–611.

Heimbach, F. & Edwards, P.J. (1983): The toxicity of 2-chloroacetamide and benomyl to earthworms under various test conditions in an artificial soil test. *Pesticide Sci.* 14, 635–636.

Heise, S. & Ahlf, W. (2005): A new microbial contact assay for marine sediments. *J Soils Sediments* 5, 9–15.

Henner, P., Schiavon, M., Druelle, V. & Lichtfouse, E. (1999): Phytotoxicity of ancient gaswork soils. *Effects of polycyclic aromatic hydrocarbons (PAHs) on plant germination. Org Geochem* 30, 963–969.

Hock, B. & Elstner, E.F. (1984): Pflanzentoxikologie. BI Wissenschaftsverlag, Mannheim. p. 346.

Hofman, J., Rhodes, A. & Semple, K.T. (2008): Fate and behaviour of phenanthrene in natural and artificial soils. *Environ. Pollut.* 152, 468–475.

Hollert, H., Keiter, S., Koenig, N., Rudolf, M., Ulrich, M. & Braunbeck, T. (2003): A new sediment contact assay to assess particle-bound pollutants using zebrafish (*Danio rerio*) embryos. *J. Soils Sediments* 3, 197–207.

Hopkin, S.P. (1997): Biology of the springtails (Insecta: Collembola). Oxford University Press, Oxford, UK.

Huebert, D.B. & Shay, J.M. (1993): Considerations in the assessment of toxicity using duckweed. *Environ. Toxicol. Chem.* 12, 481–483.

Hund-Rinke, K. & Wiechering, H. (2001): Earthworm avoidance test for soil assessment. *J. Soils Sediments* 1, 15–20.

Hund-Rinke, K., Koerdel, W., Hennecke, D., Achazi, R., Warnecke, D., Wilke, B-M., Winkel, B. & Heiden, S. (2002): Bioassays for the ecotoxicological and genotoxicological assessment of contaminated soils (results of a round-Robin test). Part II: Assessment of the habitat function of soils – tests with soil microflora and fauna. *J. Soils Sediments* 2, 83–90

Hund-Rinke, K., Achazi, R., Roembke, J. & Warnecke, D. (2003): Avoidance tests with *Eisenia fetida* as indicator for the habitat function of soils: results of a laboratory comparison test. *J. Soils Sediments* 1, 7–12.

Hund-Rinke K., Lindemann M. & Simon M. (2005): Experiences with novel approaches in earthworm testing alternatives. *J. Soils Sediments* 5, 233–239.

Ibanez, R., Viguri, J., Andrés, A. & Irabien, A. (2001): Characterization of solid wastes from non-ferrous metals re cycling. *Fresenius Environ. Bull.* 10(1), 84–88.

ISO (International Organization for Standardization) (1993a): Soil quality - Effects of pollutants on earthworms (*Eisenia fetida*) - Part 1: Determination of acute toxicity using artificial soil substrate. ISO 11268. Geneve, Switzerland.

ISO (International Organization for Standardization) (1993b): Soil quality - Determination of dry matter and water content on a mass basis - Gravimetric method. ISO 11465. Geneva, Switzerland.

ISO (International Organization for Standardization) (1994): Accuracy (trueness and precision) of measurement methods and results – Part 1: General principles and definitions. ISO 5725-1. Geneva, Switzerland.

ISO (International Organization for Standardization) (1995a): Water quality – *Pseudomonas putida* Growth inhibition test. ISO 10712. Geneva, Switzerland.

ISO (International Organization for Standardization) (1995b): Soil quality – Determination of the effects of pollutants on soil flora - Part 2: Effects of chemicals on the emergence and growth of higher plants. ISO 11269-2. Geneve, Switzerland.

ISO 11466 (International Organization for Standardization) (1995): Soil quality – Extraction of trace elements soluble in aqua regia. ISO, Geneva, Switzerland.

ISO (International Organization for Standardization) (1996): Water quality - Determination of the inhibition of mobility of *Daphnia magna* Straus (Cladocera, Crustacea) - Acute toxicity test. ISO 6341. Genève, Switzerland.

ISO (International Organization for Standardization) (1998a): Technical Corrigendum 1 to ISO 6341 (1996). Genève, Switzerland.

ISO 11047 (International Organization for Standardization) (1998): Soil quality – Determination of cadmium, chromium, cobalt, copper, lead, manganese, nickel and zinc – Flame and electrothermal atomic absorption spectrometric methods. ISO, Geneva, Switzerland.

ISO (International Organization for Standardization) (1998c): Soil Quality - Effects of pollutants on earthworms (*Eisenia fetida*). Part 2: Determination of effects on reproduction. ISO 11268-2. Geneva, Switzerland.

ISO (International Organization for Standardization) (1998d): Water quality -- Determination of the inhibitory effect of water samples on the light emission of *Vibrio fischeri* (Luminescent bacteria test) - Part 2: Method using liquid-dried bacteria. ISO 11348-2. Geneve, Switzerland.

ISO 13877 (International Organization for Standardization) (1998): Soil quality – Determination of polynuclear aromatic hydrocarbons – Method using high -performance liquid chromatography. ISO, Geneva, Switzerland.

ISO (International Organization for Standardization) (1999a): Water quality – Determination of turbidity. ISO 7027. Geneve, Switzerland.

ISO (International Organization for Standardization) (1999b): Soil Quality - Inhibition of reproduction of Collembola (*Folsomia candida*) by soil pollutants. ISO 11267. Geneva, Switzerland.

ISO (International Organization for Standardization) (1999c): Water quality – Determination of the inhibitory effect on the light emission of *Vibrio fischeri* (Luminescent bacteria test). 11348-1/2/3. Geneva, Switzerland.

ISO (International Organization for Standardization) (2000): Water quality - Determination of the genotoxicity of water and waste water using the umu-test. ISO 13829. Geneve, Switzerland.

ISO (International Organization for Standardization) (2002): Accuracy (Trueness and Precision) of Measurement Methods and Results - Part 2: Basic Method for the Determination of Repeatability and Reproducibility of a Standard Measurement. ISO 5725-2. Geneva, Switzerland.

ISO (International Organization for Standardization) (2004a): Water quality — Freshwater algal growth inhibition test with *Desmodesmus subspicatus* and *Pseudokirchneriella subcapitata*. ISO 8692. Geneva, Switzerland.

ISO (International Organization for Standardization) (2004b): Soil quality – Determination of the Effects of Pollutants on Soil Flora. Part II: Effects of Chemicals on the Emergence and Growth of Higher Plants. ISO 11269-2. Geneva, Switzerland.

ISO (International Organization for Standardization) (2004c): Soil Quality - Effects of pollutants on Enchytraeidae (*Enchytraeus* sp.). Determination of effects on reproduction and survival. ISO 16387. Geneve, Switzerland.

ISO 16703 (International Organization for Standardization) (2004): Soil quality – Determination of polynuclear aromatic hydrocarbons – Method using high -performance liquid chromatography. ISO, Geneva, Switzerland.

ISO (International Organization for Standardization), (2004e): Soil quality – Biological methods – Chronic toxicity in higher plants. ISO 22030. Geneve, Switzerland.

ISO 10390 (2005): Soil quality – Determination of pH. ISO, Geneva, Switzerland.

ISO (International Organization for Standardization) (2005b): Water quality – Determination of toxic effect of water constituents and waste water to duckweed (*Lemna minor*) – Duckweed growth inhibition test. ISO 20079. Geneva, Switzerland.

ISO (International Organization for Standardization) (2006a): Information Technology - Topic Maps - Part 2 - Data Model. ISO 13250-2. Genève, Switzerland.

ISO (International Organization for Standardization) (2006b): Water quality — Determination of chronic toxicity to *Ceriodaphnia dubia*. ISO 20665. Genève, Switzerland.

ISO (International Organization for Standardization) (2007a): Water quality - Determination of selected elements by inductively coupled plasma optical emission spectrometry (ICP-OES). ISO 11885. Geneva, Switzerland.

ISO (International Organization for Standardization). (2007b). Information Technology - Topic Maps - Part 3: XML Syntax. ISO 13250-3. Genève, Switzerland.

ISO (International Organization for Standardization) (2008a): Water quality — Determination of the inhibition of dehydrogenase activity of *Arthrobacter globiformis* — Solid contact test using the redox dye resazurine. ISO 10871. Draft. Geneve, Switzerland.

ISO (International Organization for Standardization) (2008b): Soil Quality - Avoidance test for evaluating the quality of soils and the toxicity of chemicals. Test with Earthworms (*Eisenia fetida/andrei*). ISO 17512-1. Geneve, Switzerland.

ISO (International Organization for Standardization) (2008c): Soil quality – Guidance on the assessment of tests applied in the field of ecotoxicological characterization of soils and soil materials. ISO 17616. Geneve, Switzerland.

ISO (International Organisation for Standardisation) (2008d): Water Quality – Determination of the chronic toxicity to *Brachionus calyciflorus* in 48h. ISO 20666. Geneve, Switzerland.

Jaensch, S., Amorim, M.J.B. & Roembke, J. (2005): Identification of the ecological requirements of important terrestrial ecotoxicological test species. *Environ. Rev.* 13, 51–83.

Juvonen, R., Martikainen, E., Schultz, E., Joutti, A., Ahtiainen, J. & Lehtokari, M. (2000): A battery of toxicity tests as indicators of decontamination in composting oily waste. *Ecotoxicol Environ Saf.* 47, 156–166.

Kalbe, U., Berger W., Eckardt, J. & Christoph, G. (2006). Evaluierung von Methoden zur Sickerwasserprognose. Final Report in German. Federal Institute for Materials research and Testing (BAM). Umweltbundesamt FKZ 20374278.

Kalsch, W., Junker, T. & Roembke, J. (2006a): Modifications of a chronic plant test for the assessment of contaminated soils Part 1: Method development. *J. Soils Sediments* 6, 37–45.

Kalsch, W., Junker, T. & Roembke, J. (2006b): Modifications of a chronic plant test for the assessment of contaminated soils Part II: Testing of contaminated soils. *J. Soils Sediments* 6, 92–101.

Kamiske, G.F. & Brauer, J.P. (1999): Qualitaetsmanagement von A bis Z: Erlaeuterungen moderner Begriffe des Qualitaetsmanagements (3. ed.), Hanser Ltd., Wien, Muenchen.

Kaneko, H. (1996): Evaluation of municipal waste incinerator fly ash toxicity and the role of cadmium by two aquatic toxicity tests. *Waste Management,* 16(5–6), 555–559

Katz, S.A. & Salem H. (2005): Chemistry and toxicology of building timbers pressure-treated with chromated copper arsenate: a review. *J. Appl. Toxicol.* 25, 1–7.

Keddy, C., Greene, J.C. & Bonnell, M.A. (1994): A review of whole organism bioassays for assessing the quality of soil, freshwater sediment, and freshwater in Canada. Environment Canada, Ottawa, 185.

Kinniburgh, D.G., van Riemsdijk, W.H., Koopal, L.K., Borkovec, M., Benedetti, M.F. & Avena, M.J. (1999): Ion binding to natural organic matter: competition, heterogeneity, stoichiometry and thermodynamic consistency, J. *Colloids Surf., A* 151, 147–166.

Kiss, I. & Bakonyi, G. (1992): Guideline for testing the effects of pesticides on *Folsomia candida* Willem (Collembola): laboratory test. *Bulletin of the International Organization for Biological and Integrated Control of Noxious Animals and Plants, West Palaearctic Regional Section (IOBC/WPRS)* 15, 131–137.

Knie, J. (1992): Biologische Testverfahren in der Vollzugspraxis des AbwAG und WHG. In: Biologische Testverfahren. Steinhaeuser, K.G. & Hansen P.D. (eds.), Gustav Fischer Verlag, Stuttgart/New York, 523–526.

Kostanjek, R., Lapanje, A., Drobne, D., Perovic, S., Perovic, A., Zidar, P., Strus, J., Hollert, H. & Karaman, G. (2005): Bacterial community structure analyses to assess pollution of water and sediments in the Lake Shkodra/Skadar, Balkan Peninsula. Environ. *Sci. Pollut. Res.* 12, 361–368, 2005

Kostka, J.E. & LutherIII, G.W. (1994): Partitioning and speciation of solid phase iron in saltmarsh sediments, Geochim. *Cosmochim. Acta*, 58:. 1701–1710.

Kostka-Rick, R. (2004): Oekotoxikologische Charakterisierung von Abfall – Literaturstudie. Oekologische Umweltbeobachtung 3. LfU Baden-Wuerttemberg, Karlsruhe. 108 S.

Kuemmerer, K, Al-Ahmad, A. & Mersch-Sundermann, V. (2000): Biodegradability of some antibiotics, elimination of the genotoxicity and affection of wastewater bacteria in a simple test. *Chemosphere* 40, 701–710.

Kuemmerer, K. & Alexy, R. (2006): Arzneimittel in der Umwelt: Sind klassische Tests zur Erfassung der Wirkung geeignet? In: Spurenstoffe in Gewaessern. Track, Th. & Kreysa, G. (eds.), Wiley, Weinheim.

Kula, C. (1998): Endpoints in laboratory testing with earthworms: experience with regard to regulatory decisions for plant protection products. In: Advances in Earthworm Ecotoxicology. Sheppard, S., Bembridge, J., Holmstrup, M. & Posthuma, L. (eds.), SETAC Press, Pensacola, pp. 3–14.

Kuperman, R.G., Amorim, M.J.B., Roembke, J., Lanno, R., Checkai, R.T., Dodard, S.G., Sunahara, G.I. & Scheffczyk, A. (2006a): Adaptation of the Enchytraeid toxicity test for use with natural soil types. *Eur. J. Soil Biol.* 42, S234–S243.

Kuperman, R.G., Checkai, R.T., Simini, M., Phillips, C.T., Kolakowski, J.E. & Kurnas, C.W. (2006b): Toxicities of dinitrotoluenes and trinitrobenzene freshly amended or weathered and aged in a sandy loam soil to *Enchytraeus crypticus*. *Environ. Toxicol. Chem.* 25, 1368–1375.

Kuperman, R.G., Checkai, R.T., Simini, M., Phillips, C.T., Anthony, J.S., Kolakowski, J.E. & Davis, E.A. (2006c): Toxicity of emerging energetic soil contaminant CL-20 to potworm *Enchytraeus crypticus* in freshly amended or weathered and aged treatments. *Chemosphere* 62, 1282–1293.

Lachmund, C., Koecher, B., Manz, W. & Heininger, P. (2003): Chemical and microbiological in situ characterization of benthic communities in sediments with different contamination levels. *J. Soils Sediments* 3, 188–196, 2003

Landolt, E. & Kandeler, R. (1987): The family of *Lemnaceae* – a monographic study. Biosystematic investigations in the family of duckweeds (*Lemnaceae*), Vol. 4, Publications of the Geobotanical Institute, Federal University of Technology, Foundation Ruebel, Zurich, Switzerland, 638 S.

LANUV-NRW (2006): Landesumweltamt Nordrhein-Westfalen, 2. Laenderuebergreifender Biotest-Ringversuch (Daphnientest/Leuchtbakterientest), http://www.lanuv.nrw.de/analytik/ringversuche/pdf/luebiotest_rv_2.pdf.

Lapa, N., Santos Oliveira, J.F. & Camacho, S.L. (2002a) An ecotoxic risk assessment of residue materials produced by the plasma pyrolysis/vitrification process. *Waste Manage.* 22(3), pp. 335–342.

Lapa, N., Barbosa, R., Morais, J., Mendes, B., Méhu, J. & Santos Oiveira J.F. (2002b) Ecotoxiological assessment of leachates from MSWI bottom ashes. *Waste Manage.* 22(6), pp. 583–293.

Latif, M. & Zach, A. (2000): Toxicity studies on treated residual wastes in Austria using different types of conventional assays and cost-effective microbiotests. In: New microbiotests for routine toxicity screening and biomonitoring. Persoone, G., Janssen, C. & De Coen, W. (eds.), Kluwer/Plenum, New York, NY, pp. 367–384.

Lavelle, P., Bignell, D. & Lepage, M. (1997): Soil function in a changing world: the role of invertebrate ecosystem engineers. *Eur. J. Soil Biol.* 33, 159–193.

Lee, S.Z., Allen, H.E., Huang, C.P., Sparks, D.L., Sanders, P.F. & Peijnenburg, W.J.G.M. (1996): Predicting soil–water partition coefficients for cadmium. *Environ. Sci. Technol.* 30, 3418–3424.

Lewis, M.A. (1995): Use of freshwater plants for phytotoxicity testing: A review. *Environ. pollut.* 87, 319–336.

LFU (Landesanstalt fuer Umweltschutz Baden-Wuerttemberg) (2004): Ecotoxicological Characterization of Waste – Method Development for Determining the "Ecotoxicological" (H14) Risk Criterion, Karlsruhe, Germany. 121 pp

LFUG (1995): Saechsisches Landesamt fuer Umwelt und Geologie: Materialien zur Altlastenbehandlung 7/95 – Altlastenerkundung mit biologischen Methoden. URL: http://www.umwelt.sachsen.de/de/wu/umwelt/lfug/lfug-internet/documents/mza7_95.pdf (December 2007)

Liber, K. & Solomon, K.R. (1994): Acute and chronic toxicity of 2,3,4,6-Tetrachlrophenol and Pentachlorophenol to *Daphnia* and rotifers. *Arch. Environ. Contam. Toxicol.* 26, 212–221.

Lin, C.F., Wu, C.H. & Liu, Y.C. (2007): Long-term leaching test of incinerator bottom ash: Evaluation of Cu partition. *Waste Manage.* 27, 954–960.

Lock, K. & Janssen, C.R. (2001a): Effect of clay and organic matter type on the ecotoxicity of zinc and cadmium to the potworm *Enchytraeus albidus. Chemosphere* 44, 1669–1672.

Lock, K. & Janssen, C.R. (2001b): Modeling zinc toxicity for terrestrial invertebrates. *Environ. Toxicol. Chem.* 20, 1901–1908.

Løkke, H. & Van Gestel, C.A.M. (1998): Handbook of Soil Invertebrate Toxicity Testing. Wiley, London, UK.

Loureiro, S., Soares, A.M.V.M. & Noguira A.J.A. (2005): Terrestrial avoidance behaviour tests as screening tool to assess soil contamination. *Environ. Pollut.* 138, 121–131.

Lucivjanska, V., Lucivjanska, M. & Cizek, V. (2000): Sensitivity comparison of the ISO *Daphnia* and algal test procedures with Toxkit microbiotests. In: New microbiotests for routine toxicity screening and biomonitoring. Persoone, G., Janssen, C. & De Coen, W. (eds.), Kluwer/Plenum, New York, NY, pp. 243–246.

Lukkari, T. & Haimi, J. (2005): Avoidance of Cu- and Zn-contaminated soil by three ecologically different earthworm species. *Ecotoxicol. Environ. Saf.* 62, 35–41.

Mallau, M., Eisentraeger, A., Dott, W. (2001): Gefriertrocknung von *Vibrio fisherii* und *Pseudomonas putida* in 96-well Mikrotiterplatten zur oekotoxikologischen Charakterisierung von Umweltproben. In: GDCh-Jahrestagung Chemie, Wuerzburg 23.-29. September 2001, Media Process Management GmbH, Mainz, ISBN 3-924763-98-4, 29.

Manusadzianas, L, Balkelyte, L., Sadauskas, K., Blinova, I., Pollumaa, L. & Kahru, A. (2003): Ecotoxicological study of Lithuanian and Estonian wastewaters: selection of the biotests, and correspondence between toxicity and chemical-based indices. *Aquat. Toxicol.* 63: 27–41.

Marques, S. & Ramos, J.L. (1993): Transcriptional control of the *Pseudomonas putida* TOL plasmid catabolic pathways. *Mol. Microbiol.* 9, 923–929.

Martikainen, E. (1996): Toxicity of Dimethoate to some soil animal species in different soil types. *Ecotoxicol. Environ. Saf.* 33, 128–136.

Meeussen, J.C.L. (2003): ORCHESTRA: An object-oriented framework for implementing chemical equilibrium models. *Environ. Sci. Technol.* 37, 1175–1182.

Milne, C.J., Kinniburgh, D.G., van Riemsdijk, W.H. & Tipping, E. (2003): Generic NICA-Donnan model parameters for metal–ion binding by humic substances. *Environ. Sci. Technol.* 37, 958–971.

Naddy, R.B., Stubblefield, W.A., Scott, J.R.M. & Hockett, J.R. (2002): The effect of calcium and magnesium ration on the toxicity of copper to five aquatic species in freshwater. *Environ. Toxicol. Chem.* 21(2), 347–352.

Natal-da-Luz, T., Ribeiro, R. & Sousa, J.P. (2004): Avoidance tests with Collembola and Earthworms as early screening tools for site-specific assessment of polluted soils. *Environ. Toxicol. Chem.* 23, 2188–2193.

Natal-da-Luz, T., Amorim, M.J.B., Roembke, J. & Sousa, J.P. (2008a): Avoidance tests with earthworms and springtails: Defining the minimum exposure time to observe a significant response. *Ecotoxicol. Environ. Saf.* 71(2), 545–551.

Natal-da-Luz, T., Roembke, J. & Sousa, J.P. (2008b): Avoidance tests in site-specific risk assessment – Influence of soil properties on the avoidance response of collembola and earthworms. *Environ. Toxicol. Chem.* 27(5), 1112–1117.

Neumann-Hensel, H. & Melbye, K. (2006): Optimisation of the solid-contact test with *Arthrobacter globiformis. J Soils Sediments* 6, 201–207.

Neumann-Hensel, H. (2004): Entwicklung eines Testkits fuer den Feststoffkontakttest mit *Arthrobacter globiformis. BfG Veranstaltungen Statusseminar- Sedimentkontakttests.* 5, 36–45.

NIES (National Institute for Environmental Studies) (2005): Report on the pre-validation test of the enhanced Test Guideline, *Daphnia magna* reproduction test. 83 pp.

Nyholm, N. (1990): Expression of results from growth inhibition toxicity tests with algae. *Arch. Environ. Contam. Toxicol.* 19(4), 518–522.

Oda, S., Tatarazako, N., Watanabe, H., Morita, M. & Iguchi, T. (2005): Production of male neonates in *Daphnia magna* (Cladocera, Crustacea) exposed to juvenile hormones and their analogs. *Chemosphere* 61, 1168–1174.

OECD (Organisation for Economic Co-operation and Development) (1984): Earthworm, acute toxicity tests. Guideline for Testing Chemicals No. 207. Paris, France.

OECD (Organisation for Economic Co-operation and Development) (1998): *Daphnia magna* reproduction test. Guideline for the testing of Chemicals N°211, Paris, France.

OECD (Organisation for Economic Co-operation and Development) (2000): Adsorption - Desorption Using a Batch Equilibrium Method. Guideline for Testing Chemicals No. 106. Paris, France.

OECD (Organisation for Economic Co-operation and Development) (2004a): *Daphnia* sp. acute immobilisation test. Guideline for the testing of Chemicals N°202, Paris, France.

OECD (Organisation for Economic Co-operation and Development) (2004b). Guidelines for the testing of chemicals No. 220. Enchytraeid Reproduction Test. Paris, France.

OECD (Organisation for Economic Co-operation and Development) (2004c). Guidelines for the testing of chemicals No. 222. Earthworm Reproduction Test (*Eisenia fetida/Eisenia andrei*). Paris, France.

OECD (Organisation for Economic Co-Operation and Development) (2006): Guideline for Testing of Chemicals No. 208. Terrestrial Plant Test: Seedling Emergence and Seedling Growth Test. Paris, France.

OECD (Organisation for Economic Co-operation and Development) (2006) Guideline for testing of chemicals No. 201: Freshwater Alga and Cyanobacteria: Growth Inhibition Test., Paris.

Olmstead A.W. & LeBlanc G.A. (2000): Effects of endocrine-active chemicals on the development of sex characteristics of *Daphnia magna. Environ. Toxicol. Chem.* 19, 2107–2113.

OSPAR (2000): OSPAR background document concerning the elaboration of programmes and measures relating to Whole Effluent Assessment. Ospar Commission, The Executive Secretary, London 2000, ISBN 0-946955-98-0, 142 p.

Pandard, P., Devillers J., Charissou, A.M., Poulsen, V., Jourdain, M.J., Férard, J-F., Grand, C. & Bispo, A (2006): Selecting a battery of bioassays for ecotoxicological characterization of wastes. *Sci. Total Environ.* 363, 114–125.

Pepper, S. (2000): The TAO of topic maps: finding the way in the age of infoglut. In: *Proceedings of XML Europe 2000*, Paris, 12–16 June 2000; http://www.org/papers/xmleurope2000/pdf/s11-01.pdf.

Pandard, P., Chancerelle, L. & Thybaud, E. (2007): Assessment of the repeatability of biological tests performed with waste eluates. Abstract book of the SETAC Europe 17th annual meeting, 20-24 may 2007, Porto, Portugal. Brussels: SETAC Europe, 2007, p. 133.

Persoone, G. (1998a): Development and first validation of a "stock-culture free" algal microbiotest: the Algaltoxkit. In: Microscale testing in aquatic toxicology. P.G. Wells, K. Lee and C. Blaise (eds). CRC Press, Boca Raton, Fl, USA. pp. 311–319.

Persoone, G. (1998b) : Development and validation of Toxkit microbiotests with invertebrates, in particular crustaceans : In: Microscale testing in aquatic toxicology. P.G. Wells, K. Lee and C. Blaise (eds). CRC Press, Boca Raton, Fl, USA. pp. 437-449.

Persoone, G., Janssen, C. (2006): International Intercalibration Exercise on the Algaltoxkit Microbiotest – Comparison with conventional algal assays. www.microbiotests.be, Extensive Reports on Toxkit Microbiotests. 21 pp.

Persoone, G., Janssen, C. & De Coen, W. (2000): New microbiotests for routine toxicity screening and biomonitoring. Kluwer/Plenum, New York, NY, p 550.

Phillips, C.T., Kuperman, R.G. & Checkai, R.T. (2002): Toxicity of chemical-warfare agent HD to Folsomia candida in different soil types. *Eur. J. Soil Biol.* 38, 281–285.

Phytotoxkit. (2004): Seed germination and early growth microbiotest with higher plants. Standard Operational Procedure. 24 pp.

Postma, J.F., de Valk, S., Dubbeldam, M., Maas, J.L., Tonkes, M., Schipper, C.A. & Kater, B.J. (2002): Confounding factors in bioassays with freshwater and marine organisms. *Ecotoxicol. Environ. Saf.* 53: 226–236.

Potter, C.L., Gkaser, J.A., Chang, L.W., Meier, J.R., Dosani, M.A. & Herrmann, R.F. (1999): Degradation of polynuclear aromatic hydrocarbons under bench-scale compost conditions. *Environ. Sci. Technol.* 33, 1717–1725.

Prokop, Z., Cupr, P., Zlevorova-Zlamalikova, V., Komarek, J., Dusek, L. & Holoubek, I. (2003): Mobility, bioavailability, and toxic effects of cadmium in soil samples. *Environ. Res.* 91, 119–126

ProLab 2006®, version 2.8.0.19, quo data Gesellschaft fuer Qualitaetsmanagement und Statistik mbH, Dresden (Germany)

Quilici, L., Praud-Tabaries, A., Tabaries, F. & Siret, B. (2004): Integration of ecotoxicity index and carboxylic of MSW incineration bottom ashes. *Ecotoxicology* 13, 503–509.

Ren, S. & Frymier, P.D. (2003): The use of a genetically engineered *Pseudomonas* species (Shk1) as a bioluminescent reporter for heavy metal toxicity screening in wastewater treatment plant influent. *Water Environ. Res.* 75, 21–29.

Ren, S. & Frymier, P.D. (2005): Toxicity of metals and organic chemicals evaluated with bioluminescence assays. *Chemosphere* 58, 543–550.

Riepert, F. & Kula, C. (1996): Development of laboratory methods for testing effects of chemicals and pesticides on Collembola and earthworms. *Mittl. Biologischen Bundesanstalt fuer Land- und Forstwirtschaft* 320, 8–49.

Rila, J-P. & Eisentraeger, A. (2003): Application of bioassays for risk characterization and remediation control of soils polluted with nitroaromatics and PAHs. *Water Air Soil Pollut.* 148, 223–242.

Robertson J.A., de Monredon F.D. & Dysseler P. (2000): Hydration properties of dietary fibre and resistant starch: A European collaborativ study. *Lebensmittel, Wissenschaft und Technologie-Food Sci. Technol.* 33, 72–79.

Rojickova-Padrtova, R., Marsálek, B. & Holoubek, I. (1998): Evaluation of alternative and standard toxicity assays for screening of environmental samples: Selection of an optimal test battery, *Chemosphere* 37, 495–507.

Roembke, J. & Ahtiainen, J. (2007): The search for the "ideal" soil toxicity test reference substance. *Integrated Environmental Assessment and Management* 3, 464–466.

Roembke, J. & Moser, T. (2002): Validating the enchytraeid reproduction test: organisation and results of an international ringtest. *Chemosphere* 46: 1117–1140.

Roembke, J. & Moser, H. (2007): OEkotoxikologische Charakterisierung von Aschen aus der Muellverbrennung. *VGB PowerTech* 12/2007: 62–68.

Roembke, J. (2003): Ecotoxicological laboratory tests with enchytraeids: A review. *Pedobiologia* 47: 606–616.

Roembke, J.; Bauer, C.; Brodesser, J.; Brodsky, J.; Danneberg, G.; Heimann, D.; Renner, I.; Schallnass, H.-J. (1995): Grundlagen fuer die Beurteilung des oekotoxikologischen Gefaehrdungspotentials von Altstoffen im Medium Boden. UBA-Texte, 53/95: 1-369.

Roembke, J., Vickus, P. & Bauer, C. (1992): Experiences and problems with the OECD-Earthworm Acute Test in routine testing. In: Ecotoxicology of Earthworms. Greig-Smith, P., et al. (eds.), Intercept, Andover, pp. 209–212.

Rotoxkit F chronic. (2006): Chronic Toxicity Test for Freshwater. Standard Operational Procedure. MicroBioTests Inc. 28 pp.

Rutherford, D.W., Chiou, C.T. & Kile, D.E. (1992): Influence of soil organic matter composition on the partition of organic compounds. *Environ. Sci. Technol.* 26, 336–340.

Schaefer, M. (2003): Behavioural endpoints in earthworm ecotoxicology. *J. Soils Sediments* 3, 79–84.

Schmitz, R.P.H., Eisentraeger, A. & Dott, W. (1998): Miniaturized kinetic growth inhibition assays with *Vibrio fisheri* and *Pseudomonas putida* (application, validation and comparison). *J. Microbiol. Methods* 31, 159–166.

Schmitz, R.P.H., Eisentraeger, A. & Dott, W. (1999): Agonistic and antagonistic toxic effects observed with miniaturized growth and luminescence inhibition assays. *Chemosphere* 38, 79–95.

Schoknecht, U., & Gunschera, J., Marx, H.N., Marx, G., Peylo, A., & Schwarz, G. (1998): Holzschutzmittelanalytik – Daten- und Literaturzusammenstellung fuer Wirkstoffe in geprueften Holzschutzmitteln. BAM-Forschungsbericht 225. 156 p., ISBN 3-89701-160-3 (in German).

ScienceLab (2005): Material Safety Data Sheet (MSDS) 2-Chloroacetamide. http://www.sciencelab.com/xMSDS-2_Chloroacetamide-9923402.

Shaw, L., Beaton, Y., Glover, L., Killham, K. & Meharg, A. (2000): Interactions between soil, toxicant, and a lux-marked bacterium during solid phase – Contact toxicity testing. *Environ. Toxicol. Chem.* 19, 1247–1252.

Sickerwasser Prognose (2007) Der Förderschwerpunkt "SiWaP" des Bundesministeriums für Bildung und Forschung: "Prognose des Schadstoffeintrages in das Grundwasser mit dem Sickerwasser" FZK-PtWT+E, Karlsruhe, Germany.

Slabbert, J.L. & Grabow, W.O.K. (2006): A rapid water toxicity screening test based on oxygen uptake of *Pseudomonas putida*. *Toxic. Assess.* 1(1), 13–26.

Smith, B. (2003): Ontology. In: Blackwell Guide to the Philosophy of Computing and Information. Floridi, L. (ed.), Blackwell, Oxford, pp. 155–166.

Spiller, E. (2006): Development of a novel electronic 1536 microtiter plate for quantifying the biomass in high throughput devices. Thesis at the University of Freiburg, Institut fuer Mikrosystemtechnik, Germany (urn:nbn:de:bsz:25-opus-25270; URL: http://www.freidok.uni-freiburg.de/volltexte/2527/ (December 2007)

Spurgeon, D.J., Weeks, J.M. & Van Gestel, C.A.M. (2003): A summary of eleven years progress in earthworm ecotoxicology. *Pedobiologia* 47, 588–606.

Spurgeon, D.J. & Hopkin, S.P. (1996): Effects of variations of the organic matter content and pH of soils on the availability and toxicity of zinc to the earthworm *Eisenia fetida*. *Pedobiologia* 40, 80–96.

Stantec (2004): Developmental Studies in Support of Environment Canada's Biological Test Methods for Measuring Soil Toxicity to Earthworms. Report prepared for Environmental Technology Center, Ottawa, Canada.

Stephenson, G.L., Solomon, K.R., Hale, B., Greenberg, B.M. & Scroggins, R.P. (1997): Development of suitable test methods for evaluating the toxicity of contaminated soils to a battery of plant species relevant to soil environments in Canada. In: Dwyer, F.J., Doane, T.T. & Hinman, M.L. (eds.). Environmental Toxicology and Risk Assessment: Modeling and Risk Assessment. Vol. 6. American Society for Testing and Materials, Philadelphia, PA, USA. pp. 474-489.

Suetterlin, H., Alexy, R., Kuemmerer, K. (2008): The toxicity of the quaternary ammonium compound benzalkonium chloride alone and in mixtures with other anionic compounds to bacteria in test systems with *Vibrio fischeri* and *Pseudomonas putida*. Ecotox Environ Safe: 71: 498–505.

Sverdrup, L.E., Hagen, S.B., Krog, P.H. & van Gestel, C.A.M. (2007): Benzo(a)pyrene shows low toxicity to three species of terrestrial plants, two soil invertebrates, and soil-nitrifying bacteria. *Ecotoxicol. Environ. Saf.* 66, 362–368.

SYSTAT® Software, Inc., Point Richmond, CA, USA

Thamnotoxkit F. (2001): Crustacean Toxicity Screening Test for Freshwater. Standard Operational Procedure. MicroBioTests Inc. 23 pp.

Tidepool (1995). ToxCalc programme, version 5.0.12N. Tidepool Scientific Software.

ToxRat® (2003): Software for the statistical analysis of biotests. Copyright: ToxRat Solutions GmbH, Alsdorf, Germany.

ToxRat® (2006): Software for the statistical analysis of biotests. Copyright: ToxRat Solutions GmbH, Alsdorf, Germany

Ulm, L., Vrzina, J., Schiesl, V., Puntaric, D. & Smit, Z. (2000): Sensitivity comparison of the conventional *Daphnia magna* immobilisation test with the Daphtoxkit microbiotest for household products. In: New microbiotests for routine toxicity screening and biomonitoring. Persoone, G., Janssen, C.& De Coen, W. (eds.), Kluwer/Plenum, New York, NY, pp. 247–252.

Van der Sloot, H.A., van Zomeren, A., Meeuwsen, H.C.L., Seignette, P. & Bleyerveld, R. (2007): Interpretation of test method selection, validation against field data, and predictive modeling for impact evaluation of stabilised waste disposal. *J. Hazard. Mater.* 141, 354–369.

Van der Veen, A.M.H. & Nater, D.A.G. (1993): Sample preparation from bulk samples: An overview. *Fuel Process. Technol.* 36, 1–7.

Van der Wielen, C. & Halleux, I. (2000): Shifting from the conventional ISO 8692 algal growth inhibition test to the Algaltoxkit microbiotest ? In: New microbiotests for routine toxicity screening and biomonitoring. Persoone, G., Janssen, C. & De Coen, W. (eds.), Kluwer/Plenum, New York, NY, pp. 269–272.

van Gestel, C.A.M. (1992): The influence of soil characteristics on the toxicity of chemicals for earthworms: A review. In: Ecotoxicology of Earthworms. Greig-Smith, P.W., Becker, H., Edwards, P.J. & Heimbach, F. (eds.), Intercept, Andover, pp. 44–54.

Van Zomeren, A. & Comans, R.N.J. (2004): Contribution of natural organic matter to copper leaching from municipal solid waste incinerator bottom ash. *Environ. Sci. Technol.* 38, 3927–3932.

Van Zomeren, A. & Comans, R.N.J. (2007): Measurement of humic and fulvic acid concentrations and dissolution properties by a rapid batch method. *Environ. Sci. Technol.* 41, 6755–6761.

Vangheluwe, M., Janssen C. & Van Sprang, P. (2000): Selection of bioassays for sediment toxicity screening. In: New microbiotests for routine toxicity screening and biomonitoring. Persoone, G. Janssen, C. & De Coen, W. (eds.), Kluwer/Plenum, New York, NY, pp. 449–458.

Venturin P., Ros, M. (2003): Interlaboratory comparison on simulated waste waters made up with mixed chemicals. Intercalibration Exercise on the acute *Daphnia magna* toxicity test. Slovenian National Institute of Chemistry (Internal Report in Slovenian).

VICH (2005): Environmental impact assessment (EIAs) for veterinary medicinal products (VMPs) – Phase II Guidance. VICH Guideline 38 (Ecotoxicity Phase II), Bruxelles (Belgium), October 2004.

Vindimian, E., Garric, J., Flammarion, P., Thybaud, E. & Babut, M. (1999): An index of effluent aquatic toxicity testing designed by partial least squares regression using acute and chronic tests and expert judgements. *Environ. Toxicol. Chem.* 18, 2386–2391.

VMM (1996): Ecotoxicological assessment of industrial effluents: review of methods and protocols for implementation in effluent management. Report for the Flemish Environmental Agency.

Wadhia, K, Dando, T. & Thompson, K.C. (2007): Intra-laboratory evaluation of microbial assay for risk assessment (MARA) for potential application in the implementation of the water framework directive (WFD). *J. Environ. Monit.* 9, 1–6.

Wang, W. (1991): Literature review on higher plants for toxicity testing. *Water Air Soil Pollut.* 59, 381–400.

Weber, J., Kreutzmann, J., Plantikow, A., Claus, E., Manz, W. & Heininger, P. (2006): A novel particle contact assay with the yeast *Saccharomyces cerevisiae*. *J. Soils Sediments* 6, 84–91.

Wiles, J.A. & Krogh, P.H. (1998): Tests with the Collembolans *Isotoma viridis, Folsomia candida* and *Folsomia fimetaria*. In: Handbook of Soil Invertebrate Toxicity Tests. Løkke, H. & Van Gestel, C.A.M. (eds.), Wiley, Chichester, pp. 131–156.

Wilke, B.-M., Riepert, F., Koch, C., Kühne, T. (2007): Ecotoxicological characterization of hazardous wastes. Ecotoxicol. Environm. Safety, 70: 283-293.

Wilke, B.-M., Winkel, B., Fleischmann, S. & Gong, P. (1998): Higher Plant Growth and Microbial Toxicity Tests for the Evaluation of the Ecotoxic Potential of Soils. Contaminated Soil' 98, Thomas Telford, London, UK, 345–354.

Wong, M.H. & Wong, J.W.C. (1989): Germination and seedling growth of vegetable crops in fly ash-amended soils. *Agric. Ecosyst. Environ.* 26, 23–35.

Wundram, M. & Bahadir, M. (1999): Ecotoxicological test systems for prediction of environmental behaviour of toxic compounds in underground disposals. *Fresenius Envir. Bull.* 8, 280–287.

Wundram, M., Selmar, D., Wilken, G., Larink, O., Mueller, M. & Bahadir, M. (1996): Oekotoxikologische Bewertung von Deponiesickerwaessern und Abfalleluaten. *EP* 7–9, 62–65.

Xu, Q., Barrios, C.A., Cutright, T. & Zhang Newby, B.M. (2005): Evaluation of toxicity of capsaicin and zosteric acid and their potential application as antifoulants. *Environ. Toxicol.* 20(5), 467–474.

Yeardley, J.R.B., Lazorchak, J.M. & Gast, L.C. (1996): The Potential of an earthworm avoidance test for evaluation of hazardous waste sites. *Environ. Toxicol. Chem.* 15, 1532–1537.

Zar, J.H. (1999): Biostatistical Analysis. Prentice-Hall, London, UK.

Zounková, R., Odráska, P., Dolezalová, L., Hilscherová, K., Marsálek, B. & Bláha, L. (2007): Ecotoxicity and genotoxicity assessment of cytostatic pharmaceuticals. *Environ. Toxicol. Chem.* 26, 2208–2214.

References

[illegible]

Index

GPSR Compliance

The European Union's (EU) General Product Safety Regulation (GPSR) is a set of rules that requires consumer products to be safe and our obligations to ensure this.

If you have any concerns about our products, you can contact us on ProductSafety@springernature.com

In case Publisher is established outside the EU, the EU authorized representative is:

Springer Nature Customer Service Center GmbH
Europaplatz 3
69115 Heidelberg, Germany

Batch number: 10755284

Printed by Printforce, the Netherlands